STEM CELLS AND COVID-19

STEM CELLS AND COVID-19

Edited by

CHANDRA P. SHARMA

Biomedical Technology Wing, Sree Chitra Tirunal Institute for Medical Sciences & Technology (SCTIMST), Trivandrum, India
Department of Pharmaceutical Biotechnology, Manipal College of Pharmaceutical Sciences, Manipal University, Manipal, India
College of Biomedical Engineering and Applied Sciences, Purbanchal University, Kathmandu, Nepal

DEVENDRA K. AGRAWAL

Senior Vice President for Research and Professor, Translational Research, Western University of Health Sciences, Pomona, CA, United States

FINOSH G. THANKAM

Assistant Professor, Tissue Engineering and Regenerative Medicine, Department of Translational Research, Western University of Health Sciences, Pomona, CA, United States

Academic Press is an imprint of Elsevier
125 London Wall, London EC2Y 5AS, United Kingdom
525 B Street, Suite 1650, San Diego, CA 92101, United States
50 Hampshire Street, 5th Floor, Cambridge, MA 02139, United States
The Boulevard, Langford Lane, Kidlington, Oxford OX5 1GB, United Kingdom

ISBN 978-0-323-89972-7

For information on all Academic Press publications
visit our website at https://www.elsevier.com/books-and-journals

Publisher: Stacy Masucci
Acquisitions Editor: Elizabeth Brown
Editorial Project Manager: Samantha Allard
Production Project Manager: Selvaraj Raviraj
Cover Designer: Greg Harris

Typeset by STRAIVE, India

Contents

Contributors

Devendra K. Agrawal Department of Translational Research, Western University of Health Sciences, Pomona, CA, United States

S.R. Aravind Advanced Centre for Tissue Engineering, Department of Biochemistry, University of Kerala, Thiruvananthapuram, Kerala, India

Sunitha Chandran Technology Business Incubator for Medical Devices and Biomaterials, Sree Chitra Tirunal Institute for Medical Sciences and Technology, Trivandrum, India

Soumya Chandrasekher Department of Zoology, KKTM Government College, Pullut, Trichur, India

Smitha Chenicheri PMS College of Dental Sciences and Research; Biogenix Research Center for Molecular Biology and Applied Sciences, Thiruvananthapuram, India

Bodhisatwa Das Department of Biomedical Engineering, Indian Institute of Technology Ropar, Punjab, India

Francis Boniface Fernandez Bioceramics Laboratory, Biomedical Technology Wing, Sree Chitra Tirunal Institute for Medical Sciences & Technology, Thiruvananthapuram, India

Annie John Advanced Centre for Tissue Engineering, Department of Biochemistry, University of Kerala, Thiruvananthapuram, Kerala, India

Jessy John Division of Hematology and Oncology, Hillman Cancer Center, University of Pittsburgh, Pittsburgh, PA, United States

Josna Joseph Department of Clinical Immunology and Rheumatology, Christian Medical College, Vellore, Tamil Nadu, India

Remya Kommeri McGowan Institute for Regenerative Medicine, University of Pittsburgh, Pittsburgh, PA, United States

P. Lekshmi Division of Thrombosis Research, Department of Applied Biology, Sree Chitra Tirunal Institute for Medical Sciences and Technology, Trivandrum, India

Bernadette K. Madathil Advanced Centre for Tissue Engineering, Department of Biochemistry, University of Kerala, Thiruvananthapuram, Kerala, India

Krupa Ann Mathew Advanced Centre for Tissue Engineering, Department of Biochemistry, University of Kerala, Thiruvananthapuram, Kerala, India

S. Mini Advanced Centre for Tissue Engineering, Department of Biochemistry, University of Kerala, Thiruvananthapuram, Kerala, India

Daniel Miranda, Jr. Pharmaceutical Sciences Department, College of Pharmacy, Western University of Health Sciences, Pomona, CA, United States

Sujata Mohanty Stem Cell Facility (DBT-Centre of Excellence for Stem Cell Research), All India Institute of Medical Sciences, New Delhi, India

Anwesha Mukherjee Department of Biomedical Engineering, Indian Institute of Technology Ropar, Punjab, India

Renjith P. Nair Division of Thrombosis Research, Department of Applied Biology, Sree Chitra Tirunal Institute for Medical Sciences and Technology, Trivandrum, India

E. Pranshu Rao Stem Cell Facility (DBT-Centre of Excellence for Stem Cell Research), All India Institute of Medical Sciences, New Delhi, India

David Jesse Sanchez Pharmaceutical Sciences Department, College of Pharmacy, Western University of Health Sciences, Pomona, CA, United States

Chandra P. Sharma Biomedical Technology Wing, Sree Chitra Tirunal Institute for Medical Sciences & Technology (SCTIMST), Trivandrum; Department of Pharmaceutical Biotechnology, Manipal College of Pharmaceutical Sciences, Manipal University, Manipal, India; College of Biomedical Engineering and Applied Sciences, Purbanchal University, Kathmandu, Nepal

Yashvi Sharma Stem Cell Facility (DBT-Centre of Excellence for Stem Cell Research), All India Institute of Medical Sciences, New Delhi, India

Finosh G. Thankam Department of Translational Research, Western University of Health Sciences, Pomona, CA, United States

Ancy Thomas Department of Dermatology, Northwestern University; Robert H. Lurie Comprehensive Cancer Center, Chicago, IL, United States

Jane Joy Thomas Division of Nephrology and Hypertension, Feinberg Cardiovascular and Renal Research Institute, Northwestern University, Chicago, IL, United States

Mereena George Ushakumary Division of Pulmonary Biology, Cincinnati Children's Hospital Medical Center, Cincinnati, OH, United States

Daniel R. Wilson McGowan Institute for Regenerative Medicine, University of Pittsburgh, Pittsburgh, PA; Department of Translational Research, Western University of Health Sciences, Pomona, CA, United States

Preface

The pandemic spread of COVID-19 infection throughout 190 countries and territories has created health crisis and huge economic burden across the globe. This is due to SARS-CoV-2 that induces aggravated and sustained inflammation with cytokine storm compared to other coronavirus strains, leading to multiorgan pathology. Tremendous efforts have been devoted to learn the underlying cellular and molecular mechanisms and developing effective vaccines.

Stem cells pose immense potential in developing better diagnostic, treatment, and preventive strategies in SARS-CoV-2 infection. Interestingly, hematopoietic and mesenchymal stem cells are critical to better understand the response of immune system to SARS-CoV-2 infection in both healthy and comorbid conditions, promising the development of effective vaccines and immunotherapies. Indeed, blood-forming type 1 dendritic cells and other antigen-presenting cells could present SARS-CoV-2 protein fragments to T-lymphocytes that undergo clonal expansion with the development of memory T cells to induce strong immune response and thus long-lasting immunity. Stem cells may also be helpful in disease modeling and drug screening to fight against viral infection.

In this book, our goal is to bring up-to-date knowledge on the effect of hematopoietic and mesenchymal stem cells to combat SARS-CoV-2 infection in its diagnosis, treatment, and prevention; critically discuss the challenges; and highlight outstanding questions and future perspectives for both preclinical and clinical practitioners. The book consists of 12 multiauthored chapters. All contributors are internationally recognized experts in their respective specialties and have reviewed the latest thoughts, concepts, and their implications.

We thank all the contributors for their excellent contributions, Ms. Elizabeth A. Brown for encouraging the project development at its initial stage, and Ms. Allard Samantha for her effective coordination.

We take this time to express our gratitude and thanks to our family members for sustained support during this project.

Chandra P. Sharma thanks and very much appreciates his wife Aruna Sharma for her continuous support during this entire project.

Finosh Thankam expresses his sincere gratitude for the continuous support and heartening from his wife Soumya and his son Rohin for the preparation of this book. Also, Finosh acknowledges the constant support from the editors and the excellent contribution by the authors and the perpetual blessings and inspiration from his parents, teachers, friends, and well-wishers for the successful completion of this project.

Devendra Agrawal expresses many thanks to his coeditors, Dr. Finosh G. Thankam and Prof. Chandra P. Sharma, the authors of the chapters for their exceptional contribution with the latest information in the field to complete this project, and the Elsevier editorial team for their patience and outstanding support.

Chandra P. Sharma
Devendra K. Agrawal
Finosh G. Thankam

1

Introduction

Finosh G. Thankam[a], Devendra K. Agrawal[a], and Chandra P. Sharma[b,c,d]

[a]Department of Translational Research, Western University of Health Sciences, Pomona, CA, United States, [b]Biomedical Technology Wing, Sree Chitra Tirunal Institute for Medical Sciences & Technology (SCTIMST), Trivandrum, India, [c]Department of Pharmaceutical Biotechnology, Manipal College of Pharmaceutical Sciences, Manipal University, Manipal, India, [d]College of Biomedical Engineering and Applied Sciences, Purbanchal University, Kathmandu, Nepal

Pandemic outbreak of the novel coronavirus disease (COVID-19) infection caused by SARS-CoV-2 remains to be a serious threat to human population across the globe. As of July 22, 193,361,389 clinically confirmed cases with 4,150,618 deaths shook the world. The pandemic has globally impacted all dimensions of life, especially the health care system, economy, and education [1]. COVID-19 primarily infects respiratory system causing pneumonia-like clinical presentation; however, it affects heart, kidney, liver, and central nervous system leading to multiple organ failure and subsequent death in severe cases [2]. Indeed, the angiotensin-converting enzyme 2 (ACE2) receptor in the alveolar epithelial cells and macrophages is crucial for the viral entry to human system. The accelerated replication of COVID-19 in these cells leads to the activation of apoptosis, increasing the surge of proinflammatory cytokines potentially damaging the alveoli and reducing diffusion perfusion and ventilation leading to low level of oxygen in the blood [3]. Evidently, the severely infected patients displayed hypersecretion of interleukin (IL)-6 and tumor necrosis factor α (TNFα) along with decreased density of CD4 + and CD8 + T cells which significantly declined the recovery rate [4]. In addition, the compromised alveolar epithelium facilitates the influx of neutrophils and inflammatory macrophages contributing to the pool of proinflammatory cytokines resulting in cytokine burst sustaining the alveolar damage and potential development of multiorgan dysfunction [1].

Advancements in PCR and antibody technologies have made early detection of COVID-19 infection. Moreover, the diagnostic imaging (especially CT scan) coupled with artificial intelligence has proven to be

Stem Cells and COVID-19. https://doi.org/10.1016/B978-0-323-89972-7.00013-1

advantageous [5]. Generally, the management of COVID-19 depends on the clinical symptoms where the patients with mild symptoms are treated without hospitalization. However, critical clinical care including oxygen support is required for patients with severe/aggravated COVID-19 infection. In addition, the commonly administered therapeutics include hydroxychloroquine, azithromycin, remdesivir, favipiravir, tocilizumab, and itolizumab and convalescent plasma. Despite the promising outcomes in the attenuation of symptoms, the overall clinical outcomes are minimally effective [6]. Also, a novel approach of binding the virus with soluble ACE2 receptor which neutralizes the S-protein of COVID-19 by preventing the entry to the cells and giving the opportunity for immune system to clear the virus has been attempted [7]. Currently, the global medical community is striving to tame the virus and to bring back the world as it was in pre-COVID-19 era.

As the world strives to control COVID-19, multiple approved vaccines are in clinical practice, and several are on the cusp of approval. Despite the concerns in the efficacy and safety, the COVID-19 vaccines are beneficial in preventing the pandemic spread [8]. However, potential concerns regarding the dose, effectiveness, efficiency, and duration of effect persist which warrant the next-generation vaccines with improved and long-lasting performance [8]. Multiple viral elements have been tried as target to vaccine development; however, the vaccines designed against S-protein have gained maximum significance owing to its crucial role in virus binding to the host cell. Interestingly, diverse strategies have been employed for the vaccines which include inactivated virus, viral-like nanoparticles, protein subunits, virus-vectored, DNA, mRNA, and live attenuated virus [9]. Unfortunately, the uncertainty in the performance of vaccines and the extent of tissue damage and comorbidities following the COVID-19 infection warrant effective regenerative/management strategies adjuvant to vaccine-based therapies.

Additionally, it is becoming extremely challenging because of constant mutation of the virus and the emergence and circulation of genetic variants of SARS-CoV2 around the globe during this COVID-19 pandemic. The major variants of COVID-19 that are circulating include B.1.1.7 (Alpha), B.1.351 (Beta), B.1.617.2 (Delta), and P.1 (Gamma). Among these, the cases of gamma variant (B.1.617.2) are surging in India, United States, and other countries, and appears to be 10–15 times more deadly than the earlier predominant strain (B.1.1.7-Alpha). In a very recent study, there was a modest difference in the effectiveness of the two vaccine doses between the delta variant and the alpha variant [10]. These findings support the effectiveness of the currently available vaccine against both alpha and delta variants. However, better therapies with longer duration of effect are warranted.

Several cell-based therapies have been attempted in the management of COVID-19 which revealed promising outcomes. Specifically, the contribution of stem cells in medicine and clinical practice is central which has been applied to COVID-19 as well. For instance, the intravenous administration of mesenchymal stromal cells (MSCs) in critically ill patients demonstrated the improvements in clinical symptoms, including the resolution of cough, fever, respiratory distress, and reversal of oxygen saturation within 2–4 days [7]. Importantly, MSCs have been an ideal choice for cell-based regenerative therapies owing to their lower immunogenicity, differentiation potential, immunomodulatory effects, and ease of harvest and manipulation. In addition, MSCs are abundant in most tissues and can be harvested using minimally invasive techniques. Owing to their potent immunomodulatory and prohealing effects, MSCs have been hailed to be promising for the management of COVID-19 as evident from the growing numbers of clinical trials [11]. Also, the MSCs harvested from adipose tissues, bone marrow, umbilical cord, blood, and other tissues offer flexibility in manipulation, in vitro expansion and storage which is further beneficial for COVID-19 management.

Since COVID-19 is associated with cytokine-driven multiple organ damage, MSC therapy has been innovative in the management of organ damage following COVID-19 infection. However, the potential healing mechanisms elicited by MSCs in COVID-19 mediated organ damage remain unanswered [12]. Route of administration, dose, biomarker definition of MSCs, and expansion status warrant further optimization. Despite these challenging questions, secretome from MSCs released as exosomes offers promising translational potential addressing the hurdles of MSC therapy [13]. The MSC-derived exosomes have been superior in preventing lung injury and lung fibrosis, offering the flexibility of injection, inhalation, and infusion suggesting their potential application in COVID-19 management. Interestingly, several clinical trials have been registered focusing on the application of exosomes in COVID-19 therapy [11,12]. In addition, the advantages of exosomes including safer than parent cells, easily diffusible nature, and ability to cross cellular/tissue barriers and the immunomodulatory contents (proteins, growth factors, miRNAs, and lncRNAs) upgrade the stem cell-derived exosomes to be the superior biological therapeutics for COVID-19 management.

The advancements in cell technologies and vesicle engineering have significantly contributed to clinical medicine and disease management. Apart from MSCs, induced pluripotent stem cells (iPSC) and programmed cells are ideal for the management of COVID-19 [1,14]. The combinatorial approaches employing stem cell technologies and advanced gene technologies including genome editing, CRISPR-cas9, and single cell genomics unveil the ideal stem cell population customized for COVID-19 management. Significant research commitment is required to translate the stem cells as a promising therapeutic for the management of COVID-19 infection and associated complications. Current knowledge regarding the prevention of COVID-19 virus entry and replication by stem cells in the respiratory and/or other tissues is limited; however, such possibility cannot be neglected owing to the healing effects of stem cells against viral infections. Taken together, stem cell biology leaves behind a promising opportunity as novel management strategy for COVID-19 and associated organ damage.

As the COVID-19 pandemic continues to challenge the global scientific community, the information presented in this book primarily focuses on the perspectives in the potential of stem cells in developing better diagnostic, treatment, and preventive strategies in COVID-19 infection. The 12 chapters contributed by highly qualified investigators critically review the therapeutic potential of stem cells in response of immune system to coronavirus infection in both healthy and comorbid conditions. In this book, our goal is to bring up-to-date knowledge on the effect of diverse stem cells and stem-cell derived entities to combat COVID-19 infection in its diagnosis, treatment, and prevention; critically discuss the challenges; and highlight outstanding questions and future perspectives for both preclinical and clinical practitioners.

References

[1] Djidrovski I, Georgiou M, Hughes GL, et al. SARS-CoV-2 infects an upper airway model derived from induced pluripotent stem cells. Stem Cells 2021. https://doi.org/10.1002/stem.3422.

[2] Zhu N, Zhang D, Wang W, et al. A novel coronavirus from patients with pneumonia in China, 2019. N Engl J Med 2020;382:727–33. https://doi.org/10.1056/NEJMoa2001017.

[3] Thankam FG, Agrawal DK. Molecular chronicles of cytokine burst in patients with coronavirus disease 2019 (COVID-19) with cardiovascular diseases. J Thorac Cardiovasc Surg 2020. https://doi.org/10.1016/j.jtcvs.2020.05.083.

[4] Diao B, Wang C, Tan Y, et al. Reduction and functional exhaustion of T cells in patients with coronavirus disease 2019 (COVID-19). Front Immunol 2020. https://doi.org/10.3389/fimmu.2020.00827.

[5] Alsharif W, Qurashi A. Effectiveness of COVID-19 diagnosis and management tools: a review. Radiography (Lond, Engl: 1995) 2021;27:682–7. https://doi.org/10.1016/j.radi.2020.09.010.

[6] Mahendiratta S, Bansal S, Sarma P, et al. Stem cell therapy in COVID-19: pooled evidence from SARS-CoV-2, SARS-CoV, MERS-CoV and ARDS: a systematic review. Biomed Pharmacother 2021;137. https://doi.org/10.1016/j.biopha.2021.111300, 111300.

[7] Anka AU, Tahir MI, Abubakar SD, et al. Coronavirus disease 2019 (COVID-19): An overview of the immunopathology, serological diagnosis and management. Scand J Immunol 2020. https://doi.org/10.1111/sji.12998, e12998.

[8] Soiza RL, Scicluna C, Thomson EC. Efficacy and safety of COVID-19 vaccines in older people. Age Ageing 2020;afaa274. https://doi.org/10.1093/ageing/afaa274.

[9] Dai L, Gao GF. Viral targets for vaccines against COVID-19. Nat Rev Immunol 2020;1–10. https://doi.org/10.1038/s41577-020-00480-0.

[10] Lopez Bernal J, Andrews N, Gower C, et al. Effectiveness of Covid-19 vaccines against the B.1.617.2 (Delta) Variant. N Engl J Med 2021. https://doi.org/10.1056/NEJMoa2108891.

[11] Riedel RN, Pérez-Pérez A, Sánchez-Margalet V, et al. Stem cells and COVID-19: are the human amniotic cells a new hope for therapies against the SARS-CoV-2 virus? Stem Cell Res Ther 2021;12:155. https://doi.org/10.1186/s13287-021-02216-w.

[12] Afarid M, Sanie-Jahromi F. Mesenchymal stem cells and COVID-19: cure, prevention, and vaccination. Stem Cells Int 2021;2021. https://doi.org/10.1155/2021/6666370, e6666370.

[13] Gennai S, Monsel A, Hao Q, et al. Microvesicles derived from human mesenchymal stem cells restore alveolar fluid clearance in human lungs rejected for transplantation. Am J Transplant Off J Am Soc Transplant Am Soc Transpl Surg 2015;15:2404–12. https://doi.org/10.1111/ajt.13271.

[14] Fang W, Agrawal DK, Thankam FG. 'Smart-exosomes': a smart approach for tendon regeneration. Tissue Eng Part B Rev 2021. https://doi.org/10.1089/ten.TEB.2021.0075.

2

Characteristics and immunobiology of COVID-19

Remya Kommeri[a], Finosh G. Thankam[b], Devendra K. Agrawal[b], and Daniel R. Wilson[a,b]

[a]McGowan Institute for Regenerative Medicine, University of Pittsburgh, Pittsburgh, PA, United States, [b]Department of Translational Research, Western University of Health Sciences, Pomona, CA, United States

Introduction

COVID-19 (coronavirus disease-2019) has been identified as a serious respiratory infection by the SARS-CoV2 virus (severe acute respiratory syndrome coronavirus 2), was first reported in Wuhan, Hubei Province, China in December 2019. Since then, COVID-19 has exponentially spread worldwide as a pandemic eventually declared by WHO on March 11, 2020. It continues currently as the utmost urgent/critical ailment of global public health [1]. Moreover, the COVID-19 pandemic has not only affected global health scenarios but also has heavily burdened the economic, financial, political, educational, and other dimensions of humanity across the world. As of June 20, 2021, according to Worldometer, 179,245,386 cases with 3,881,890 deaths and 163,798,559 recovered cases were reported globally. Data collection varies across nations, but officially the USA has reported the most cases (34,406,001 cases), followed by India (29,934,361 cases), and Brazil (17,927,928 cases; [2]). That these statistics continue to gather rapidly reflects how this severe global health challenge requires effective strategies for the diagnosis, prevention, and management of COVID-19. These, in turn, rely on understanding the vector structures, modes of infection, molecular pathologies, and optimal translational targets.

Generally, COVID-19 infection is characterized by lung dysfunction and respiratory difficulties leading to pneumonia-like clinical presentation causing death in severe cases. The major pathological mechanism for aggravated respiratory pathology in COVID-19 infection has been attributed to increased pro-inflammatory milieu due to cytokine storm [3]. Cytokine storm severely impairs gas exchange, induces edema and trauma in the lungs, elicits acute respiratory

Stem Cells and COVID-19. https://doi.org/10.1016/B978-0-323-89972-7.00008-8

distress, and increases susceptibility to secondary infections. In addition, preexisting comorbidities including type II diabetes, cardiovascular diseases, hypertension, and older age are major aggravating factors in COVID-19 infection. Unfortunately, failure of proper management and lack of effective treatment modalities further complicate clinical presentations resulting in increased mortality [4]. Generally, postinfection manifestations and symptoms arise within 14 days, with a mean incubation time of 5 days. Despite the multiple clinical presentations, fever, cough, difficulty in breathing, loss of smell and taste, and fatigue are the typical initial symptoms of COVID-19 infection [5]. Alarmingly, asymptomatic presentations, extremely higher infectivity, and lack of effective drugs or vaccines are major impediments to management and prevention of disease [6]. Hence, understanding the basic biology of COVID-19 and molecular mechanisms underlying the cytokine burst are crucial to develop effective management strategies. With this as background, the focus of this chapter is to delineate current knowledge regarding the basic immunobiology of COVID-19 with a decidedly translational perspective.

Structure and etiology

International Committee on Taxonomy of Viruses has classified coronaviruses under the family Coronaviridae, subfamily Coronavirinae, and further classified into four genera such as Alphacoronavirus, Betacoronavirus, Gammacoronavirus, and Deltacoronavirus. In the family, 6 types of viruses infect humans, of which four cause the common cold, whereas the remaining two lead to fatal diseases. The fatal strains are SARS-CoV and MERS-CoV, identified in 2003 and 2012, respectively [7]. That the new variant, SARS-CoV2, has sequence similarities to bat coronavirus and human SARS-CoV indicates homology to the betacoronavirus family (Ref. [8], p. 2). SARS-CoV2 has a single-stranded positive-sense RNA genome of approximately 29.9 kb [9] packed in a nucleocapsid (N) and further surrounded by an envelope composed of three major structural proteins: membrane protein (M), spike protein (S), and envelope protein (E) [10]. Rather than structural proteins, 16 nonstructural proteins (NSP) are identified in SARS-CoV2 as essential for modulation and survival of the virus [11]. The spike protein is one of the decisive components for viral entry to host cells. Spike protein is composed of two subunits—an external S1 subunit with receptor-binding domain to bind with the angiotensin-converting enzyme 2 receptors on the human cells, and an S2 subunit comprised of transmembrane and internal domains that facilitate fusion of viral membrane with host cells [12]. Being an inevitable structural protein, S proteins are mostly targeted in therapeutic approaches [13], as well as easily recognized by the host immune system to develop antiviral immunity. The major functions of the identified structural and nonstructural proteins are presented in Table 1.

Pathology of infection

The onset of infection is typically followed by three stages of infection: an asymptomatic phase, upper and conducting airway infection, and—most perilous—hypoxia followed by Acute Respiratory Distress Syndrome (ARDS) [30] (Fig. 1).

The initial 1–2 days of SARS-CoV2 viral infection is mostly asymptomatic. The densely glycosylated S1 spike protein on the inhaled SARS-CoV2 virus binds with angiotensin-converting enzyme receptor 2 (ACE2R) expressed on cell linings of the nasal cavity and conducting airway. That the affinity of SARS-CoV2 spike protein to the ACE2R is 10–20 times more than in other SARS CoV viruses underlies the highly infectious profile of COVID-19. The transmembrane protease serine (TMPRSSs) on host cells cleaves the S1/S2 protease site and exposes S2 fusion protein that enables fusion of viral membrane with host cell membranes [31]. Coexpression of ACE2R and TMPRSSs has been observed in alveolar type 2 pneumocytes, epithelium of the upper esophagus, and absorptive enterocytes that ease the entry of the virus through the alveolar, esophageal, and intestinal epithelium [32]. Once the viral genome enters host cells, the lytic cycle begins with initial transcription and translation of viral genome and proteins using host machinery, assembly of the viral genome into the nucleocapsid, and lysis of host cells in the systemic circulation [33].

In few days, SARS-CoV2 infects upper and conducting airways enough to activate the innate immune system, thereby manifesting clinical signs and symptoms such as cough and fever. Approximately 80% of patients show these mild symptoms and stay in stage 2, with infection restricted to the upper and conducting airways; however, this elicits minimal migration toward the lungs [34]. Unfortunately, migration of the SARS-CoV2 virus to the lungs and alveolar cells complicates the clinical manifestations and prognosis in at least 20% of patients, causing more than 2% mortality as compounded by comorbidity factors [34]. SARS-CoV2 virus preferably infects alveolar type II cells, the precursor cells for type 1 pneumocytes in the lungs [35]. Infected peripheral and subpleural alveolar cells die and release pleural effusions to form hyaline membranes that critically limit alveolar gaseous exchange [36]. The endogenous healing response initiates epithelial regeneration; however, this fails to heal wound caused by infection and results in fibrosis—the most commonly observed pathological manifestation associated with ARDS [37]. Failure of gaseous exchange rapidly reduces blood oxygenation and induces tachypnea. Tachypnea

Table 1 Major functions of structural and nonstructural proteins in SARS-CoV2 virus.

Structural proteins in COVID-19		
Protein	**Functional domains**	**Significance in pathology**
Spike protein (S protein)	Type I transmembrane N-linked glycosylated protein (150–200 kDa) consists of S1 and S2 subunit Major domains: receptor-binding domain (RBD) of S1: facilitate host cell recognition and ACE2 receptor binding. Fusion peptide (FP) in S2: integration between viral and host cell membrane [14]	Entry of virus to the host cells Major determinant of host immune response [13]
Membrane (M) protein	O-linked glycoprotein (25–30 kDa) Possesses three distinct transmembrane domains [15]	Facilitate molecular assembly of virus particles Glycosylation determines organ tropism and IFN signaling by virus [16]
Envelope (E) protein	Five-helix bundle surrounding a dehydrated narrow pore with bipartite channel (8–12 kDa)	Plays a major role in pathogenesis, virus assembly, and release [17]
N (nucleocapsid) protein	RNA-binding domain, N terminal, and C terminal domains Facilitate the affinity for viral versus nonviral RNA on assembly [18]	Facilitates virion assembly High immunogenic nature and target for vaccine design [19]
Nonstructural proteins in COVID-19		
NSP1	N-terminal product of the viral replicase Leader protein	Host translation inhibitor Degrade host mRNAs [20]
NSP2		Binds to prohibitin 1 and prohibitin 2 Alters the mitochondrial biogenesis in the host cell [21]
NSP3	Papain-like proteinase	Responsible for release of NSP1, NSP2, and NSP3 from the N-terminal region of pp1a and 1ab [22]
NSP4	Membrane-spanning protein	Viral replication-transcription complex Helps modify ER membranes [23]
NSP5	Proteinase and main proteinase	Cleaves at multiple distinct sites to yield mature and intermediate nonstructural proteins
NSP6	Putative transmembrane domain	Induces the formation of ER-derived autophagosomes Induces double-membrane vesicles [24]

Table 1 Major functions of structural and nonstructural proteins in SARS-CoV2 virus—cont'd

Structural proteins in COVID-19		
Protein	**Functional domains**	**Significance in pathology**
NSP7	RNA-dependent RNA polymerase	Forms complex with NSP8 and NSP12 to yield the RNA polymerase activity of NSP8 [25]
NSP8	Multimeric RNA polymerase; replicase	Makes heterodimer with NSP8 and 12 [26]
NSP9	Single-stranded RNA-binding viral protein	May bind to helicase
NSP10	Growth-factor-like protein possessing two zinc-binding motifs	Unknown
NSP11	Consists of 13 amino acids and identical to the first segment of Nsp12	Unknown
NSP12	RNA-dependent RNA polymerase	Replication and methylation [27]
NSP13	RNA-dependent RNA polymerase	Replication and transcription [28]
NSP14	Proofreading Exoribonuclease domain	3′–5′ direction and N7-guaninemethyltransferase activity
NSP15	EndoRNAse; nsp15-A1 and nsp15B-NendoU	Mn(2 +)- dependent endoribonuclease activity
NSP16	2′-*O*-ribose methyltransferase	Methyltransferase that mediates mRNA cap 2′-O-ribose methylation to the 5′-cap structure of viral mRNAs [29]

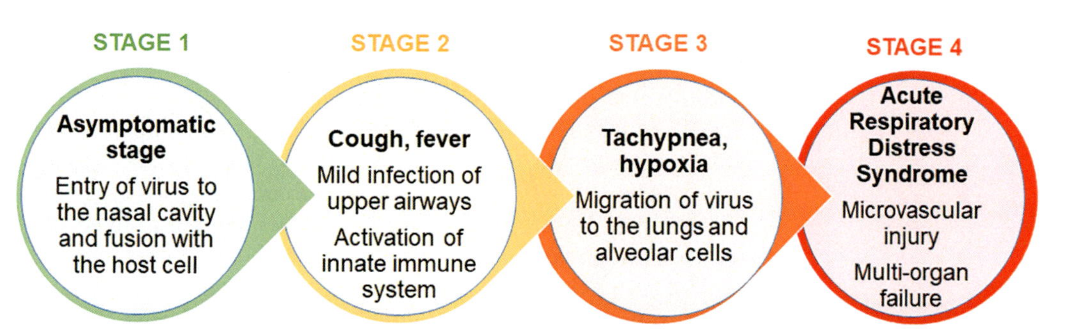

Fig. 1 Pathological stages in COVID-19.

further complicates the pulmonary efficiency resulting in low oxygenation, multiorgan failure, microvascular injury, and ischemic stroke, among other end organ sequelae [32].

Mortality in COVID-19 infection is mostly associated with age and comorbidities in the patients. Elder patients are more susceptible to reduced mucociliary clearance [38], weak immune system response, poor epithelial regeneration, and comorbid conditions such as cardiac disorders, kidney failure, and cancers [39].

Immunology and signaling mechanisms

The immunopathology of the patient has intimate and unexpected correlations with the COVID-19 infection. The unpredictable dysregulation of immune cell balance and clever evasion of the immune system have been identified in patients with severe symptoms including ARDS. Infections at Stage 3 are invariably associated with the reduction of immune cells, especially the $CD3^+$, $CD4^+$, $CD8^+$ T lymphocytes, B cells, and NK cells [40,41]. Moreover, a higher number of naïve T cells and reduced memory T cells are observed, followed by an infection which causes high level of inflammatory responses and more cases of relapse [41,42]. Naïve T cells are dedicated to defending the new pathogens with tightly controlled cytokine release that is essential to counteract the infection; however, this may cause hyperinflammation at the site of infection or even more dispersed systemic damage. Interestingly, the influence of SARS-CoV2 on immune cells is beyond the existing infectious course, as the immune cells are generally low in ACE2 receptors [43]. Importantly, the B cells have an important role in defense; although naïve B cells are reduced, a higher number of antibody-producing plasma cells have been observed in recovered patients. Cross-reactivity of serum indicates the presence of antibodies against the nucleocapsid antigen of the virus within 17–19 days of onset of symptoms [44].

The recovery or severity of infection strictly depends on the balance of the immune system in the host. The entry of virus to host cells follows pyroptosis of cells and release of damage-associated molecular patterns (DAMPs), including ATP, nucleic acids, as well as pathogen-associated molecular patterns (PAMPS) in viral RNA and dsDNA [40]. These patterns are recognized by neighboring cells by either endosomal Toll-like receptors TLR3 and TLR7/8, and the cytosolic RNA sensor, retinoic acid-inducible gene (RIG-I)/MDA5 [45]. Detecting dangerous viral pattern initiates several interlinked signaling cascade, including IRF3 (IFN regulatory factor-3), nuclear factor κB (NF-κB), JAK (Janus kinase)/STAT (signal transducer and activator of transcription) signaling pathways, that ultimately leads to the release of pro-inflammatory cytokines [46]. Cytokines are the alarm

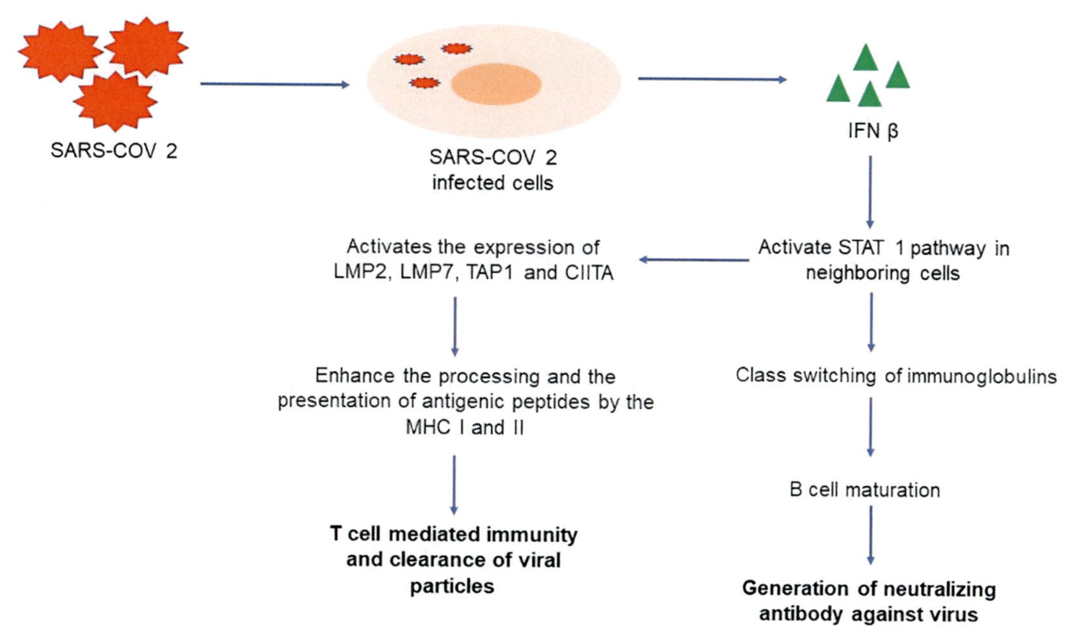

Fig. 2 Role of IFN-β in the healthy immune response against SARS-COV2.

signals for the immune cells to detect the threat and recruit T cells, monocytes, and macrophages.

The choice of the immune system depends on the balance of cytokines at the site of infection. IFNβ produced by infected cells plays an inevitable role [47] (Fig. 2). The surrounding cells bind IFNβ protein via IFNα/β receptors (INFAR1) and activate the STAT1 signaling pathway that elicits an incredible antiviral immunity [48]. Activated STAT binds ISRE (Interferon Stimulated Response Element) in the genome to promote a battery of genes, such as proteasome subunits LMP2 [49], LMP7, TAP1 [50], and CIITA (class II transactivator) [51]. These are necessary to process and present antigenic peptides of the major histocompatibility complexes (MHC) I and II. In addition, controlled expression of immunoglobulins facilitates class switching for B cell maturation [52], promotes BCR stimulation [53], and induces expression of protein T-bet, which thereby activates class switching of immunoglobulins [54].

Generally, healthy immune system function correlates with progress toward STAT1 signaling that activates the virus-specific NK cells and antibody-producing B cells. These then phagocytose or neutralize the virus which is immediately cleared by macrophages, thereby reducing pro-inflammatory cytokines and hamper disease progress. However, the failure of such mechanisms allows immune evasion by the virus, cytokine storm, and severe symptoms sometimes leading to multiorgan damage [55].

Cytokine storm

Uncontrolled release of pro-inflammatory cytokine constitutes the cytokine storm, which manifests in COVID-19 patients who suffer severe complications and often die. Considerably higher levels of pro-inflammatory IL-2, IL-7, IL-10, G-CSF, TNF, CXCL10, MCP1, and MIP1α cytokines are identified in the serum of hospitalized COVID-19 patients [56]. Moreover, a significantly higher concentration of IL-6 is observed in nonsurvivors compared to surviving patients. This is due to cytokine bursts that are the major cause of mortality associated with SARS-CoV2 infection [57].

The role of IL-6 in cytokine storm and associated complications is more important than other pro-inflammatory mediators as it is considered a major cause of mortality (Fig. 3) [58]. Entry of SARS-CoV2 via ACE2R dysregulates the balance of the RAAS system and rapidly increases serum angiotensin 2 level sufficient to initiate pro-inflammatory cascades through angiotensin 2 type 1 receptors (AT1R) [59]. In addition, Ang 2-AT1R signaling activates ADAM metallopeptidase domain 17 (ADAM17) that digests several membrane proteins such as EGF, epiregulin, amphiregulin, transforming growth factor-alpha (TNF-α) [60], causing hyperactivation of NF-κB pathway to release more pro-inflammatory cytokines including IL-6. Apart from the NF-κB pathway, coactivation of STAT3 signaling [61] adds to the cytokine load. Simultaneously, ADAM17 peptidases convert membrane-bound IL-6Rα to the soluble form (SIL-6Rα), which is capable of initiating intracellular signaling in neighboring nonimmune cells, including endothelial cells, epithelial cells, and fibroblasts, which often elicits a positive feedback loop. This cascade involving Ang 2-AT1R signaling and NF-κB/STAT3 pathway is an IL-6 amplifier [58] that has been identified in nonsurviving patients. This cytokine storm increases tissue inflammation, irreversible lung injury, blood coagulation, microvascular damage, and multiorgan failure causing death in many COVID-19 patients.

Unfortunately, the exact molecular mechanisms underlying COVID-19 infection are largely unknown and warrant further research. In addition, comprehensive definition and distinguishing strategy of cytokine storm phenomena such as physiological inflammatory responses are currently incomplete. However, elevated levels of circulatory cytokines, severe and acute systemic inflammation, and/or cytokine-mediated organ damage are key features of cytokine storm caused by COVID-19 [62] (Ref. [63], p. 19). Hence, cytokine storm represents an immune system hyperactivation inducing disparate collateral pathology and the viral burden often leads to secondary organ damage. Importantly, elucidation of cytokine storm phenomena is a significant avenue of research for management of COVID-19 pathology.

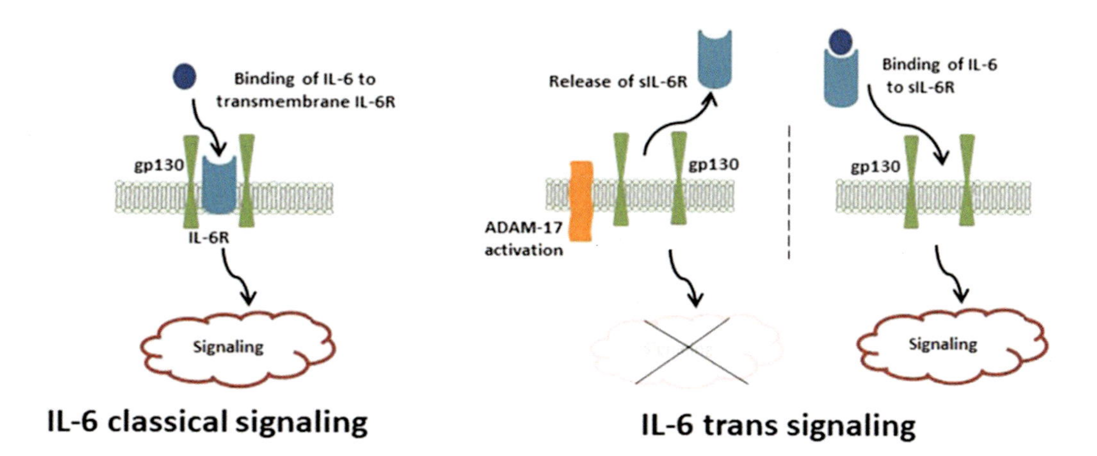

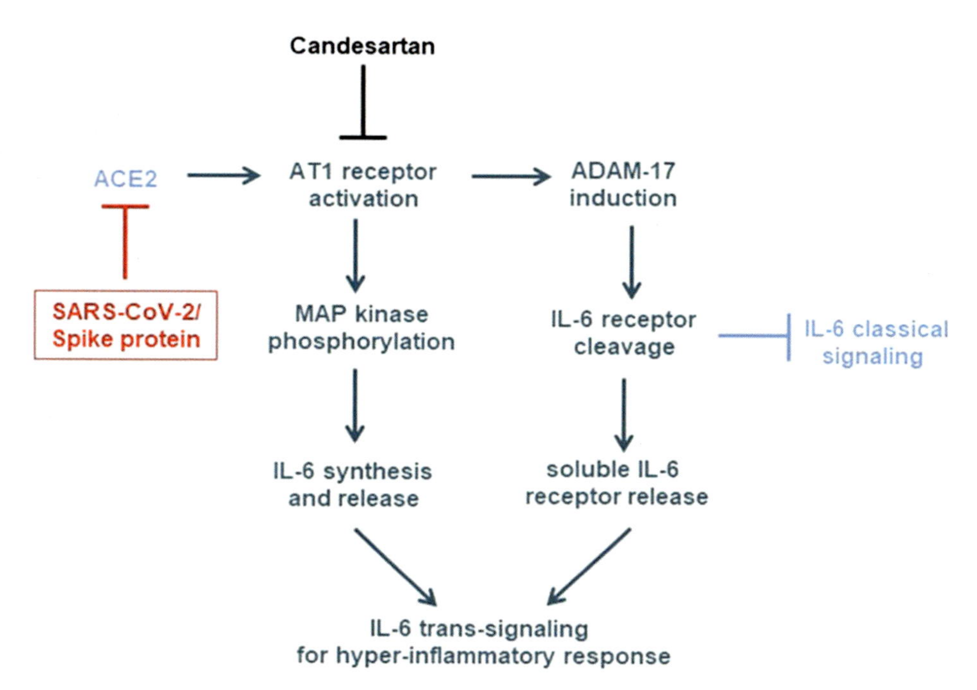

Fig. 3 IL-6 trans-signaling and amplifier cascade. The image was taken from the article by Patra, T., Meyer, K., Geerling, L., Isbell, T.S., Hoft, D.F., Brien, J., Pinto, A.K., Ray, R.B., Ray, R., 2020. SARS-CoV-2 spike protein promotes IL-6 trans-signaling by activation of angiotensin II receptor signaling in epithelial cells. PLoS Pathog 16, e1009128. doi:10.1371/journal.ppat.1009128.

Current treatment approaches

Treatment approaches for COVID-19 that target the cytokine storm have not yet been successful. Presently, development of vaccines has substantially prevented COVID-19 spread, despite the emergence of multiple virulent mutant resistant strains. Moreover, reports of post-vaccination infection warrant further progress in vaccine therapy. Current management focuses on general supportive and critical care, and relief of symptoms and complications owing to the lack of standard therapeutic approaches. Moreover, clinical data on COVID-19 infection are surprisingly scarce in a global scenario [64]. Oxygen support has been recommended for patients with hypoxemia with mechanical ventilation in severe cases. Antibiotic treatments have been administered to prevent secondary infections and glucocorticoids have been prescribed for severe clinical symptoms [65]. Interestingly, several protease inhibitors used to prevent viral replication in HIV patients have been found to be effective for COVID-19 infection. Also, several drugs (individually and in combination) including darunavir, atazanavir-cobicistat, lopinavir, ritonavir, favipiravir, and hydroxychloroquine are reportedly superior in the management of symptoms [66]. Furthermore, experimental and clinical evidence revealed improvements in symptoms following treatment with immunomodulators. For instance, monoclonal antibodies against key cytokines including tocilizumab (antibody directed against the Il-6 receptor), anakinra (inhibitors IL-1α and IL-1β), baricitinib (inhibitor of Janus kinases), and chloroquine and hydroxychloroquine (antiviral drugs) revealed appreciably better outcomes in management of cytokine storm (Refs. [66,67], p. 19). Similarly, anticoagulant therapy using heparin has significantly increased survival rates and life expectancy of COVID-19 patients as reported by a seminal retrospective study conducted in Wuhan [68]. Importantly, convalescent plasma containing specific antibodies against COVID-19 from survivors has benefited some critically ill patients. Increased neutralization by IgG antibodies and decreased T cell-mediated immunity improve humoral response and can prevent cytokine storm [65].

Mesenchymal stem cell therapy (MSC) for management of cytokine storm and other COVID-19 complications is emerging owing to their exceptional immunomodulatory and repair potential. In addition, the ability of MSCs to regulate hyperactivity of immune cells and secretion of abundant growth factors and regenerative mediators may explain their therapeutic potential for COVID-19 pathology [69]. Interestingly, administration of MSCs in critically ill COVID-19 patients correlates with appreciable clinical improvements resulting in increased demand for stem cell-based strategies evident in the surge of the reports to the stem cell clinical trial registry. Challenges including possible cellular sources, dosage, in vitro manipulation, diversity of target patients, and clinical monitoring warrant further attention [70]. Nonetheless, MSCs form a quite promising management strategy for aggravated COVID-19 pathology, and chapters in this book focus on diverse perspectives of stem cell-based applications in the management of COVID-19.

Translational avenues and perspectives

The intimidating explosion of the COVID-19 pandemic reverberates across the globe with drastic morbidity and mortality. Unfortunately, proper understanding regarding the structure, mutation rate, regulation of replication, and molecular pathology underlying COVID-19 infection and exact mechanisms of morbidity and mortality remain obscure. The majority of patients presenting with aggravated pathology have comorbidities, including diabetes, cancer, asthma, CVDs, and other inflammatory diseases. Thus many of them surrendered to death due to adverse events associated with these comorbidities. Investigations of molecular events associated with aggravated COVID-19 pathology secondary to comorbidities warrant customized translational strategies for patient care. In addition, in vitro, in vivo, in silico, and clinical assessments are warranted on multiple aspects of cytokine storm, including possible sterile inflammatory milieu to open novel translational avenues for the management of COVID-19 pathology and spread. Moreover, deciphering potential therapeutic targets to address the severity of symptoms in COVID-19 infection and comorbidity is urgent. Unfortunately, lack of proper animal models and unavailability of postmortem/biopsy samples are barriers to translational research. Therefore further investigations are needed. As COVID-19 continues to *'mask'* humanity, global attention is on translational medicine in hopes for effective preventive and managerial strategies to control this pandemic. Importantly, thorough understanding of COVID-19 immunobiology is the key to successful management opportunities. Hence, a multidisciplinary approach that bridges basic and clinical knowledge is an immediate priority to tame feral COVID-19 pandemic and revive global healthcare system.

Acknowledgments

This work was supported by the research funds of Western University of Health Sciences to FGT, and research grants R01 HL144125 and R01HL147662 to DKA from the National Institutes of Health, USA. The contents of this chapter are solely the responsibility of the authors and do not necessarily represent the official views of the National Institutes of Health. RK acknowledges the United States-India Educational Foundation (USIEF) for Fulbright-Nehru postdoctoral fellowship.

Conflict of Interests

All authors have read the Elsevier's policy on disclosure of potential conflicts of interest. Author C (DKA) has received grants from the National Institutes of Health. Author B (FGT) received start-up funds from Western University of Health Sciences. Author D (DRW) has no relevant affiliations or financial or nonfinancial involvement with any organization or entity with financial or nonfinancial interest or conflict with the subject matter or materials discussed in the manuscript apart from those disclosed. Author A (RK), Author B (FGT), Author C (DKA, and Author D (DRW), declare that they have no conflict of interest. No writing assistance was utilized in the production of this manuscript.

References

[1] Riedel RN, Pérez-Pérez A, Sánchez-Margalet V, Varone CL, Maymó JL. Stem cells and COVID-19: are the human amniotic cells a new hope for therapies against the SARS-CoV-2 virus? Stem Cell Res Ther 2021;12:155. https://doi.org/10.1186/s13287-021-02216-w.

[2] COVID Live Update. 181,758,960 cases and 3,936,714 deaths from the coronavirus—worldometer [WWW document], https://www.worldometers.info/coronavirus/#countries; 2021. [Accessed 27 June 2021].

[3] Zhao Y, Nie H-X, Hu K, Wu X-J, Zhang Y-T, Wang M-M, Wang T, Zheng Z-S, Li X-C, Zeng S-L. Abnormal immunity of non-survivors with COVID-19: predictors for mortality. Infect Dis Poverty 2020;9:108. https://doi.org/10.1186/s40249-020-00723-1.

[4] McPadden J, Warner F, Young HP, Hurley NC, Pulk RA, Singh A, Durant TJ, Gong G, Desai N, Haimovich A, Taylor RA, Gunel M, Cruz CSD, Farhadian SF, Siner J, Villanueva M, Churchwell K, Hsiao A, Torre CJ, Velazquez EJ, Herbst RS, Iwasaki A, Ko AI, Mortazavi BJ, Krumholz HM, Schulz WL. Clinical characteristics and outcomes for 7,995 patients with SARS-CoV-2 infection (preprint). Infect Dis (except HIV/AIDS) 2020. https://doi.org/10.1101/2020.07.19.20157305.

[5] Liu J, Liu S. The management of coronavirus disease 2019 (COVID-19). J Med Virol 2020;92:1484–90. https://doi.org/10.1002/jmv.25965.

[6] Habas K, Nganwuchu C, Shahzad F, Gopalan R, Haque M, Rahman S, Majumder AA, Nasim T. Resolution of coronavirus disease 2019 (COVID-19). Expert Rev Anti Infect Ther 2020;18:1201–11. https://doi.org/10.1080/14787210.2020.1797487.

[7] Cui J, Li F, Shi Z-L. Origin and evolution of pathogenic coronaviruses. Nat Rev Microbiol 2019;17:181–92. https://doi.org/10.1038/s41579-018-0118-9.

[8] Coronaviridae Study Group of the International Committee on Taxonomy of Viruses. The species severe acute respiratory syndrome-related coronavirus: classifying 2019-nCoV and naming it SARS-CoV-2. Nat Microbiol 2020;5:536–44. https://doi.org/10.1038/s41564-020-0695-z.

[9] Lu R, Zhao X, Li J, Niu P, Yang B, Wu H, Wang W, Song H, Huang B, Zhu N, Bi Y, Ma X, Zhan F, Wang L, Hu T, Zhou H, Hu Z, Zhou W, Zhao L, Chen J, Meng Y, Wang J, Lin Y, Yuan J, Xie Z, Ma J, Liu WJ, Wang D, Xu W, Holmes EC, Gao GF, Wu G, Chen W, Shi W, Tan W. Genomic characterisation and epidemiology of 2019 novel coronavirus: implications for virus origins and receptor binding. Lancet 2020;395:565–74. https://doi.org/10.1016/S0140-6736(20)30251-8.

[10] Brian DA, Baric RS. Coronavirus genome structure and replication. In: Enjuanes L, editor. Coronavirus replication and reverse genetics, current topics in microbiology and immunology. Berlin, Heidelberg: Springer Berlin Heidelberg; 2005. p. 1–30. https://doi.org/10.1007/3-540-26765-4_1.

[11] Wang M-Y, Zhao R, Gao L-J, Gao X-F, Wang D-P, Cao J-M. SARS-CoV-2: structure, biology, and structure-based therapeutics development. Front Cell Infect Microbiol 2020;10. https://doi.org/10.3389/fcimb.2020.587269.

[12] Bosch BJ, van der Zee R, de Haan CAM, Rottier PJM. The coronavirus spike protein is a class I virus fusion protein: structural and functional characterization of the fusion core complex. J Virol 2003;77:8801–11. https://doi.org/10.1128/jvi.77.16.8801-8811.2003.

[13] Huang Y, Yang C, Xu X, Xu W, Liu S. Structural and functional properties of SARS-CoV-2 spike protein: potential antivirus drug development for COVID-19. Acta Pharmacol Sin 2020;41:1141–9. https://doi.org/10.1038/s41401-020-0485-4.

[14] Casalino L, Gaieb Z, Goldsmith JA, Hjorth CK, Dommer AC, Harbison AM, Fogarty CA, Barros EP, Taylor BC, McLellan JS, Fadda E, Amaro RE. Beyond shielding: the roles of Glycans in the SARS-CoV-2 spike protein. ACS Cent Sci 2020;6:1722–34. https://doi.org/10.1021/acscentsci.0c01056.

[15] Sturman LS, Holmes KV, Behnke J. Isolation of coronavirus envelope glycoproteins and interaction with the viral nucleocapsid. J Virol 1980;33:449–62. https://doi.org/10.1128/jvi.33.1.449-462.1980.

[16] Laude H, Gelfi J, Lavenant L, Charley B. Single amino acid changes in the viral glycoprotein M affect induction of alpha interferon by the coronavirus transmissible gastroenteritis virus. J Virol 1992;66:743–9. https://doi.org/10.1128/jvi.66.2.743-749.1992.

[17] Nieto-Torres JL, DeDiego ML, Verdiá-Báguena C, Jimenez-Guardeño JM, Regla-Nava JA, Fernandez-Delgado R, Castaño-Rodriguez C, Alcaraz A, Torres J, Aguilella VM, Enjuanes L. Severe acute respiratory syndrome coronavirus envelope protein Ion Channel activity promotes virus fitness and pathogenesis. PLoS Pathog 2014;10. https://doi.org/10.1371/journal.ppat.1004077, e1004077.

[18] McBride R, van Zyl M, Fielding B. The coronavirus nucleocapsid is a multifunctional protein. Viruses 2014;6:2991–3018. https://doi.org/10.3390/v6082991.

[19] Chang C, Sue S-C, Yu T, Hsieh C-M, Tsai C-K, Chiang Y-C, Lee S, Hsiao H, Wu W-J, Chang W-L, Lin C-H, Huang T. Modular organization of SARS coronavirus nucleocapsid protein. J Biomed Sci 2006;13:59–72. https://doi.org/10.1007/s11373-005-9035-9.

[20] Huang C, Lokugamage KG, Rozovics JM, Narayanan K, Semler BL, Makino S. SARS coronavirus nsp1 protein induces template-dependent endonucleolytic cleavage of mRNAs: viral mRNAs are resistant to nsp1-induced RNA cleavage. PLoS Pathog 2011;7. https://doi.org/10.1371/journal.ppat.1002433, e1002433.

[21] Cornillez-Ty CT, Liao L, Yates JR, Kuhn P, Buchmeier MJ. Severe acute respiratory syndrome coronavirus nonstructural protein 2 interacts with a host protein complex involved in mitochondrial biogenesis and intracellular signaling. J Virol 2009;83:10314–8. https://doi.org/10.1128/JVI.00842-09.

[22] Lei J, Kusov Y, Hilgenfeld R. Nsp3 of coronaviruses: structures and functions of a large multi-domain protein. Antiviral Res 2018;149:58–74. https://doi.org/10.1016/j.antiviral.2017.11.001.

[23] Sakai Y, Kawachi K, Terada Y, Omori H, Matsuura Y, Kamitani W. Two-amino acids change in the nsp4 of SARS coronavirus abolishes viral replication. Virology 2017;510:165–74. https://doi.org/10.1016/j.virol.2017.07.019.

[24] Cottam EM, Whelband MC, Wileman T. Coronavirus NSP6 restricts autophagosome expansion. Autophagy 2014;10:1426–41. https://doi.org/10.4161/auto.29309.

[25] te Velthuis AJW, van den Worm SHE, Snijder EJ. The SARS-coronavirus nsp7+nsp8 complex is a unique multimeric RNA polymerase capable of both de novo initiation and primer extension. Nucleic Acids Res 2012;40:1737–47. https://doi.org/10.1093/nar/gkr893.

[26] Shi Z, Gao H, Bai X, Yu H. Cryo-EM structure of the human cohesin-NIPBL-DNA complex. Science 2020;368:1454–9. https://doi.org/10.1126/science.abb0981.

[27] Subissi L, Posthuma CC, Collet A, Zevenhoven-Dobbe JC, Gorbalenya AE, Decroly E, Snijder EJ, Canard B, Imbert I. One severe acute respiratory syndrome coronavirus protein complex integrates processive RNA polymerase and exonuclease activities. Proc Natl Acad Sci 2014;111:E3900–9. https://doi.org/10.1073/pnas.1323705111.
[28] Jang K-J, Jeong S, Kang DY, Sp N, Yang YM, Kim D-E. A high ATP concentration enhances the cooperative translocation of the SARS coronavirus helicase nsP13 in the unwinding of duplex RNA. Sci Rep 2020;10:4481. https://doi.org/10.1038/s41598-020-61432-1.
[29] Decroly E, Debarnot C, Ferron F, Bouvet M, Coutard B, Imbert I, Gluais L, Papageorgiou N, Sharff A, Bricogne G, Ortiz-Lombardia M, Lescar J, Canard B. Crystal structure and functional analysis of the SARS-coronavirus RNA cap 2′-O-methyltransferase nsp10/nsp16 complex. PLoS Pathog 2011;7. https://doi.org/10.1371/journal.ppat.1002059, e1002059.
[30] Mason RJ. Pathogenesis of COVID-19 from a cell biology perspective. Eur Respir J 2020;55. https://doi.org/10.1183/13993003.00607-2020.
[31] Meng T, Cao H, Zhang H, Kang Z, Xu D, Gong H, Wang J, Li Z, Cui X, Xu H, Wei H, Pan X, Zhu R, Xiao J, Zhou W, Cheng L, Liu J. The insert sequence in SARS-CoV-2 enhances spike protein cleavage by TMPRSS. bioRxiv 2020. https://doi.org/10.1101/2020.02.08.926006.
[32] Shanmugam C, Mohammed AR, Ravuri S, Luthra V, Rajagopal N, Karre S. COVID-2019 – a comprehensive pathology insight. Pathol Res Pract 2020;216. https://doi.org/10.1016/j.prp.2020.153222, 153222.
[33] Ryu W-S. Virus life cycle. In: Molecular Virology of Human Pathogenic Viruses; 2017. p. 31–45. https://doi.org/10.1016/B978-0-12-800838-6.00003-5.
[34] Wu Z, McGoogan JM. Characteristics of and important lessons from the coronavirus disease 2019 (COVID-19) outbreak in China: summary of a report of 72 314 cases from the Chinese center for disease control and prevention. JAMA 2020;323:1239. https://doi.org/10.1001/jama.2020.2648.
[35] Mossel EC, Wang J, Jeffers S, Edeen KE, Wang S, Cosgrove GP, Funk CJ, Manzer R, Miura TA, Pearson LD, Holmes KV, Mason RJ. SARS-CoV replicates in primary human alveolar type II cell cultures but not in type I-like cells. Virology 2008;372:127–35. https://doi.org/10.1016/j.virol.2007.09.045.
[36] Qian Z, Travanty EA, Oko L, Edeen K, Berglund A, Wang J, Ito Y, Holmes KV, Mason RJ. Innate immune response of human alveolar type II cells infected with severe acute respiratory syndrome–coronavirus. Am J Respir Cell Mol Biol 2013;48:742–8. https://doi.org/10.1165/rcmb.2012-0339OC.
[37] Xu Z, Shi L, Wang Y, Zhang J, Huang L, Zhang C, Liu S, Zhao P, Liu H, Zhu L, Tai Y, Bai C, Gao T, Song J, Xia P, Dong J, Zhao J, Wang F-S. Pathological findings of COVID-19 associated with acute respiratory distress syndrome. Lancet Respir Med 2020;8:420–2. https://doi.org/10.1016/S2213-2600(20)30076-X.
[38] Ho JC, Chan KN, Hu WH, Lam WK, Zheng L, Tipoe GL, Sun J, Leung R, Tsang KW. The effect of aging on nasal Mucociliary clearance, beat frequency, and ultrastructure of respiratory cilia. Am J Respir Crit Care Med 2001;163:983–8. https://doi.org/10.1164/ajrccm.163.4.9909121.
[39] Bernabeu-Wittel M, Ternero-Vega JE, Díaz-Jiménez P, Conde-Guzmán C, Nieto-Martín MD, Moreno-Gaviño L, Delgado-Cuesta J, Rincón-Gómez M, Giménez-Miranda L, Navarro-Amuedo MD, Muñoz-García MM, Calzón-Fernández S, Ollero-Baturone M. Death risk stratification in elderly patients with covid-19. A comparative cohort study in nursing homes outbreaks. Arch Gerontol Geriatr 2020;91. https://doi.org/10.1016/j.archger.2020.104240, 104240.
[40] Catanzaro M, Fagiani F, Racchi M, Corsini E, Govoni S, Lanni C. Immune response in COVID-19: addressing a pharmacological challenge by targeting pathways triggered by SARS-CoV-2. Signal Transduct Target Ther 2020;5:1–10. https://doi.org/10.1038/s41392-020-0191-1.

[41] Qin C, Zhou L, Hu Z, Zhang S, Yang S, Tao Y, Xie C, Ma K, Shang K, Wang W, Tian D-S. Dysregulation of immune response in patients with coronavirus 2019 (COVID-19) in Wuhan, China. Clin Infect Dis 2020;71:762–8. https://doi.org/10.1093/cid/ciaa248.
[42] Tan L, Wang Q, Zhang D, Ding J, Huang Q, Tang Y-Q, Wang Q, Miao H. Lymphopenia predicts disease severity of COVID-19: a descriptive and predictive study. Signal Transduct Target Ther 2020;5:33. https://doi.org/10.1038/s41392-020-0148-4.
[43] Zhu N, Zhang D, Wang W, Li X, Yang B, Song J, Zhao X, Huang B, Shi W, Lu R, Niu P, Zhan F, Ma X, Wang D, Xu W, Wu G, Gao GF, Tan W. A novel coronavirus from patients with pneumonia in China, 2019. N Engl J Med 2020;382:727–33. https://doi.org/10.1056/NEJMoa2001017.
[44] Long Q-X, Liu B-Z, Deng H-J, Wu G-C, Deng K, Chen Y-K, Liao P, Qiu J-F, Lin Y, Cai X-F, Wang D-Q, Hu Y, Ren J-H, Tang N, Xu Y-Y, Yu L-H, Mo Z, Gong F, Zhang X-L, Tian W-G, Hu L, Zhang X-X, Xiang J-L, Du H-X, Liu H-W, Lang C-H, Luo X-H, Wu S-B, Cui X-P, Zhou Z, Zhu M-M, Wang J, Xue C-J, Li X-F, Wang L, Li Z-J, Wang K, Niu C-C, Yang Q-J, Tang X-J, Zhang Y, Liu X-M, Li J-J, Zhang D-C, Zhang F, Liu P, Yuan J, Li Q, Hu J-L, Chen J, Huang A-L. Antibody responses to SARS-CoV-2 in patients with COVID-19. Nat Med 2020;26:845–8. https://doi.org/10.1038/s41591-020-0897-1.
[45] de Marcken M, Dhaliwal K, Danielsen AC, Gautron AS, Dominguez-Villar M. TLR7 and TLR8 activate distinct pathways in monocytes during RNA virus infection. Sci Signal 2019;12. https://doi.org/10.1126/scisignal.aaw1347, eaaw1347.
[46] Thankam FG, Agrawal DK. Molecular chronicles of cytokine burst in patients with coronavirus disease 2019 (COVID-19) with cardiovascular diseases. J Thorac Cardiovasc Surg 2021;161:e217–26. https://doi.org/10.1016/j.jtcvs.2020.05.083.
[47] Molaei S, Dadkhah M, Asghariazar V, Karami C, Safarzadeh E. The immune response and immune evasion characteristics in SARS-CoV, MERS-CoV, and SARS-CoV-2: vaccine design strategies. Int Immunopharmacol 2021;92. https://doi.org/10.1016/j.intimp.2020.107051, 107051.
[48] Najjar I, Fagard R. STAT1 and pathogens, not a friendly relationship. Biochimie 2010;92:425–44. https://doi.org/10.1016/j.biochi.2010.02.009.
[49] Chatterjee-Kishore M, Wright KL, Ting JP-Y, Stark GR. How Stat1 mediates constitutive gene expression: a complex of unphosphorylated Stat1 and IRF1 supports transcription of the LMP2 gene. EMBO J 2000;19:4111–22. https://doi.org/10.1093/emboj/19.15.4111.
[50] Chatterjee-Kishore M, Kishore R, Hicklin DJ, Marincola FM, Ferrone S. Different requirements for signal transducer and activator of transcription 1alpha and interferon regulatory factor 1 in the regulation of low molecular mass polypeptide 2 and transporter associated with antigen processing 1 gene expression. J Biol Chem 1998;273:16177–83. https://doi.org/10.1074/jbc.273.26.16177.
[51] Min W, Pober JS, Johnson DR. Kinetically coordinated induction of TAP1 and HLA class I by IFN-gamma: the rapid induction of TAP1 by IFN-gamma is mediated by Stat1 alpha. J Immunol (Baltim Md: 1950) 1996;156:3174–83.
[52] Yoshimoto T, Okada K, Morishima N, Kamiya S, Owaki T, Asakawa M, Iwakura Y, Fukai F, Mizuguchi J. Induction of IgG2a class switching in B cells by IL-27. J Immunol (Baltim Md) 2004;1950(173):2479–85. https://doi.org/10.4049/jimmunol.173.4.2479.
[53] Xu W, Nair JS, Malhotra A, Zhang JJ. B cell antigen receptor signaling enhances IFN-gamma-induced Stat1 target gene expression through calcium mobilization and activation of multiple serine kinase pathways. J Interferon Cytokine Res Off J Int Soc Interferon Cytokine Res 2005;25:113–24. https://doi.org/10.1089/jir.2005.25.113.
[54] Xu W, Zhang JJ. Stat1-dependent synergistic activation of T-bet for IgG2a production during early stage of B cell activation. J Immunol (Baltim Md) 2005;1950(175):7419–24. https://doi.org/10.4049/jimmunol.175.11.7419.

[55] Tay MZ, Poh CM, Rénia L, MacAry PA, Ng LFP. The trinity of COVID-19: immunity, inflammation and intervention. Nat Rev Immunol 2020;20:363–74. https://doi.org/10.1038/s41577-020-0311-8.
[56] Huang C, Wang Y, Li X, Ren L, Zhao J, Hu Y, Zhang L, Fan G, Xu J, Gu X, Cheng Z, Yu T, Xia J, Wei Y, Wu W, Xie X, Yin W, Li H, Liu M, Xiao Y, Gao H, Guo L, Xie J, Wang G, Jiang R, Gao Z, Jin Q, Wang J, Cao B. Clinical features of patients infected with 2019 novel coronavirus in Wuhan, China. Lancet (Lond Engl) 2020;395:497–506. https://doi.org/10.1016/S0140-6736(20)30183-5.
[57] Zhang Y, Li J, Zhan Y, Wu L, Yu X, Zhang W, Ye L, Xu S, Sun R, Wang Y, Lou J. Analysis of serum cytokines in patients with severe acute respiratory syndrome. Infect Immun 2004;72:4410–5. https://doi.org/10.1128/IAI.72.8.4410-4415.2004.
[58] Hojyo S, Uchida M, Tanaka K, Hasebe R, Tanaka Y, Murakami M, Hirano T. How COVID-19 induces cytokine storm with high mortality. Inflamm Regen 2020;40:37. https://doi.org/10.1186/s41232-020-00146-3.
[59] Eguchi S, Kawai T, Scalia R, Rizzo V. Understanding angiotensin II type 1 receptor signaling in vascular pathophysiology. Hypertension (Dallas, Tex: 1979) 2018;71:804–10. https://doi.org/10.1161/HYPERTENSIONAHA.118.10266.
[60] Scheller J, Chalaris A, Garbers C, Rose-John S. ADAM17: a molecular switch to control inflammation and tissue regeneration. Trends Immunol 2011;32:380–7. https://doi.org/10.1016/j.it.2011.05.005.
[61] Murakami M, Kamimura D, Hirano T. Pleiotropy and specificity: insights from the interleukin 6 family of cytokines. Immunity 2019;50:812–31. https://doi.org/10.1016/j.immuni.2019.03.027.
[62] Fajgenbaum DC, June CH. Cytokine Storm. N Engl J Med 2020;383:2255–73. https://doi.org/10.1056/NEJMra2026131.
[63] Henderson LA, Canna SW, Schulert GS, Volpi S, Lee PY, Kernan KF, Caricchio R, Mahmud S, Hazen MM, Halyabar O, Hoyt KJ, Han J, Grom AA, Gattorno M, Ravelli A, De Benedetti F, Behrens EM, Cron RQ, Nigrovic PA. On the alert for cytokine storm: immunopathology in COVID-19. Arthritis Rheumatol (Hoboken, NJ) 2020;72:1059–63. https://doi.org/10.1002/art.41285.
[64] Rajendran K, Krishnasamy N, Rangarajan J, Rathinam J, Natarajan M, Ramachandran A. Convalescent plasma transfusion for the treatment of COVID-19: systematic review. J Med Virol 2020;92:1475–83. https://doi.org/10.1002/jmv.25961.
[65] Li Y, Liu S, Zhang S, Ju Q, Zhang S, Yang Y, Wang H. Current treatment approaches for COVID-19 and the clinical value of transfusion-related technologies. Transfus Apher Sci Off J World Apher Assoc Off J Eur Soc Haemapheresis 2020;59. https://doi.org/10.1016/j.transci.2020.102839, 102839.
[66] Stasi C, Fallani S, Voller F, Silvestri C. Treatment for COVID-19: an overview. Eur J Pharmacol 2020;889. https://doi.org/10.1016/j.ejphar.2020.173644, 173644.
[67] Hu B, Huang S, Yin L. The cytokine storm and COVID-19. J Med Virol 2021;93:250–6. https://doi.org/10.1002/jmv.26232.
[68] Tang N, Bai H, Chen X, Gong J, Li D, Sun Z. Anticoagulant treatment is associated with decreased mortality in severe coronavirus disease 2019 patients with coagulopathy. J Thromb Haemost 2020;18:1094–9. https://doi.org/10.1111/jth.14817.
[69] Li Z, Niu S, Guo B, Gao T, Wang L, Wang Y, Wang L, Tan Y, Wu J, Hao J. Stem cell therapy for COVID-19, ARDS and pulmonary fibrosis. Cell Prolif 2020;53. https://doi.org/10.1111/cpr.12939, e12939.
[70] Choudhery MS, Harris DT. Stem cell therapy for COVID-19: possibilities and challenges. Cell Biol Int 2020;44:2182–91. https://doi.org/10.1002/cbin.11440.

3

An insight into the molecular mechanisms of mesenchymal stem cells and their translational approaches to combat COVID-19

Yashvi Sharma, E. Pranshu Rao, and Sujata Mohanty
Stem Cell Facility (DBT-Centre of Excellence for Stem Cell Research), All India Institute of Medical Sciences, New Delhi, India

Abbreviations

ACE-2	angiotensin convertase enzyme 2
ALI	acute lung injury
ARDS	acute respiratory distress syndrome
COPD	chronic obstructive pulmonary disorder
COVID-19	coronavirus disease 2019
EV	extracellular vesicles
MSC	mesenchymal stem cells
nCoV	novel coronavirus
RAS	renin-angiotensin system
SARS-CoV-2	severe acute respiratory syndrome coronavirus 2

Introduction and background

SARS-CoV-2 is a novel strain in a family of viruses known as *Coronaviridae*, recognized to cause the deadly pandemic disease COVID-19 worldwide. The cases of this virus were first identified in December 2019, in the city of Wuhan, China, and were further spread all over the world, resulting in a mass public health emergency [1]. It was initially considered a pneumonia virus majorly causing damage to lungs, but eventually it was observed that it caused a broad spectrum of complications resulting in whole body disabilities [2]. The incidence of this disease declined around November 2020; however, another wave of COVID-19 hit the world in early 2021, which was likely caused by a more infectious mutant strain of SARS-CoV-2 thereby causing worsened devastation of public health and healthcare systems [3]. This strain of virus is extremely contagious and is known to spread by direct

Stem Cells and COVID-19. https://doi.org/10.1016/B978-0-323-89972-7.00003-9

contact and via aerosols. The incubation period reported for this virus is 2–14 days and the common symptoms experienced during this infection included fever, dry cough, cold, headache, loss of smell and taste, drop in the oxygen saturation-pO_2, dyspnea, diarrhea, and many more [4]. Many of the affected were able to survive this disease without the requirement of hospitalization but in few cases extreme complications were caused by ARDS and multiple organ failure resulting in death. This virus has high transmission and infection rates with 216 million people affected and 4.5 million deaths as of August 2021 [5]. This RNA virus enters via the respiratory tract and works by causing uncontrolled release of cytokines and excessive inflammation of the lungs. This cytokine squall further spreads to the rest of the body attacking other organs, resulting in a multiple organ collapse situation [6]. Currently the treatment options are majorly symptomatic, and some approaches being considered for the severely affected individuals include administration of drugs like dexamethasone, remdesivir, baricitinib, tocilizumab, favipiravir, monoclonal antibody treatment, and convalescent plasma therapy [7–10]. Apart from that, several vaccines have come up including Oxford-AstraZeneca vaccine sold under the brand name Vaxzevria and Covishield, Pfizer-BioNTech vaccine also known as Comirnaty, Sputnik V, Covaxin, Johnson & Johnson COVID-19 vaccine, Moderna, etc. [11–13]. While the search for the perfect cure is still ongoing, some hope in the war against this virus has come up via stem cell therapy as it has shown astounding outcomes in preclinical and clinical trials by reduction of inflammation and uncontrolled immune response of the body. Mesenchymal stem cells, out of all other types of stem cells, have been extensively considered as an alternative treatment option for COVID-19 apart from the standard therapies [14]. Mesenchymal stem cells are adult stem cells which can easily be derived from various sources like bone marrow, adipose tissue, umbilical cord, etc. and possess less ethical concerns in their utility as compared to the embryonic stem cells, which is a primary reason for their enhanced popularity in regenerative medicine. Mesenchymal stem cells have an evidenced immunomodulatory and curative potential which is demonstrated via various different clinical trials utilizing them [15]. They operate in part through a paracrine mechanism via the release of extracellular vesicles containing functional factors like miRNAs, mRNAs, proteins, cytokines, enzymes, etc. [16]. They have even been found to rescue cells via the transport of mitochondria packaged inside extracellular vesicles called exosomes. Mesenchymal stem cells are thereby diverse in their mechanism of action and mode of repair. In this chapter, we will briefly describe the pathogenesis of SARS-CoV-2 and the development of COVID-19, followed by the conventional and translational approaches being considered as therapeutics. We will also specifically highlight the molecular mechanisms and role of mesenchymal stem cells in targeting the pathophysiology of SARS-CoV-2 Infection.

Pathophysiology of SARS-CoV-2

SARS-CoV-2 is a large enveloped RNA virus. Its genome is sized around 29.9 kb and comprises of a single-stranded RNA in a lipid bilayer [17]. The 4 proteins contributing to its structure include Nucleocapsid protein (N), Membrane protein (M), Envelope protein (E), and Spike protein (S) along with some other accessory proteins [18] (Fig. 1.). The nucleocapsid protein is comparatively conserved and is bound to the genome of this virus. It is involved in the stability of the viral genome along with operations like viral replication and host response to infection, for example, by eliciting the proliferation of T-cells specific to the virus [19]. Membrane protein is the structural protein which gives shape to the virus and is present in high quantities. It forms a scaffold-like structure that helps in the assembly of virus [20]. Envelope protein is the smallest protein in the viral structure. It is transmembrane in organization and is involved in the viral release and assembly [21]. Spike protein is the key protein in the functioning of this virus as it serves as the primary point of contact in the infection. These are crown shaped protrusions arising from the membrane of the virus which have a role in identification and recognition of the host receptors and fusion with target cells [22]. It enhances the affinity of infected cells to attract and adhere to noninfected cells in order to

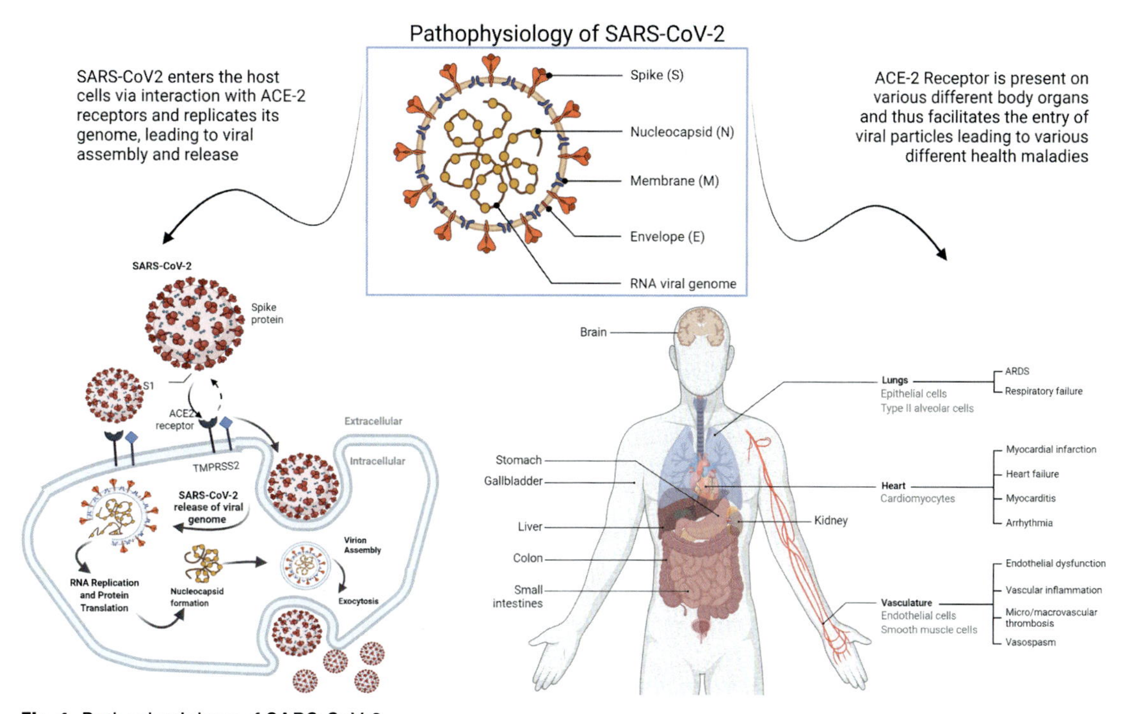

Fig. 1 Pathophysiology of SARS-CoV-2.

facilitate the spread of infection [22]. This spike protein interacts with a homolog of the Angiotensin convertase enzyme, specifically ACE-2 receptors, to expedite the entry of nCoV into the cells [22]. These receptors are present majorly on the type II alveolar epithelial cells and in some minute quantity on the epithelial cells in the oral and nasal mucosa and nasopharynx, inferring that this virus focally targets the lungs [23]. ACE2 is highly expressed on various other cells and tissues as well including cardiac muscle cells, neurons, proximal convoluted tubules in kidney, epithelium of the lower urinary tract, intestinal absorptive cells, etc. [24]. After interaction of the ACE-2 receptors with the spike protein, the transmembrane protease serine protease-2 (TMPRSS-2) and ADAM metallopeptidase domain 17 (ADAM17) in the host cell prime the spike proteins and lead to the membrane fusion of the virus and the host [25]. This results in entry to the cells where the RNA is released and viral replication and transmission begins. This is accompanied by the hyperactivation of the Renin-Angiotensin system (RAS) which is known to be involved in the progression of various other clinical implications like cardiac failure, high blood pressure, pulmonary hypertension, pneumonia, fibrotic scarring, and sepsis [25]. Apart from this, another major mechanism that this RNA virus relies on to cause destruction is the building up of excessive inflammation and a heightened immune response via the release of inflammation promoting cytokines, chemokines, and mass recruitment of inflammatory cells, also complemented with the downregulation of retaliation by the interferons [26]. Such mechanisms are lethal to the body and can ultimately result in septic shock which endangers the life of the affected [27].

Conventional and contemporary translational approaches to combat COVID-19

The treatment for COVID-19 has basically been symptomatic and supportive. For asymptomatic and mild cases, home isolation along with generic medication is suggested for the maintenance of the vitals. In the absence of a definitive cure and a directed approach, there are several treatment options that have been explored and employed as standard treatment protocols (Fig. 2).

Drugs

There are various different antiviral and antiinflammatory drugs which have been used symptomatically to gain relief against COVID-19 infection. Certain drugs like Ribavirin, Remdesivir, and Favipiravir which sought effective during the previous SARS sequela were allowed

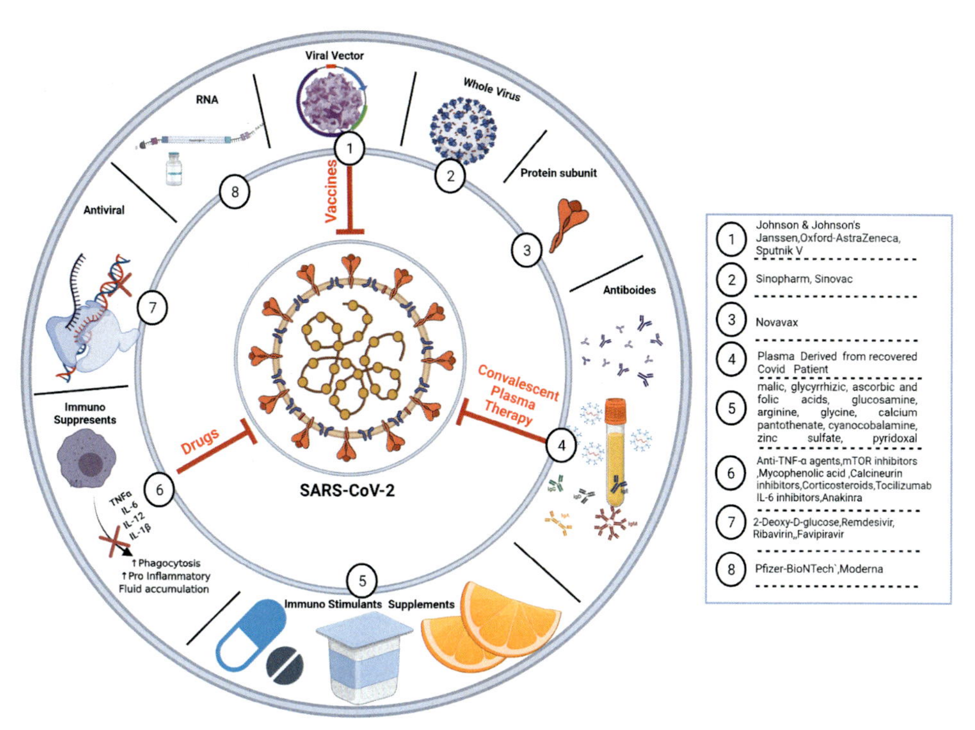

Fig. 2 Current treatment modalities employed for the treatment of COVID-19.

to be used against this virus. They basically function on a molecular level via the inhibition of the RNA polymerase of this virus; however, it has many side effects ranging from mild ailments like rashes and diarrhea, to seriously damaging effects on kidneys, muscles, and brain [28]. Some antiviral drugs including Ritonavir and Lopinavir basically work by inhibition of the viral protease and are commonly used combination against HIV infection [29]. These were also approved for use against COVID-19 and proved effective; however, they were also extremely toxic to the body in higher doses and were recommended for urgent and over cautious use only. Chloroquine and hydroxychloroquine are drugs which are commonly and widely used against malarial infection and also confer an immunomodulatory effect in their translational mechanism [29]. These drugs also exert a certain antiviral effect by inhibiting the viral entry into the cells and were thereby found to help in improvement of SARS-CoV-2 infection by impairing the ACE-2 receptor [30]. This drug is also affordable and effective with minimal side effects. In order to target the hyperinflammatory conditions created in vivo by this RNA virus, the use of steroids has been suggested in case of patients requiring mechanical ventilation [31]. The National Institute of Health and Care Excellence along with WHO has recommended the use of Dexamethasone and Hydrocortisone in order to manage COVID-19 infection; however, its side effects like hyperglycemia need to be majorly addressed.

Convalescent plasma therapy

Administration of convalescent plasma or pooled plasma containing the immunoglobulins from recovered patients has been another translational approach to combat the SARS-CoV-2 infection (Fig. 3 and Supplementary Table 1 in the online version at https://doi.org/10.1016/B978-0-323-89972-7.00003-9). It works on the principle of passive immunization and has been used in the past as a treatment option for various other viral outbreak diseases, e.g., Measles, Severe influenza, MERS [10,32]. This therapy has shown results in terms of faster recovery and better survival rates [33,34]. Even in case of SARS-CoV-2 infection, convalescent plasma therapy was found to be safe, tolerated, and effective in terms of neutralizing the viral load but only when high titer plasma is administered [35–37]. It does not have standalone significance in terms of recovery but can be used in an additional and supportive manner to the conventional treatment protocol for patients who aren't showing marked improvement despite use of steroids.

Vaccines

Vaccines are concordantly being considered as the most effective translation approach to prevent and control the SARS-CoV-2 infection. Ever since late 2020, many vaccines had entered the preclinical and clinical stages. The spike protein of SARS-CoV-2 has been in primary focus for the development of vaccines [38–40]. There are many different types of vaccines being tried; however, the leading ones that have surfaced worldwide are the Pfizer-BioNTech BNT162b2 and Oxford-AstraZeneca ChAdOx1-S vaccines [41]. Pfizer-BioNTech BNT162b2, popularly known as Comirnaty, is an mRNA vaccine comprising an optimized form of the mRNA of Spike protein of SARS-CoV-2, while Oxford-AstraZeneca ChAdOx1-S sold under the brand names Vaxzevria and Covishield is a viral vector vaccine wherein the vector is a chimpanzee adenovirus [42–44]. Clinical trials for both the vaccines have shown them to improve the symptoms and severity of the disease [45–47]. With a very few off-the-shelf vaccines and clinical trials still underway, there is yet an ever increasing and unmet need for vaccines which can be produced, stored, and allocated in a timely manner, and further can successfully cater to the diversified sets of population in terms of socioeconomic strata.

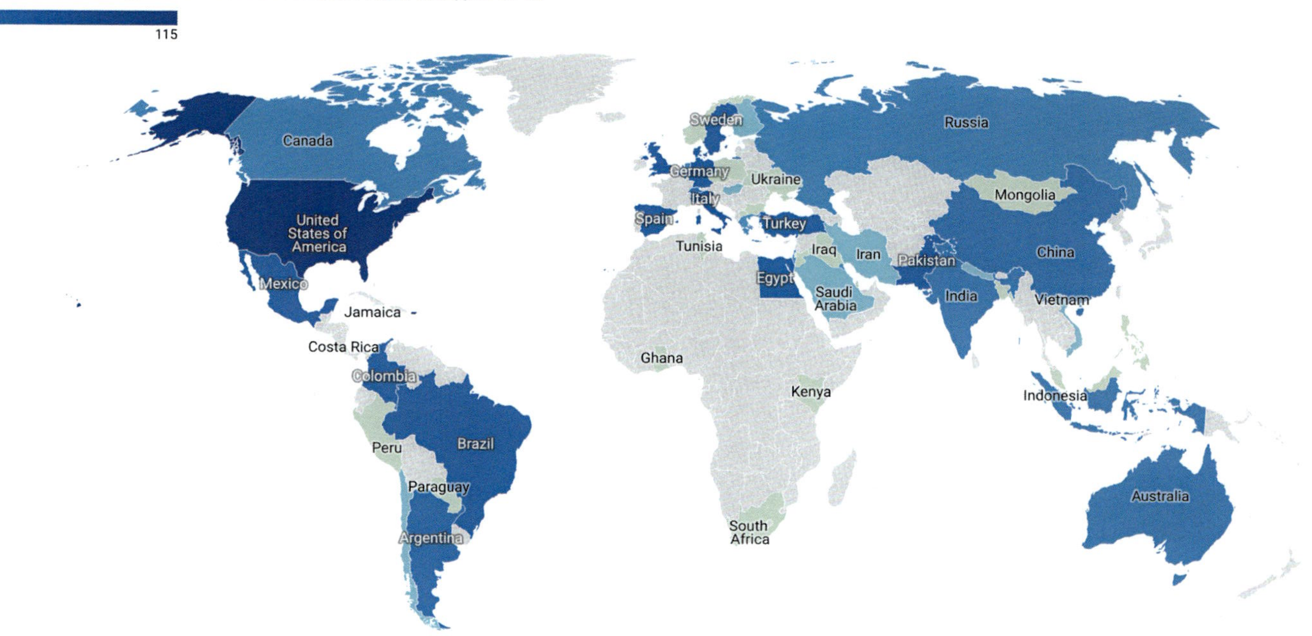

Fig. 3 A global status of Convalescent Plasma Therapy being employed for COVID-19 Management as per clinicaltrials.gov (Supplementary Table 1 in the online version at https://doi.org/10.1016/B978-0-323-89972-7.00003-9).

With the rise and fall of conventional and allopathic drug-based therapies, there is still no convincing therapeutic approaches that could holistically target the virus and its symptomatology. Here in the current pandemic situation, the search for therapies to be used in Emergency Use Authorization (EUA), have led to the rise of trials utilizing stem cell therapy against SARS-CoV-2. Stem cell therapy has emerged as a novel translation approach majorly due to their multimodal approaches toward combating the symptomatology of this virus.

Stem cell therapy as a novel translation approach against the COVID-19 pandemic

Stem cells are unspecialized cells which can self-renew to form identical stem cells or can differentiate into specialized cell types in the presence of appropriate cues. They function as an internal repair system in vivo, and thereby form the core of regenerative medicine [48]. Stem cells can be embryonic or adult in their source. While embryonic-derived stem cells possess many ethical concerns in their use, adult stem cells are relatively easier to obtain. A specific type of stem cells, called the mesenchymal stem cells, is predominantly being considered for therapeutics due to their plasticity, ability to differentiate into different kinds of specialized cells, and their unmatched immunomodulatory potential. The prospects of mesenchymal stem cells as a therapeutic aid have been explored in various different diseases at clinical and preclinical stages, including cardiovascular diseases, neurodegenerative diseases, hepatic diseases, renal diseases, cancer, and the outcomes have been astounding [49–55]. They are found to be evidently beneficial in various kinds of lung disorders as well. Due to their extensive curative potential in different diseases and disorders, mesenchymal stem cells are being explored as an unconventional translational approach against COVID-19 [56].

Molecular mechanisms exhibited by mesenchymal stem cells to combat SARS-CoV-2 infection

The potential of MSCs has been explored in a wide range of lung disorders. They have been found to improve lung functioning and recovery in both acute and chronic lung diseases, including pulmonary edema, ALI, COPD, Asthma, silicosis, etc. The therapeutic effects of MSCs are known to be unaffected upon varying the route of administration but it has been suggested that intravenous administration results in more effective immunomodulation whereas the local administration, e.g., via the intratracheal route in case of respiratory disorders like emphysema, results in more intense reparative effects locally [57,58]. However, the pathophysiology of COVID-19 makes the affected individuals specifically prone to develop ARDS-like symptomatology, which is one of the characteristic features which leads to mortality in COVID-19 patients [59]. Wilson et al. tested the safety profile for the infusion of allogeneic BM-MSCs in patients with moderate-to-severe ARDS and was found to be well tolerated with no adverse events [60]. This was also evidenced by another study by Matthay et al. which suggested that intravenous infusion of MSCs was also safe and well tolerated in patients of moderate-to-severe ARDS [61]. Chen et al. studied the outcome of MSC treatment against ARDS induced by influenza A (H7N9) which caused an epidemic in 2013 [62]. They suggested that the transplantation of MSCs significantly improved the survival of the experimental group as compared to the control group without any deteriorative repercussions in the follow-up period [62]. As H7N9 share a relevant symptomatology with COVID-19, it was hinted that MSC transplantation could be a success in improving the survival for COVID-19 patients as well. The prime mechanisms by which MSCs can combat SARS-CoV-2 include regeneration, differentiation, immunomodulation, mitochondrial transfer, and paracrine action mediated via extracellular vesicles (Fig. 4).

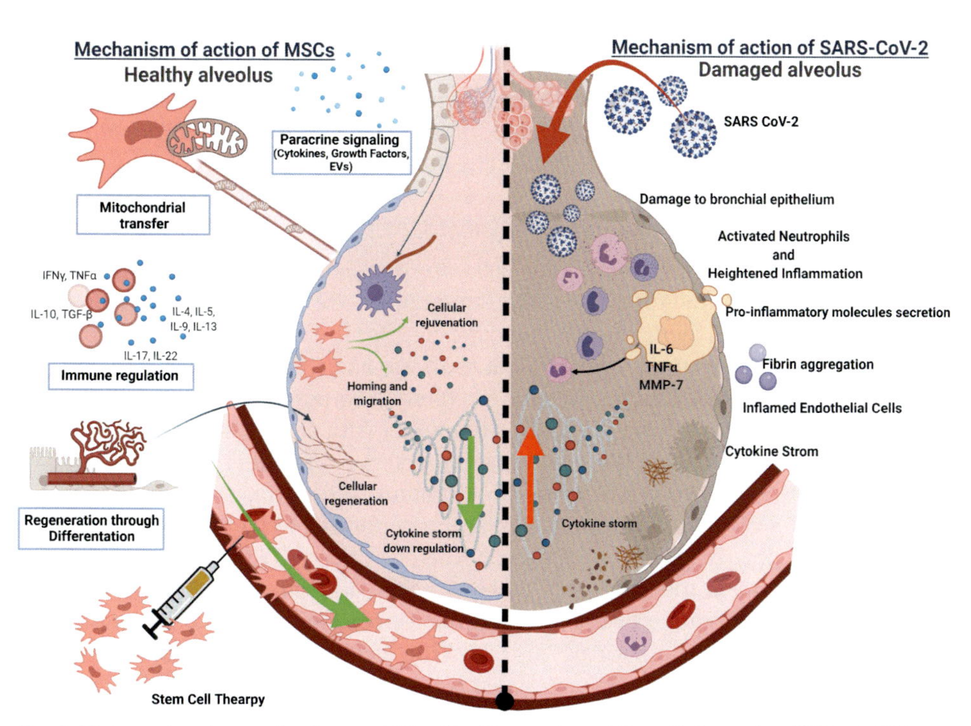

Fig. 4 Major molecular mechanisms exhibited by Mesenchymal Stem Cells against SARS-CoV-2 infection.

Regeneration by differentiation

Mesenchymal stem cells are known to differentiate into various cell types like chondrocytes, adipocytes, osteoblasts, etc. but there is barely any study which suggests the differentiation of MSCs into the cells of the epithelial lineage which primarily constitute the resident tissue cells of the lungs and respiratory system [57]. However, MSCs do help in secretion and upregulation of various other growth factors, including hepatocyte, keratinocyte, and vascular endothelial growth factor, which ultimately bestow protective, regenerative, and proliferative effects on the epithelial and endothelial cells, thereby strengthening the epithelial-alveolar lung barriers [63]. Xin et al. demonstrated that the expression of HOX9A, an essential factor for lung endothelial and epithelial cells, was increased more than 10-fold within 24 h of MSC administration [64]. This fastened the restoration against lung damage, upregulated E-cadherin, and downregulated the MMP-9, thereby protecting the lung endothelium [64]. In another study Li et al. also suggested that mechanically stretched MSCs helped in strengthening of the endothelial cell barrier of the lung in part by altering the expression of Vascular Endothelial-Cadherin and Connexin-43 [65]. MSCs exert a protective effect on alveolar-capillary barrier and it was found that UC-MSCs were better in clearing the dysregulated alveolar fluid and improved penetrability in in vitro airway epithelial assays as compared to BM-MSCs [66]. Therefore we can state that MSCs, by themselves, confer their therapeutic effects in case of COVID-19 majorly via the upregulation of antiapoptotic and proliferative factors rather than by exhibiting regeneration by differentiation.

Immunomodulatory and antiinflammatory effect of MSCs

Mesenchymal stem cells are popularly known to exhibit immunomodulatory and antiinflammatory effects. This property of MSCs makes them an excellent candidate for therapy against COVID-19 and thereby numerous clinical trials are listed around the globe utilizing different sources of MSCs for therapeutic purposes (Fig. 5 and Table 1).

MSCs have the ability to interact with almost all kinds of immune cells via secretion of mediatory factors or via direct cell-to-cell contact [67]. They regulate the proliferation, activation, and functionality of various immune cells, including dendritic cells, lymphocytes, macrophages, granulocytes, etc. MSCs even possess the ability to cause anergy in the subsets of T-helper cells. MSCs are found to inhibit T-cell proliferation by upregulating the expression of CDK (cyclin-dependent kinase) inhibitor p27kip1 and the cyclin-D2, along with the interaction of the programmed death 1 receptor (PD 1) with its ligands PD L1 and PD L2, thereby alleviating the T-cell directed inflammation [63]. Furthermore, MSCs can also repress the T-cell directed inflammation

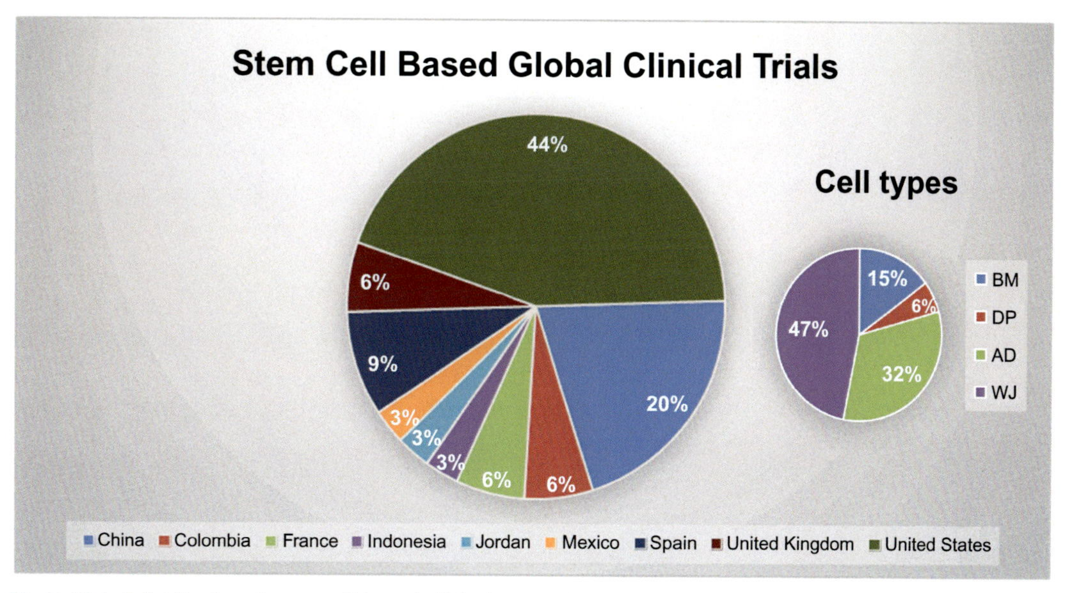

Fig. 5 Global distribution of stem cell-based clinical research and MSC source used as per data from clinicaltrials.gov.

via the action of secretory factors like PGE-2, IDO, NO, TGF-β, etc. which counter the proliferation and activation of T-cells [63]. MSCs also modulate the immune response via their action over the macrophages, by changing their secretory profile and phenotype from pro-inflammatory toward pro-resolution [68,69]. MSCs need to be first activated via the action of pro-inflammatory cytokines in order to exhibit immunomodulation by secretion of immunosuppressive molecules [70]. These features of MSCs can make them a strong competitor in therapeutics for the SARS-CoV-2 infection, as the central feature of COVID-19 pathophysiology is the extreme cytokine response, which results in excessive inflammation targeting multiple organs in the body in an unregulated manner. The flag holders of this cytokine storm are the pro-inflammatory chemokines and interleukins, including IL-1, IL-7, IL-1β, TNFα, IFN-γ, TGF-β [71]. MSCs secrete soluble mediators in their extracellular milieu for a localized effect which increases the levels of antiinflammatory cytokines, especially IL-10 and helps in the decline of pro-inflammatory factors including C-reactive protein by enhancing the lymphocyte number along with the regulatory dendritic cells and T-regulatory cells [71]. Many preclinical studies have shown astounding results for MSCs in countering the overinflammatory responses in respiratory disorders, which has paved way for clinical trials as well [66]. Lately, even the US FDA has allowed the use of MSCs for COVID-19 treatment for compassionate use [67].

Table 1 Clinical trials registered to test the application of MSCs as a treatment option in COVID-19.

<table>
<tr><th>S. no.</th><th>NCT number</th><th>Title</th><th>Interventions</th><th>Phase</th><th>Enrolment</th><th>Study type</th><th>Location</th></tr>
<tr><td>1.</td><td>NCT04346368</td><td>Bone marrow-derived mesenchymal stem cell treatment for severe patients with coronavirus disease 2019 (COVID-19)</td><td>Bone marrow MSCs</td><td>Phase1 | Phase 2</td><td>20</td><td>Interventional</td><td>Guangzhou, Guangdong, China</td></tr>
<tr><td>2.</td><td>NCT04397471</td><td>A study to collect bone marrow for process development and production of BM-MSC to treat severe COVID19 pneumonitis</td><td>Bone marrow MSCs</td><td>Phase 1 | Phase 2</td><td>10</td><td>Observational</td><td>Cambridgeshire, United Kingdom</td></tr>
<tr><td>3.</td><td>NCT04336254</td><td>Safety and efficacy study of allogeneic human dental pulp mesenchymal stem cells to treat severe COVID-19 patients</td><td>Dental pulp MSCs</td><td>Phase 1 | Phase 2</td><td>20</td><td>Interventional</td><td>Wuhan, Hubei, China</td></tr>
<tr><td>4.</td><td>NCT04302519</td><td>Novel coronavirus induced severe pneumonia treated by dental pulp mesenchymal stem cells</td><td>Dental pulp MSCs</td><td>Early Phase 1</td><td>24</td><td>Interventional</td><td>Shanghai, China</td></tr>
<tr><td>5.</td><td>NCT04428801</td><td>Autologous adipose-derived stem cells (AdMSCs) for COVID-19</td><td>Adipose-derived MSCs</td><td>Phase 2</td><td>200</td><td>Interventional</td><td>Huston, USA</td></tr>
<tr><td>6.</td><td>NCT04486001</td><td>Study of intravenous administration of allogeneic adipose stem cells for COVID-19</td><td>Adipose-derived MSCs</td><td>Phase 1</td><td>20</td><td>Interventional</td><td>California, United States</td></tr>
<tr><td>7.</td><td>NCT04366323</td><td>Clinical trial to assess the safety and efficacy of intravenous administration of allogeneic adult mesenchymal stem cells of expanded adipose tissue in patients with severe pneumonia due to COVID-19</td><td>Adipose-derived MSCs</td><td>Phase 1 | Phase 2</td><td>26</td><td>Interventional</td><td>Sevilla, Spain |</td></tr>
</table>

8.	NCT04348435	A randomized, double-blind, placebo-controlled clinical trial to determine the safety and efficacy of hope biosciences allogeneic mesenchymal stem cell therapy (HB-adMSCs) to provide protection against COVID-19	Adipose-derived MSCs	Phase 2	100	Interventional	Sugar Land, Texas, United States
9.	NCT04349631	A clinical trial to determine the safety and efficacy of hope biosciences autologous mesenchymal stem cell therapy (HB-adMSCs) to provide protection against COVID-19	Adipose-derived MSCs	Phase 2	56	Interventional	Sugar Land, Texas, United States
10.	NCT04909892	Study of allogeneic adipose-derived mesenchymal stem cells to treat post COVID-19 "long haul" pulmonary compromise	Adipose-derived MSCs	Phase 2	30	Interventional	California, United States
11.	NCT04905836	Study of allogeneic adipose-derived mesenchymal stem cells for treatment of COVID-19 acute respiratory distress	Adipose-derived MSCs	Phase 2	100	Interventional	California, United States
12.	NCT04728698	Study of intravenous administration of allogeneic adipose-derived mesenchymal stem cells for COVID-19-induced acute respiratory distress	Adipose-derived MSCs	Phase 2	100	Interventional	California, United States
13.	NCT04798066	Intermediate size expanded access protocol evaluating HB-adMSC's for the treatment of post-COVID-19 syndrome	Adipose-derived MSCs	Intermediate-size Population	Sugar Land, Texas, United States		
14.	NCT04362189	Efficacy and safety study of allogeneic HB-adMSCs for the treatment of COVID-19	Adipose-derived MSCs	Phase 2	100	Interventional	Houston, Texas, United States

Continued

Table 1 Clinical trials registered to test the application of MSCs as a treatment option in COVID-19—cont'd

<table>
<tr><th>S. no.</th><th>NCT number</th><th>Title</th><th>Interventions</th><th>Phase</th><th>Enrolment</th><th>Study type</th><th>Location</th></tr>
<tr><td>15.</td><td>NCT04348461</td><td>BAttLe against COVID-19 using MesenchYmal stromal cells</td><td>Adipose-derived MSCs</td><td>Phase 2</td><td>100</td><td>Interventional</td><td>Fundacion Jimenez Diaz, Madrid, Spain</td></tr>
<tr><td>16.</td><td>NCT04390139</td><td>Efficacy and safety evaluation of mesenchymal stem cells for the treatment of patients with respiratory distress due to COVID-19</td><td>Umbilical cord MSCs</td><td>Phase 1 | Phase 2</td><td>30</td><td>Interventional</td><td>Terrassa, Barcelona, Spain</td></tr>
<tr><td>17.</td><td>NCT04313322</td><td>Treatment of COVID-19 patients using Wharton's Jelly-mesenchymal stem cells</td><td>Umbilical cord MSCs</td><td>Phase 1</td><td>5</td><td>Interventional</td><td>Amman, Jordan</td></tr>
<tr><td>18.</td><td>NCT04625738</td><td>Efficacy of infusions of MSC from Wharton jelly in the SARS-Cov-2 (COVID-19) related acute respiratory distress syndrome</td><td>Umbilical cord MSCs</td><td>Phase 2</td><td>30</td><td>Interventional</td><td>Nancy, France</td></tr>
<tr><td>19.</td><td>NCT04494386</td><td>Umbilical cord lining stem cells (ULSC) in patients with COVID-19 ARDS</td><td>Umbilical cord MSCs</td><td>Phase 1 | Phase 2</td><td>60</td><td>Interventional</td><td>South Dakota, United States</td></tr>
<tr><td>20.</td><td>NCT04390152</td><td>Safety and efficacy of intravenous Wharton's jelly derived mesenchymal stem cells in acute respiratory distress syndrome due to COVID 19</td><td>Umbilical cord MSCs</td><td>Phase 1 | Phase 2</td><td>40</td><td>Interventional</td><td>Antioquia, Colombia</td></tr>
<tr><td>21.</td><td>NCT04333368</td><td>Cell therapy using umbilical cord-derived mesenchymal stromal cells in SARS-CoV-2-related ARDS</td><td>Umbilical cord MSCs</td><td>Phase 1 | Phase 2</td><td>47</td><td>Interventional</td><td>Paris, France</td></tr>
</table>

22.	NCT04429763	Safety and efficacy of mesenchymal stem cells in the management of severe COVID-19 pneumonia	Umbilical cord MSCs	Phase 2	30	Interventional	Bogotá, Colombia
23.	NCT04456361	Use of mesenchymal stem cells in acute respiratory distress syndrome caused by COVID-19	Umbilical cord MSCs	Early Phase 1	9	Interventional	Baja California, Mexico
24.	NCT04457609	Administration of allogenic UC-MSCs as adjuvant therapy for critically-Ill COVID-19 patients	Umbilical cord MSCs	Phase 1	40	Interventional	DKI Jakarta, Indonesia
25.	NCT04355728	Use of UC-MSCs for COVID-19 patients	Umbilical cord MSCs	Phase 1 | Phase 2	24	Interventional	Florida, United States
26.	NCT04452097	Use of hUC-MSC product (BX-U001) for the treatment of COVID-19 with ARDS	Umbilical cord MSCs	Phase 1 | Phase 2	39	Interventional	California, United States
27.	NCT03042143	Repair of acute respiratory distress syndrome by stromal cell administration (REALIST) (COVID-19)	Umbilical cord MSCs	Phase 1 | Phase 2	75	Interventional	Northern Ireland, United Kingdom
28.	NCT04273646	Study of human umbilical cord mesenchymal stem cells in the treatment of severe COVID-19	Umbilical cord MSCs	Not applicable	48	Interventional	Wuhan, Hubei, China
29.	NCT04288102	Treatment with human umbilical cord-derived mesenchymal stem cells for severe corona virus disease 2019 (COVID-19)	Umbilical cord MSCs	Phase 2	100	Interventional	Wuhan, Hubei, China
30.	NCT04490486	Umbilical cord tissue (UC) derived mesenchymal stem cells (MSCs) versus placebo to treat acute pulmonary inflammation due to COVID-19	Umbilical cord MSCs	Phase 1	21	Interventional	Florida, United States

Continued

Table 1 Clinical trials registered to test the application of MSCs as a treatment option in COVID-19—cont'd

S. no.	NCT number	Title	Interventions	Phase	Enrolment	Study type	Location
31.	NCT04269525	Umbilical cord(UC)-derived mesenchymal stem Cells(MSCs) treatment for the 2019-novel coronavirus (nCOV) pneumonia	Umbilical cord MSCs	Phase 2	16	Interventional	Wuhan, Hubei, China
32.	NCT04657458	Expanded access protocol on bone marrow mesenchymal stem cell derived extracellular vesicle infusion treatment for patients with COVID-19 associated ARDS	Bone marrow MSC derived extracellular vesicles	Early Phase 1	0	Expanded Access: intermediate-size population	Texas, United States
33.	NCT04276987	A pilot clinical study on inhalation of mesenchymal stem cells exosomes treating severe novel coronavirus pneumonia	Bone marrow MSC derived extracellular vesicles	Phase 1	24	Interventional	Shanghai, China
34.	NCT04798716	The use of exosomes for the treatment of acute respiratory distress syndrome or novel coronavirus pneumonia caused by COVID-19	Bone marrow MSC derived extracellular vesicles	Phase 1 | Phase 2	55	Interventional	California, United States

Mitochondrial transfer as a rescue mechanism

Mitochondria are considered to be the powerhouse of the cells. It is responsible for energy generation and homeostatic maintenance of the cells. Lately it has been suggested that mitochondrial transfer via MSCs is a molecular mechanism via which MSCs exhibit their regenerative activity. There can be various modes of mitochondrial transfer including direct cell-cell fusion, transfer through gap junctions, transfer through intermediate tunneling nanotubules, and release through packaging in extracellular vesicles [72,73]. The functional recovery by the transfer of mitochondria through MSCs was found to be characterized by improvement in oxygen utilization and enhanced intracellular ATP content [74]. Mitochondrial transfer via MSCs has been listed as one of the rescue mechanisms in lung injury, asthma, and other respiratory ailments as well [74]. In a model of LPS-induced acute lung injury, it was found that administration of MSCs was able to restore alveolar function and increased the animal survival. Furthermore, the leukocytes infiltration was reduced while the ATP function was enhanced. The underlying mechanism for the improvement was suggested to be the transfer of mitochondria from MSCs to damaged lung epithelium via the connexin 43 aided tunneling nanotubules and through packaging in extracellular vesicles [75]. In another study using an ARDS mice model, it was found that MSC-EVs were able to reinstate mitochondrial functions partially via mitochondrial transfer and were able to enhance the integrity of alveolar-capillary barrier, moreover restoring normal levels of oxidative phosphorylation [76]. In ARDS niche, MSCs have also been found to promote antiinflammatory activities and enhanced phagocytosis via macrophages through EV aided mitochondrial transfer as the transfer of mitochondria from MSCs to immune cells increases their phagocytic and antimicrobial action [77,78]. The reduction of inflammation is therefore a key effect of mitochondrial transfer via MSCs to macrophages and T-cells [79,80]. In a study it was suggested to rely over PGC-1α aided biogenesis of mitochondria and PGC-1α/TFEB aided autophagy of lysosomes [81].

The paracrine effect

The secretome of MSCs largely comprises of soluble molecules like cytokines, chemokines, growth factors, and nonsoluble, encapsulated, extracellular vesicles [82]. MSCs secrete an antibiotic, LL37, which is found to neutralize SARS-CoV-2 and it also secretes immunomodulatory molecules like PGE-2 and TGF-β, which help in neutralization of the cytokine storm [83]. Furthermore, like all other cells, MSCs also secrete some lipid membrane bound vesicles in their extracellular space which have roles ranging from excretion, intracellular homeostasis, intercellular communication, to regulation of different physiological and pathophysiological processes. They can be as small as 30 nm and go up to 10,000 nm and can be classified into 3 overlapping categories, namely exosomes (30–150 nm), microvesicles (0.1–1 μm), and apoptotic bodies (1–5 μm) [84]. The category of extracellular vesicles that has recently gained spotlight is the smallest vesicles known as exosomes. These vesicles are thought to transport molecules including DNA, mRNA, miRNA, proteins, enzymes, etc. as a contributory mechanism to mediate cellular repair, establish links for intercellular communication, and to promote disease progression [84]. Owing to the already established regenerative and reparative potential of mesenchymal stem cells, the exosomes derived from this specific cell type have sought the interest of scientific community as a potential therapeutic aid. The remedial capabilities of exosomes derived from MSCs have been tested in different diseases and disorders targeting various organs and organ systems like hepatic system, renal system, cardiovascular system, central nervous system, etc. Due to the failure of existing healthcare aids and therapies in the pandemic situation, the astounding results obtained in preclinical and clinical studies for the application of MSC exosomes in different cardiovascular diseases, lung and respiratory disorders gave strong hopes in the treatment of COVID-19. Basic and translational research investigations have suggested that extracellular vesicles, especially exosomes derived from MSCs, work in a synonymous manner to their parent cells. This per se means that they are able to elicit the regenerative and reparative phenomena in the target cells and organs as do the mesenchymal stem cells itself. There are several clinical trials which have tested their safety and efficacy in COVID-19. In a clinical trial by Sengupta et al., it was suggested that BM-MSC-derived exosomes were safe upon transplantation and demonstrated an ability to subside the cytokine storm which is a characteristic of COVID-19 along with reinstating the oxygenation ability of the lungs [85]. Furthermore, the packaging of MSC-derived extracellular vesicles constitutes factors for direct supplementation of therapeutic factors including upregulatory mRNA and proteins, mitochondrial transfer to ailing cells, agents for transcriptional and translational modifications like cytokines and chemokines which specifically interfere with biological pathways, and miRNA and siRNA causing transcriptional and translational repression to support the remedial institution of MSC-derived extracellular vesicles [86].

Conclusion and outlook for the future

The COVID-19 pandemic is one of the largest and deadliest health crises faced by the world. With the possibility of another wave striking, it is imperative to be prepared with appropriate healthcare needs and remedies. Vaccines are considered a confident approach as a preventive measure but mass vaccination is still a tedious process in largely populated countries. Furthermore, it is doubtable that the vaccines will be completely effective against the different variants of this virus. In such circumstances, the immunomodulatory potential of mesenchymal stem cells has gained recognition. However, there are also many ethical and safety concerns which are speculated in whole cell therapy like the risk of emboli and immunogenicity. Due to such concerns the scientific community is exploring the possibility of utilizing MSC-derived extracellular vesicles, which facilitate their paracrine functionality. MSC-derived extracellular vesicles have added advantages of their parent cells for being small and nonimmunogenic. Their lipid bilayer-enclosed membrane bound structure also aids to possibility of drug encapsulation and specific cell targeting for their far-reaching abilities. Building upon this capacity these could serve as mainstream candidates for therapy against SARS-CoV-2. Yet there are several hindrances limiting their therapeutic applications like inadequate measure of their large-scale production and their complete biochemical and physical characterization which is still an area incompletely explored.

In conclusion, we would like to highlight that even though there are many approaches being explored to fight the COVID-19 pandemic, the use of stem cell and its derivatives can serve as a potential therapeutic aid that can holistically manage the pandemic situation in the long term and make way for efficient management of any such health crisis in the future.

Acknowledgment

The images are created with Biorender.com.
The details of clinical data were derived from clinicaltrials.gov.

Conflict of interest

The authors declare that they have no competing interest.

Financial support

The authors declare no relevance or involvement of any funding source with the subject matter or materials discussed in the manuscript.

References

[1] Madabhavi I, Sarkar M, Kadakol N. COVID-19: a review. Monaldi Arch Chest Dis 2020;90(2):1–11.
[2] Naji H. Clinical characterization of COVID-19. Eur J Med Health Sci 2020;2(2):1–5.

[3] Asrani P, Eapen MS, Hassan MI, Sohal SS. Implications of the second wave of COVID-19 in India. Lancet Respir Med 2021;9(9):e93–4.
[4] He X, Cheng X, Feng X, Wan H, Chen S, Xiong M. Clinical symptom differences between mild and severe COVID-19 patients in China: a meta-analysis. Front Public Health 2021;8:954.
[5] Anon. Weekly epidemiological update on COVID-19—31 August 2021. 55th ed; 2021. https://www.who.int/publications/m/item/weekly-epidemiological-update-on-covid-19- - -31-august-2021.
[6] Maheshkumar K, Wankhar W, Gurugubelli KR, Mahadevappa VH, Lepcha L, kumar Choudhary A. Angiotensin-converting enzyme 2 (ACE2): COVID 19 gate way to multiple organ failure syndromes. Respir Physiol Neurobiol 2020;283:103548.
[7] Ali MJ, Hanif M, Haider MA, Ahmed MU, Sundas FN, Hirani A, Khan IA, Anis K, Karim AH. Treatment options for COVID-19: a review. Front Med 2020;7:480.
[8] Wu R, Wang L, Kuo HC, Shannar A, Peter R, Chou PJ, Li S, Hudlikar R, Liu X, Liu Z, Poiani GJ. An update on current therapeutic drugs treating COVID-19. Curr Pharm Rep 2020;6(3):56–70.
[9] Jahanshahlu L, Rezaei N. Monoclonal antibody as a potential anti-COVID-19. Biomed Pharmacother 2020;129, 110337.
[10] Chen L, Xiong J, Bao L, Shi Y. Convalescent plasma as a potential therapy for COVID-19. Lancet Infect Dis 2020;20(4):398–400.
[11] Tregoning JS, Brown ES, Cheeseman HM, Flight KE, Higham SL, Lemm NM, Pierce BF, Stirling DC, Wang Z, Pollock KM. Vaccines for COVID-19. Clin Exp Immunol 2020;202(2):162–92.
[12] Forni G, Mantovani A. COVID-19 vaccines: where we stand and challenges ahead. Cell Death Differ 2021;28(2):626–39.
[13] Belete TM. Review on up-to-date status of candidate vaccines for COVID-19 disease. Infect Drug Resist 2021;14:151.
[14] Metcalfe SM. Mesenchymal stem cells and management of COVID-19 pneumonia. Med Drug Discov 2020;1(5), 100019.
[15] Gupta S, Krishnakumar V, Sharma Y, Dinda AK, Mohanty S. Mesenchymal stem cell derived exosomes: a nano platform for therapeutics and drug delivery in combating COVID-19. Stem Cell Rev Rep 2021;17(1):33–43.
[16] Sharma Y, Gupta S, Mohanty S. Mesenchymal stem cell-derived exosome as a nano weapon to target the COVID-19 pandemic. Biocell 2021;45(3):517.
[17] Astuti I. Severe acute respiratory syndrome coronavirus 2 (SARS-CoV-2): an overview of viral structure and host response. Diabetes Metab Syndr Clin Res Rev 2020;14(4):407–12.
[18] Satarker S, Nampoothiri M. Structural proteins in severe acute respiratory syndrome coronavirus-2. Arch Med Res 2020;51(6):482–91.
[19] Dutta NK, Mazumdar K, Gordy JT. The nucleocapsid protein of SARS–CoV-2: a target for vaccine development. J Virol 2020;94(13). e00647-20.
[20] St-Germain JR, Astori A, Samavarchi-Tehrani P, Abdouni H, Macwan V, Kim DK, et al. A SARS-CoV-2 BioID-based virus-host membrane protein interactome and virus peptide compendium: new proteomics resources for COVID-19 research. bioRxiv 2020;8:1–20.
[21] Wang MY, Zhao R, Gao LJ, Gao XF, Wang DP, Cao JM. SARS-CoV-2: structure, biology, and structure-based therapeutics development. Front Cell Infect Microbiol 2020;10:1–17.
[22] Ali A, Vijayan R. Dynamics of the ACE2–SARS-CoV-2/SARS-CoV spike protein interface reveal unique mechanisms. Sci Rep 2020;10(1):1–2.
[23] Ni W, Yang X, Yang D, Bao J, Li R, Xiao Y, Hou C, Wang H, Liu J, Yang D, Xu Y. Role of angiotensin-converting enzyme 2 (ACE2) in COVID-19. Crit Care 2020;24(1):1–10.
[24] Salamanna F, Maglio M, Landini MP, Fini M. Body localization of ACE-2: on the trail of the keyhole of SARS-CoV-2. Front Med 2020;7:935.

[25] Beyerstedt S, Casaro EB, Rangel ÉB. COVID-19: angiotensin-converting enzyme 2 (ACE2) expression and tissue susceptibility to SARS-CoV-2 infection. Eur J Clin Microbiol Infect Dis 2021;3:1–5.
[26] Zhang Q, Bastard P, Bolze A, Jouanguy E, Zhang SY, Effort CH, Cobat A, Notarangelo LD, Su HC, Abel L, Casanova JL. Life-threatening COVID-19: defective interferons unleash excessive inflammation. Med 2020;1(1):14–20.
[27] Hantoushzadeh S, Norooznezhad AH. Possible cause of inflammatory storm and septic shock in patients diagnosed with (COVID-19). Arch Med Res 2020;51(4):347–8.
[28] Hanna R, Dalvi S, Sălăgean T, Pop ID, Bordea IR, Benedicenti S. Understanding COVID-19 pandemic: molecular mechanisms and potential therapeutic strategies. An evidence-based review. J Inflamm Res 2021;14:13.
[29] Raj CD, Kandaswamy DK, Danduga RC, Rajasabapathy R, James RA. COVID-19: molecular pathophysiology, genetic evolution and prospective therapeutics—a review. Arch Microbiol 2021;8:1–5.
[30] Sinha N, Balayla G. Hydroxychloroquine and covid-19. Postgrad Med J 2020;96(1139):550–5.
[31] Waterer GW, Rello J. Steroids and COVID-19: we need a precision approach, not one size fits all; 2020. p. 701–5.
[32] Cheng Y, Wong R, Soo YO, Wong WS, Lee CK, Ng MH, Chan P, Wong KC, Leung CB, Cheng G. Use of convalescent plasma therapy in SARS patients in Hong Kong. Eur J Clin Microbiol Infect Dis 2005;24(1):44–6.
[33] Duan K, Liu B, Li C, Zhang H, Yu T, Qu J, Zhou M, Chen L, Meng S, Hu Y, Peng C. Effectiveness of convalescent plasma therapy in severe COVID-19 patients. Proc Natl Acad Sci 2020;117(17):9490–6.
[34] Ye M, Fu D, Ren Y, Wang F, Wang D, Zhang F, Xia X, Lv T. Treatment with convalescent plasma for COVID-19 patients in Wuhan, China. J Med Virol 2020;92(10):1890–901.
[35] Li L, Zhang W, Hu Y, Tong X, Zheng S, Yang J, Kong Y, Ren L, Wei Q, Mei H, Hu C. Effect of convalescent plasma therapy on time to clinical improvement in patients with severe and life-threatening COVID-19: a randomized clinical trial. JAMA 2020;324(5):460–70.
[36] Al-Riyami AZ, Schäfer R, van den Berg K, Bloch EM, Estcourt LJ, Goel R, Hindawi S, Josephson CD, Land K, McQuilten ZK, Spitalnik SL. Clinical use of Convalescent Plasma in the COVID-19 pandemic: a transfusion-focussed gap analysis with recommendations for future research priorities. Vox Sang 2021;116(1):88–98.
[37] Simonovich VA, Burgos Pratx LD, Scibona P, Beruto MV, Vallone MG, Vázquez C, Savoy N, Giunta DH, Pérez LG, Sánchez MD, Gamarnik AV. A randomized trial of convalescent plasma in Covid-19 severe pneumonia. N Engl J Med 2021;384(7):619–29.
[38] Coleman CM, Liu YV, Mu H, Taylor JK, Massare M, Flyer DC, Glenn GM, Smith GE, Frieman MB. Purified coronavirus spike protein nanoparticles induce coronavirus neutralizing antibodies in mice. Vaccine 2014;32(26):3169–74.
[39] Khalaj-Hedayati A. Protective immunity against SARS subunit vaccine candidates based on spike protein: lessons for coronavirus vaccine development. J Immunol Res 2020;18:2020.
[40] Huang J, Huang H, Wang D, Wang C, Wang Y. Immunological strategies against spike protein: neutralizing antibodies and vaccine development for COVID-19. Clin Transl Med 2020;10(6):1–6.
[41] Bernal JL, Andrews N, Gower C, Robertson C, Stowe J, Tessier E, et al. Effectiveness of the Pfizer-BioNTech and Oxford-AstraZeneca vaccines on covid-19 related symptoms, hospital admissions, and mortality in older adults in England: test negative case-control study. BMJ 2021;373:1–11.

[42] Anon, https://www.pfizer.com/news/press-release/press-release-detail/pfizer-and-biontech-choose-lead-mrna-vaccine-candidate-0.
[43] Tanne JH. Covid-19: FDA panel votes to authorise Pfizer BioNTech vaccine. BMJ 2020;371. https://doi.org/10.1136/bmj.m4799. m4799.
[44] Knoll MD, Wonodi C. Oxford-AstraZeneca COVID-19 vaccine efficacy. Lancet 2021;397(10269):72–4.
[45] Eyre DW, Lumley SF, Wei J, Cox S, James T, Justice A, et al. Quantitative SARS-CoV-2 anti-spike responses to Pfizer-BioNTech and Oxford-AstraZeneca vaccines by previous infection status. Clin Microbiol Infect 2021;27:1–20.
[46] Checcucci E, Piramide F, Pecoraro A, Amparore D, Campi R, Fiori C, et al. The vaccine journey for COVID-19: a comprehensive systematic review of current clinical trials in humans. Panminerva Med 2020. https://doi.org/10.23736/S0031-0808.20.03958-0.
[47] Babaei F, Mirzababaei M, Nassiri-Asl M, Hosseinzadeh H. Review of registered clinical trials for the treatment of COVID-19. Drug Dev Res 2021;82(4):474–93.
[48] Rawat S, Srivastava P, Prabha P, Gupta S, Kanga U, Mohanty S. A comparative study on immunomodulatory potential of tissue specific hMSCs. J Dent Med Sci 2018;17(6):32–40. https://doi.org/10.9790/0853-1706143240.
[49] Qiu J, Xiao H, Zhou S, Du W, Mu X, Shi G, Tan X. Bone marrow mesenchymal stem cells inhibit cardiac hypertrophy by enhancing FoxO1 transcription. Cell Biol Int 2021;45(1):188–97.
[50] Pang LX, Cai WW, Li Q, Li HJ, Fei M, Yuan YS, Sheng B, Zhang K, An RC, Ou YW, Zeng WJ. Bone marrow-derived mesenchymal stem cells attenuate myocardial ischemia–reperfusion injury via upregulation of splenic regulatory T cells. BMC Cardiovasc Disord 2021;21(1):1–2.
[51] Magota H, Sasaki M, Kataoka-Sasaki Y, Oka S, Ukai R, Kiyose R, Onodera R, Kocsis JD, Honmou O. Intravenous infusion of mesenchymal stem cells delays disease progression in the SOD1G93A transgenic amyotrophic lateral sclerosis rat model. Brain Res 2021;15(1757), 147296.
[52] Rong X, Liu J, Yao X, Jiang T, Wang Y, Xie F. Human bone marrow mesenchymal stem cells-derived exosomes alleviate liver fibrosis through the Wnt/β-catenin pathway. Stem Cell Res Ther 2019;10(1):1.
[53] Li S, Wang Y, Wang Z, Chen L, Zuo B, Liu C, Sun D. Enhanced renoprotective effect of GDNF-modified adipose-derived mesenchymal stem cells on renal interstitial fibrosis. Curr Stem Cell Res Ther 2021;12(1):1–7.
[54] Cavaglieri RC, Martini D, Sogayar MC, Noronha IL. Mesenchymal stem cells delivered at the subcapsule of the kidney ameliorate renal disease in the rat remnant kidney model. In: Transplantation proceedings, Vol. 41 (3). Elsevier; 2009. p. 947–51.
[55] Eirin A, Zhu XY, Krier JD, Tang H, Jordan KL, Grande JP, Lerman A, Textor SC, Lerman LO. Adipose tissue-derived mesenchymal stem cells improve revascularization outcomes to restore renal function in swine atherosclerotic renal artery stenosis. Stem Cells 2012;30(5):1030–41.
[56] Coelho A, Alvites RD, Branquinho MV, Guerreiro SG, Maurício AC. Mesenchymal stem cells (MSCs) as a potential therapeutic strategy in COVID-19 patients: literature research. Front Cell Dev Biol 2020;8:1392.
[57] Antunes MA, e Silva JR, Rocco PR. Mesenchymal stromal cell therapy in COPD: from bench to bedside. Int J Chron Obstruct Pulmon Dis 2017;12:3017.
[58] Antunes MA, Abreu SC, Cruz FF, Teixeira AC, Lopes-Pacheco M, Bandeira E, Olsen PC, Diaz BL, Takyia CM, Freitas IP, Rocha NN. Effects of different mesenchymal stromal cell sources and delivery routes in experimental emphysema. Respir Res 2014;15(1):1–4.
[59] Wang W, Lei W, Jiang L, Gao S, Hu S, Zhao ZG, Niu CY, Zhao ZA. Therapeutic mechanisms of mesenchymal stem cells in acute respiratory distress syndrome reveal potentials for Covid-19 treatment. J Transl Med 2021;19(1):1–3.

[60] Wilson JG, Liu KD, Zhuo H, Caballero L, McMillan M, Fang X, Cosgrove K, Vojnik R, Calfee CS, Lee JW, Rogers AJ. Mesenchymal stem (stromal) cells for treatment of ARDS: a phase 1 clinical trial. Lancet Respir Med 2015;3(1):24–32.

[61] Matthay MA, Calfee CS, Zhuo H, Thompson BT, Wilson JG, Levitt JE, Rogers AJ, Gotts JE, Wiener-Kronish JP, Bajwa EK, Donahoe MP. Treatment with allogeneic mesenchymal stromal cells for moderate to severe acute respiratory distress syndrome (START study): a randomised phase 2a safety trial. Lancet Respir Med 2019;7(2):154–62.

[62] Chen J, Hu C, Chen L, Tang L, Zhu Y, Xu X, Chen L, Gao H, Lu X, Yu L, Dai X. Clinical study of mesenchymal stem cell treatment for acute respiratory distress syndrome induced by epidemic influenza A (H7N9) infection: a hint for COVID-19 treatment. Engineering 2020;6(10):1153–61.

[63] Harrell CR, Sadikot R, Pascual J, Fellabaum C, Jankovic MG, Jovicic N, Djonov V, Arsenijevic N, Volarevic V. Mesenchymal stem cell-based therapy of inflammatory lung diseases: current understanding and future perspectives. Stem Cells Int 2019;2:2019.

[64] Xin X, Yan L, Guangfa Z, Yan H, Keng L, Chunting W. Mesenchymal stem cells promoted lung wound repair through hox A9 during endotoxemia-induced acute lung injury. Stem Cells Int 2017;29:2017.

[65] Li JZ, Meng SS, Xu XP, Huang YB, Mao P, Li YM, Yang Y, Qiu HB, Pan C. Mechanically stretched mesenchymal stem cells can reduce the effects of LPS-induced injury on the pulmonary microvascular endothelium barrier. Stem Cells Int 2020;30:2020.

[66] Song N, Wakimoto H, Rossignoli F, Bhere D, Ciccocioppo R, Chen KS, et al. Mesenchymal stem cell immunomodulation: in pursuit of controlling COVID-19 related cytokine storm. Stem Cells 2021;39(6):707–22.

[67] Jeyaraman M, John A, Koshy S, Ranjan R, Jain R, Swati K, et al. Fostering mesenchymal stem cell therapy to halt cytokine storm in COVID-19. Biochim Biophys Acta (BBA) - Mol Basis Dis 2020;1867(2):166014.

[68] Vasandan AB, Jahnavi S, Shashank C, Prasad P, Kumar A, Prasanna SJ. Human mesenchymal stem cells program macrophage plasticity by altering their metabolic status via a PGE 2-dependent mechanism. Sci Rep 2016;6(1):1–7.

[69] Cho DI, Kim MR, Jeong HY, Jeong HC, Jeong MH, Yoon SH, Kim YS, Ahn Y. Mesenchymal stem cells reciprocally regulate the M1/M2 balance in mouse bone marrow-derived macrophages. Exp Mol Med 2014;46(1):e70.

[70] Wang M, Yuan Q, Xie L. Mesenchymal stem cell-based immunomodulation: properties and clinical application. Stem Cells Int 2018;2018, 3057624.

[71] Kavianpour M, Saleh M, Verdi J. The role of mesenchymal stromal cells in immune modulation of COVID-19: focus on cytokine storm. Stem Cell Res Ther 2020;11(1):1–9.

[72] Paliwal S, Chaudhuri R, Agrawal A, Mohanty S. Regenerative abilities of mesenchymal stem cells through mitochondrial transfer. J Biomed Sci 2018;25(1):1–2.

[73] Mohammadalipour A, Dumbali SP, Wenzel PL. Mitochondrial transfer and regulators of mesenchymal stromal cell function and therapeutic efficacy. Front Cell Dev Biol 2020;8:1519.

[74] Li C, Cheung MK, Han S, Zhang Z, Chen L, Chen J, Zeng H, Qiu J. Mesenchymal stem cells and their mitochondrial transfer: a double-edged sword. Biosci Rep 2019;39(5), BSR20182417.

[75] Islam MN, Das SR, Emin MT, Wei M, Sun L, Westphalen K, et al. Mitochondrial transfer from bone-marrow-derived stromal cells to pulmonary alveoli protects against acute lung injury. Nat Med 2012;18:759–65. https://doi.org/10.1038/nm.2736.

[76] Silva JD, Su Y, Calfee CS, Delucchi KL, Weiss D, McAuley DF, O'Kane C, Krasnodembskaya AD. MSC extracellular vesicles rescue mitochondrial dysfunction and improve barrier integrity in clinically relevant models of ARDS. Eur Respir J 2020;58:2002978.
[77] Morrison TJ, Jackson MV, Cunningham EK, Kissenpfennig A, McAuley DF, O'Kane CM, Krasnodembskaya AD. Mesenchymal stromal cells modulate macrophages in clinically relevant lung injury models by extracellular vesicle mitochondrial transfer. Am J Respir Crit Care Med 2017;196(10):1275–86.
[78] Jackson MV, Morrison TJ, Doherty DF, McAuley DF, Matthay MA, Kissenpfennig A, O'Kane CM, Krasnodembskaya AD. Mitochondrial transfer via tunneling nanotubes is an important mechanism by which mesenchymal stem cells enhance macrophage phagocytosis in the in vitro and in vivo models of ARDS. Stem Cells 2016;34(8):2210–23.
[79] Luz-Crawford P, Hernandez J, Djouad F, Luque-Campos N, Caicedo A, Carrère-Kremer S, Brondello JM, Vignais ML, Pène J, Jorgensen C. Mesenchymal stem cell repression of Th17 cells is triggered by mitochondrial transfer. Stem Cell Res Ther 2019;10(1):1–3.
[80] Court AC, Le-Gatt A, Luz-Crawford P, Parra E, Aliaga-Tobar V, Bátiz LF, Contreras RA, Ortúzar MI, Kurte M, Elizondo-Vega R, Maracaja-Coutinho V. Mitochondrial transfer from MSCs to T cells induces Treg differentiation and restricts inflammatory response. EMBO Rep 2020;21(2), e48052.
[81] Yuan Y, Yuan L, Li L, Liu F, Liu J, Chen Y, et al. Mitochondrial transfer from MSCs to macrophages restricts inflammation and alleviates kidney injury in diabetic nephropathy mice via PGC-1α activation. Stem Cells 2021;39(7):913–28.
[82] Eleuteri S, Fierabracci A. Insights into the secretome of mesenchymal stem cells and its potential applications. Int J Mol Sci 2019;20(18):4597.
[83] Chouw A, Milanda T, Sartika CR, Kirana MN, Halim D, Faried A. Potency of mesenchymal stem cell and its secretome in treating COVID-19. Regen Eng Transl Med 2021;10:1–2.
[84] Doyle LM, Wang MZ. Overview of extracellular vesicles, their origin, composition, purpose, and methods for exosome isolation and analysis. Cell 2019;8(7):727.
[85] Sengupta V, Sengupta S, Lazo A, Woods P, Nolan A, Bremer N. Exosomes derived from bone marrow mesenchymal stem cells as treatment for severe COVID-19. Stem Cells Dev 2020;29(12):747–54.
[86] Silva J, Murray LM, Su Y, Mc Auley DF, O'Kane CM, Krasnodembskaya A. Transfer of mitochondria from mesenchymal stromal cells through extracellular vesicles improves alveolar epithelial-capillary barrier in ARDS. ERJ Open Res 2020;6:98.

4

Inflammatory multisystem syndrome in COVID-19: Insights on off-target organ system in susceptible and recovering population

Ancy Thomas

Department of Dermatology, Northwestern University, Chicago, IL, United States; Robert H. Lurie Comprehensive Cancer Center, Chicago, IL, United States

Introduction

SARS-CoV-2 is a single-stranded RNA beta coronavirus that causes lethal respiratory tract infections in humans. This virus was first identified in patients with pneumonia in Wuhan, China, in late 2019 and has rapidly spread to all continents as a pandemic. Clinical manifestation of COVID-19 ranges from asymptomatic or mild respiratory illness to moderate and severe disease, rapidly progressive pneumonitis, and respiratory and multiorgan failure with fatal outcomes. The high variation of severity of infection among patients is generally associated with general health status and the severity of immune responses. Studies on the immune system of patients reveal wide variability of immune responses and clinical outcomes with severe acute respiratory syndrome coronavirus 2 (SARS-CoV-2) infection. This provides insights about the ability of SARS-CoV-2 to elicit antiviral immune responses that are different from most other human viral pathogens. This influences the longevity of immunological memory, and approaches that mediate robust protection from viral infections. These immune responses during COVID-19 recovery play a key role in determining disease severity [1].

Stem Cells and COVID-19. https://doi.org/10.1016/B978-0-323-89972-7.00004-0

Antigen processing and immune responses in COVID-19

Coronaviruses use the spike S protein on its membrane to infect by binding to the membrane of the host cell. The spike protein binds to the ACE2 receptor on the host cell surface and gets pinched inside the host cell. In general, the presentation of viral antigenic peptides is complexed with MHC (major histocompatibility complex) class I and class II molecules and presented to CD8 and CD4 T cells. The epitope prediction of COVID-19 revealed around 405 T cell epitopes which exhibit effective affinity toward MHC class I and II in addition to two potent neutralizing B cell epitopes based on Spike (S) protein [2]. Recent studies show the role of an enzyme called Furin present in the host cells, which plays a crucial role in SARS-CoV-2 entry [3]. Furin, expressed in various human organs such as the lungs, small intestine, and liver, activates SARS-CoV-2 and increases the vigorous viral proliferation, transmission and stability and causes multiorgan failure.

In humoral immune response to virus infection, T helper (Th) cells assist B cells to differentiate into plasma cells, which in return produces antibodies (Abs) specific to a viral antigen (Ag). The neutralizing antibodies block the virus from further infection. Cellular immunity is mediated by T lymphocytes—helper T cells direct the overall adaptive immune response while cytotoxic T cells play a critical role in clearance and killing of the infected cells.

Recent reports suggest a sharp reduction in CD8 + and CD4 + T cell counts of the SARS-CoV-2-infected individuals which may result in compromised T memory cell generation and persistence in SARS-CoV-2 survivors. In general, lowering of CD3 + T cells, CD4 + T cells, CD8 + T cells, and CD45 + T cells is an implication of infection establishment in severe cases. In severe cases of COVID-19, a sustained decrease in lymphocyte subsets, especially CD4 + T cells and NK cells, and proinflammatory cytokine storms lead to poor prognosis [4]. In COVID-19 patients, cytotoxic CD8 + T cells exhibit functional exhaustion patterns, such as the expression of NKG2A, PD-1, and TIM-3. Since SARS-CoV-2 restrains antigen presentation by downregulating MHC class I and II molecules and, therefore, inhibits the T cell-mediated immune responses, humoral immune responses also play a substantial role in protecting the host.

Monocytes and macrophages are two major leukocyte populations that modulate immune system in the lung. The hyperactivation of these cell populations in lung contributes to lung tissue injury and organ damage. Alveolar and other tissue resident monocytes and macrophages show a dysfunctional phenotype when infected with SARS-CoV-2 and affect their ability to generate strong adaptive immune responses. Their altered functions can lead to cytokine storm, acute inflammation, and organ damage [5]. Peripheral circulating monocytes with inflammation phenotype also contribute to cytokine storm and other adverse immune effects through its interaction with B and T lymphocytes [6]. Very high levels of IL-6 can stimulate the production of inflammatory chemokines and cytokines, such as interferon gamma (IFN-γ), IL-1β, etc. These data open the scope of targeting these cytokines and chemokines and their receptors as targets for drug development. Understanding the mechanism of T cell dysfunction is a next step in the development of targeted therapy against severe COVID-19.

Regulatory T cells (Tregs) suppress immune response and maintain homeostasis of the immune system by inhibiting T cell proliferation and cytokine production. Altered Treg phenotype with a sharp increase in Treg proportions and intracellular levels of Treg transcription factor FoxP3 is reported in severe COVID-19 patients with poor outcome. They overexpress many immune suppressive effectors along with proinflammatory molecules like IL-32 and suppress antiviral T cell responses [7]. This suggests that Tregs may play nefarious roles in COVID-19 via suppressing antiviral T cell responses during the severe phase of the disease.

Patients with severe COVID-19 are also characterized by decrease in lymphocyte count, especially CD8 + T cells and increase in neutrophil counts. The neutrophil-to-lymphocyte ratio (NLR) may be predictive of COVID-19 severity in patients. T cell restoration may reduce the inflammatory response during COVID-19 infection and thus severity [4].

Cytokine storm

Cytokine storm syndrome is a fatal blow up of cytokines due to overreaction by the host immune system in response to an antigen or pathogen. This is one of the causes of COVID-19 severity and mortality in COVID-19 patients. In response to the invaded COVID-19 virus, the immune effector cells release high amounts of chemokines and proinflammatory cytokines that cause uncontrolled systemic inflammatory response [8].

After entering respiratory epithelial cells, SARS-CoV-2 virus evokes inflammatory cytokine production accompanied by a weak interferon (IFN) response. This is followed by the infiltration of macrophages and neutrophils into the lung tissue, which results in a cytokine storm that is characterized by a high expression of IL-6 and TNF-α. The Th1 immune response leads to secretion of proinflammatory cytokines, such as granulocyte-macrophage colony-stimulating factor (GM-CSF) and interleukin-6 (IL-6). GM-CSF further activates CD14 + CD16 + inflammatory monocytes to produce large quantities of IL-6, tumor necrosis factor-α (TNF-α), and other cytokines [9]. IFN-γ induction may be an important amplifier of cytokine production. The levels of IL-1β, IL-6, IL-8, IL-12, inducible protein 10 (IP-10), MCP-1, and IFN-γ are increased during SARS-CoV-2 infection [10]. Drugs targeting IL-18, IL-1, IL-6, and Interferon-gamma would be a possibility to treat COVID-19-associated cytokine storm [12].

Inflammation in COVID-19

COVID-19 most commonly infects the lower respiratory tract of its host via attachment of the virus's Spike (S) protein to the human Angiotensin-converting enzyme (ACE2) receptors found on the alveolar epithelial cells. ACE2 receptors are also present on capillary endothelial cells, meaning organs rich in these cells are vulnerable to secondary infection from SARS-CoV-2, including the kidney, gut, and brain [13].

Diabetes patients with COVID-19 were reported to have higher susceptibility and inflammation. Blood glucose increases the levels of ACE2 present on macrophages and monocytes, helping the virus infect the very cells that should be helping to kill it. Production of reactive oxygen species by the infected immune cells also facilitates inflammation and tissue damage. Chatila and his colleagues identified the excess levels of Notch4 protein receptor found on regulatory T cells (Tregs) as a key instigating mechanism of the runaway inflammation. The Notch 4 overexpression switches Tregs from a normal tissue repair program to one permissive for tissue inflammation. Antiinflammatory drugs such as steroid dexamethasone, colchicine, etc. reduce inflammation and reduce mortality. Innate immune cells, such as macrophages, may contribute to the disease progression through the excess production of IL-6 and CRP which generates excessive inflammation.

Respiratory system failure

COVID-19 can cause multiple organ and tissue damage, particularly in the respiratory system, that leads to death due to massive alveolar damage and progressive respiratory failure [14].

Many acute COVID-19 patients develop acute respiratory distress syndrome (ARDS) that is characterized by acute progressive hypoxic respiratory failure caused by severe hypoxia and acute lung injury (ALI). This develops into interstitial pulmonary edema and pulmonary interstitial fibrosis that can lead to multiple organ failure [15,16]. COVID-19-associated pneumonia presents acute changes with diffuse alveolar damage (DAD) with vascular congestion, intraalveolar edema, hemorrhage, proteinaceous exudate, macrophages, denudation, and reactive hyperplasia of pneumocytes, patchy inflammatory cellular infiltration, and multinucleated giant cells. Many patients who develop ARDS survive the acute phase of the disease; however, many of them develop progressive pulmonary fibrosis subsequently which leads to death [17]. Dysregulation of immune inflammatory molecules results in epithelial and endothelial damage and uncontrolled fibrosis in the lung with fibroblasts exhibiting markers of stress and senescence, resistance to apoptosis, and excessive production of extracellular matrix components [18].

In the pathological anatomy of COVID-19, lungs are hyperemic with increased secretions from the tracheal and bronchial mucosa, diffuse alveolar injury and exudative inflammation with excessive white blood cells infiltration mainly monocytes and macrophages in the alveoli [19].

ACE2 receptors expressed on various host cell surfaces act as the binding domain for viral entry and evasion in COVID-19. Very low levels of ACE2 expression are reported in the ciliated epithelial cells in the upper airways, allowing SARS-CoV-2 infection at these initial sites. The antiviral immune responses elicited after the initial viral entry upregulate ACE2 expression and facilitate the viral spread across the respiratory mucosa and into the parenchyma of the lung, where it can infect type 2 alveolar epithelial cells [20]. Other factors such as smoking, preexisting lung complications, etc. may impact the levels of ACE2 expression in lung that may contribute to disease severity. Understanding the transcriptional regulation of SARS-CoV-2 infection will serve as a reference dataset for future studies of primary samples of COVID-19 patients. In vitro studies addressing the viral replication cycles would provide additional insights.

Cardiovascular manifestations in COVID-19

Respiratory illness is the dominant clinical manifestation of COVID-19; however, 8%–12% of the patients develop acute cardiac injury. This is due to the damage by the viral infections in cardiomyocytes followed by systemic inflammation that leads to myocarditis [21]. Other major cardiac manifestations include acute coronary event, heart failure, cardiac arrhythmia, etc. with high rise in myocardial biomarkers such as cardiac troponins. This directs to the importance of myocardial biomarker evaluation in patients with COVID-19 for timely intervention. The myocardial injury can cause atrial or ventricular fibrosis, and subsequent cardiac arrhythmias after recovery.

Myocarditis is caused by the effects of SARS-CoV-2 infection in the myocardium. Patients with myocardial injury are also associated with higher incidence of ARDS and in need of ventilation. Patients with an increase of troponin are also characterized by higher levels of leukocytes, D-dimer, CRP, ferritin, and IL-6, which point to inflammatory hyperactivity in myocardial injury [11]. Myocarditis related to SARS-CoV-2 is also an important acute ventricular dysfunction associated with diffuse myocardial edema [22].

In COVID-19 patients, both tachy- and brady-arrhythmias are reported which are also one of the major complications. The interplay between host immune system and the viral pathogen is proposed as a cause for arrhythmias. This causes altered intercellular coupling, interstitial edema, and cardiac fibrosis that lead to ion channel dysfunction, repolarization, and action potential abnormalities [23,24]. The increased levels of cytokines release during COVID-19 also may cause the inflammation of conduction tissue and subsequent arrhythmias. In some reports of COVID-19, bradycardia with accelerated idioventricular rhythm and intermittent high-degree AV block were seen in a patient with COVID-19 with a few cases of sinus node dysfunction [25,26]. Further studies are required to assess the risk of patient with preexisting heart conditions and develop more effective treatment strategies.

The detrimental effects of COVID-19 infection in cardiac function could also be prolonged by the severe downregulation of myocardial and pulmonary ACE2 pathways, thereby mediating myocardial inflammation, lung edema, and acute respiratory failure [27]. ACE2 expression in cardiovascular system, cytokine storm triggered by hyperactivation of T cells, respiratory dysfunction and hypoxemia caused by COVID-19, therapeutic use of corticosteroids, etc. can result in damage of myocardial cells [10,28,29].

Neurological and renal manifestation of COVID-19

The most characteristic symptom of SARS-CoV-2 infection is respiratory distress; however, many patients showed neurological abnormalities such as headache, nausea, and vomiting. This indicated that coronavirus can invade nervous system including brain. Acute neurological manifestations such as encephalitis, meningitis, acute cerebrovascular disease, and Guillain-Barré Syndrome (GBS) are associated with COVID-19 in some patients [30]. Consistent with the other organs affected by SARS-CoV-2, ACE2 receptors are expressed on glial tissues, neurons, and brain vasculature which make them a target for virus evasion [31].

Few other mechanisms of targeting CNS are direct infection injury, blood circulation pathway, neuronal pathway, immune-mediated injury, and hypoxic injury [30]. The direct entry into brain tissues is via dissemination and spread from the cribriform plate which is in close proximity to the olfactory bulb and this is supported by the presence of anosmia and hyposmia in COVID-19 patients [32]. In hematogenous spread, the virus enters into cerebral circulation where sluggish blood movement in microvessels enables interaction of the viral spike protein with ACE2 receptors of capillary endothelium. This facilitates the viral budding from capillary endothelium that results in the damage to the endothelial lining and favors viral entry into the milieu of brain and sometimes leading to hemorrhage in the brain [33]. In the neuronal pathway, anterograde and retrograde transport with the help of motor proteins kinesin and dynein via sensory and motor nerve endings or afferent nerve endings of the vagus nerve from the lungs mediates viral entry [34,35]. High immune activation and surge of proinflammatory cytokines escalate the immune-mediated injury in CNS [36]. A few case reports suggest the possibility of encephalitis due to SARS-CoV-2 [37]. Loss of the sense of smell (anosmia) is another neural manifestation of SARS-CoV-2 which is widely used for the initial screening of infection. In COVID-19, inflammation of the coverings of the brain and the spinal cord can later develop as viral meningitis or encephalitis [38], damage the peripheral nerves and leads to Guillain-Barré Syndrome. Approximately, 5.7% of patients with severe COVID-19 developed acute cerebrovascular disease [39] and it usually presents as stroke.

Many patients with acute symptoms of COVID-19 show signs of kidney damage, particularly people with preexisting kidney problems. The major signs of kidney problems in patients with COVID-19 include high levels of protein in the urine and abnormal blood parameters. Kidney has abundant expression of ACE2 receptor in various cell types including brush border of proximal tubular cells and, to a lesser extent, in podocytes and vascular endothelial cells [40,41]. American Society of Nephrology (ASN) reported 15% incidence of acute kidney injury (AKI) in patients with COVID-19 admitted to ICU. Major kidney damage was caused by varying degrees of acute tubular necrosis, luminal brush border sloughing and vacuole degeneration, and lymphocyte infiltration. The tubular damage is caused by either a direct cytotoxicity of COVID-19 or by immune-mediated tubule pathogenesis [42]. Most COVID-19 patients without previous CKD returned to baseline kidney function without the need for chronic renal replacement therapy (RRT).

GI tract manifestations

COVID-19 is mainly a pulmonary disease; however, gastrointestinal symptoms and signs are also prevalent in 11.4 to 50% of patients

[43]. Most common gastrointestinal symptoms are loss of appetite (anorexia), diarrhea, nausea, and vomiting [44], and intensity of these symptoms correlates with the severity of infection. The infection in GI tract is also accompanied by inflammation or intestinal damage with loss of intestinal barrier integrity and gut microbes [45]. It is possible that direct functional damage of the GI tract can be caused by viral interaction with the squamous and columnar epithelium mediated by ACE 2 receptor. However, the expression of ACE 2 receptors and serine protease complex TMPRSS2 is significantly lower or absent in the esophagus and stomach as compared to the intestine [46,47].

Diarrhea is the most common GI symptom associated with COVID-19, but it is usually mild; however, some patients develop severe diarrhea with electrolyte disturbances or bloody, inflammatory diarrhea during or before onset of pulmonary symptoms and develop as ICU case. Bleeding from GI tract is another frequent reason for emergency consultation. Investigations into details of these complications are limited. Research advancements in this area with advanced endoscopy would help the clinical gastroenterologists to get the best guidance through their daily practice.

Conclusion

In COVID-19 patients, the severity of infection and mortality is due to multiorgan failure, including lungs, heart, kidney, brain, etc. Our current understanding suggests that the ACE2 expression and its signaling pathway play a key role in mediating organ damage. The host immune system which tries to protect from COVID-19 also plays a critical role in disease severity and organ failure. When the immune system is functioning normally, COVID-19 infection will go unnoticed. The rapid hyperactivation of innate and adaptive immune system to combat the virus infection which is characterized by increased production of cytokines and chemokines, such as interleukin (IL)-2, IL-7, IFNy, granulocyte-colony stimulating factor, macrophage inflammatory protein 1, monocyte chemoattractant protein 1, and tumor necrosis factor-α (TNF-α), etc., generates inflammation and cytokine storm. These hyperimmune responses also cause organ damage. The high variability in these responses in different patients adds up challenges to use a standard treatment for all. More investigation and characterization of these responses in patients in diverse age groups, gender, ethnicity, etc. would help to predict the severity of infection and take the necessary steps to manage inflammation, organ damage, and death due to multiorgan failure.

Acknowledgment

We would like to acknowledge the help of Ms. Misba Majood for the creation of figures using BioRende.com.

References

[1] Hope JL, Bradley LM. Lessons in antiviral immunity. Science 2021;371(6528):464–5.
[2] Fast E, Chen B. Potential T-cell and B-cell Epitopes of 2019-nCoV. bioRxiv 2020. https://doi.org/10.1101/2020.02.19.955484.
[3] Walls AC, Park YJ, Tortorici MA, Wall A, McGuire AT, Veesler D. Structure, function, and antigenicity of the SARS-CoV-2 spike glycoprotein. Cell 2020;181(2):281–292.e6. https://doi.org/10.1016/j.cell.2020.02.058. Epub 2020 March 9. Erratum in: Cell. 2020 December 10; 183(6):1735. PMID: 32155444; PMCID: PMC7102599.
[4] Liu Y, Tan W, Chen H, Zhu Y, Wan L, Jiang K, Guo Y, Tang K, Xie C, Yi H, Kuang Y, Luo Y. Dynamic changes in lymphocyte subsets and parallel cytokine levels in patients with severe and critical COVID-19. BMC Infect Dis 2021;21(1):79. https://doi.org/10.1186/s12879-021-05792-7. PMID: 33461503; PMCID: PMC7812569.
[5] Meidaninikjeh S, Sabouni N, Marzouni HZ, Bengar S, Khalili A, Jafari R. Monocytes and macrophages in COVID-19: friends and foes. Life Sci 2021;269. https://doi.org/10.1016/j.lfs.2020.119010, 119010.
[6] Zhou Y, Fu B, Zheng X, Wang D, Zhao C, Qi Y, Sun R, Tian Z, Xu X, Wei H. Aberrant pathogenic GM-CSF+ T cells and inflammatory CD14+CD16+ monocytes in severe pulmonary syndrome patients of a new coronavirus. bioRxiv 2020. 2020.02.12.945576.
[7] Galvan-Pena S, Leon J, Chowdhary K, Michelson DA, Vijaykumar B, Yang L, Magnuson A, Manickas-Hill Z, Piechocka-Trocha A, Worrall DP, Hall KE, Ghebremichael M, Walker BD, Li JZ, Yu XG, Mathis D, Benoist C. Profound Treg perturbations correlate with COVID-19 severity. bioRxiv [Preprint] 2020. https://doi.org/10.1101/2020.12.11.416180. PMID: 33330871; PMCID: PMC7743083.
[8] Yao X, Ye F, Zhang M, Cui C, Huang B, Niu P, Liu X, Zhao L, Dong E, Song C, Zhan S, Lu R, Li H, Tan W, Liu D. In vitro antiviral activity and projection of optimized dosing design of hydroxychloroquine for the treatment of severe acute respiratory syndrome coronavirus 2 (SARS-CoV-2). Clin Infect Dis 2020. https://doi.org/10.1093/cid/ciaa237. pii: ciaa237.
[9] Haiming W, Xiaoling X, Yonggang Z, et al. Aberrant pathogenic GM-CSF+ T cells and inflammatory CD14+CD16+ monocytes in severe pulmonary syndrome patients of a new coronavirus. BioRXiv 2020.
[10] Wong CK, Lam CW, Wu AK, et al. Plasma inflammatory cytokines and chemokines in severe acute respiratory syndrome. Clin Exp Immunol 2004;136(1):95–103.
[11] Shi S, Qin M, Shen B, Cai Y, Liu T, Yang F, et al. Association of cardiac injury with Mortality in hospitalized patients with COVID-19 in Wuhan, China. JAMA Cardiol 2020;5:802–10.
[12] Cameron MJ, Ran L, Xu L, Danesh A, Bermejo-Martin JF, Cameron CM, Muller MP, Gold WL, Richardson SE, Poutanen SM, Willey BM, DeVries ME, Fang Y, Seneviratne C, Bosinger SE, Persad D, Wilkinson P, Greller LD, Somogyi R, Humar A, Keshavjee S, Louie M, Loeb MB, Brunton J, McGeer AJ, Canadian SARS Research Network, Kelvin DJ. Interferon-mediated immunopathological events are associated with atypical innate and adaptive immune responses in patients with severe acute respiratory syndrome. J Virol 2007;81(16):8692–706.
[13] Samavati L, Uhal B. ACE2, much more than just a receptor for SARS-COV-2. Front Cell Infect Microbiol 2020;10(317).

[14] Xu Z, Lei S, Wang Y, et al. Pathological findings of COVID-19 associated with acute respiratory distress syndrome. Lancet Resp Med 2020;8:420–2.
[15] Xu Z, Shi L, Wang Y, Zhang J, Huang L, Zhang C, Liu S, Zhao P, Liu H, Zhu L, Tai Y, Bai C, Gao T, Song J, Xia P, Dong J, Zhao J, Wang FS. Pathological findings of COVID-19 associated with acute respiratory distress syndrome. Lancet Respir Med 2020;8(4):420–2. https://doi.org/10.1016/S2213-2600(20)30076-X. Epub 2020 February 18. Erratum in: Lancet Respir Med. 2020 February 25; PMID: 32085846; PMCID: PMC7164771.
[16] Ware LB, Matthay MA. The acute respiratory distress syndrome. N Engl J Med 2000;342:1334–49.
[17] George PM, Wells AU, Jenkins RG. Pulmonary fibrosis and COVID-19: the potential role for antifibrotic therapy. Lancet Resp Med 2020;8:807–15.
[18] Barratt SL, Creamer A, Hayton C, Chaudhuri N. Idiopathic pulmonary fibrosis (IPF): an overview. J Clin Med 2018;7:201.
[19] Rabaan AA, Al-Ahmed S, Haque S, et al. SARS-CoV-2, SARS-CoV, and MERS-COV: a comparative overview. Infez Med 2020;28:174–84.
[20] Nawijn MC, Timens W. Can ACE2 expression explain SARS-CoV-2 infection of the respiratory epithelia in COVID-19? Mol Syst Biol 2020;16(7). https://doi.org/10.15252/msb.20209841, e9841.
[21] Bansal M. Cardiovascular disease and COVID-19. Diabetes Metab Syndr 2020;14(3):247–50. https://doi.org/10.1016/j.dsx.2020.03.013. Epub 2020 March 25. PMID: 32247212; PMCID: PMC7102662.
[22] Hua A, O'Gallagher K, Sado D, Byrne J. Life-threatening cardiac tamponade complicating myo-pericarditis in COVID-19. Eur Heart J 2020;41:2130. https://doi.org/10.1093/eurheartj/ehaa253.
[23] Gaaloul I, Riabi S, Harrath R, Evans M, Salem NH, Mlayeh S, Huber S, Aouni M. Sudden unexpected death related to enterovirus myocarditis: histopathology, immunohistochemistry and molecular pathology diagnosis at post-mortem. BMC Infect Dis 2012;12:212.
[24] Tse G, Yeo JM, Chan YW, Lai ETHL, Yan BP. What is the arrhythmic substrate in viral myocarditis? Insights from clinical and animal studies. Front Physiol 2016;7:308.
[25] Kir D, Mohan C, Sancassani R. Heart brake: an unusual cardiac manifestation of coronavirus disease 2019 (COVID-19). JACC Case Rep 2020;2:1252–5.
[26] Peigh G, Leya MV, Baman JR, Cantey EP, Knight BP, Flaherty JD. Novel coronavirus 19 (COVID-19) associated sinus node dysfunction: a case series. Eur Heart J—Case Rep 2020. https://doi.org/10.1093/ehjcr/ytaa132.
[27] Oudit GY, Kassiri Z, Jiang C, et al. SARS-coronavirus modulation of myocardial ACE2 expression and inflammation in patients with SAR. Eur J Clin Invest 2009;39:618–25.
[28] Huang C, Wang Y, Li X, et al. Clinical features of patients infected with 2019 novel coronavirus in Wuhan, China. Lancet 2020;395:497–506.
[29] Cameron MJ, Bermejo-Martin JF, Danesh A, Muller MP, Kelvin DJ. Human immunopathogenesis of severe acute respiratory syndrome (SARS). Virus Res 2008;133:13–9.
[30] Ahmed MU, Hanif M, Ali MJ, et al. Neurological manifestations of COVID-19 (SARS-CoV-2): a review. Front Neurol 2020;11:518. Published 2020 May 22 https://doi.org/10.3389/fneur.2020.00518.
[31] Turner AJ, Hiscox JA, Hooper NM. ACE2: from vasopeptidase to SARS virus receptor. Trends Pharmacol Sci 2004;25:291–4. https://doi.org/10.1016/j.tips.2004.04.001.
[32] Baig AM, Khaleeq A, Ali U, Syeda H. Evidence of the COVID-19 virus targeting the CNS: tissue distribution, host–virus interaction, and proposed neurotropic mechanisms. ACS Chem Nerosci 2020;11:995–8. https://doi.org/10.1021/acschemneuro.0c00122.

[33] Wrapp D, Wang N, Corbett KS, Goldsmith JA, Hsieh CL, Abiona O, et al. Cryo-EM structure of the 2019-nCoV spike in the prefusion conformation. Science 2020;367:1260–3. https://doi.org/10.1126/science.abb2507.
[34] Li YC, Bai WZ, Hashikawa T. The neuroinvasive potential of SARS-CoV2 may play a role in the respiratory failure of COVID-19 patients. J Med Virol 2020;92:552–5. https://doi.org/10.1002/jmv.25728.
[35] Wong SH, Lui RN, Sung JJ. Covid-19 and the digestive system. J Gastroenterol Hepatol 2020;35:744–8. https://doi.org/10.1111/jgh.15047.
[36] Bohmwald K, Gálvez NMS, Ríos M, Kalergis AM. Neurologic alterations due to respiratory virus infections. Front Cell Neurosci 2018;12:386. https://doi.org/10.3389/fncel.2018.00386.
[37] Poyiadji N, Shahin G, Noujaim D, Stone M, Patel S, Griffith B. COVID-19-associated acute hemorrhagic necrotizing encephalopathy: CT and MRI features. Radiology 2020. https://doi.org/10.1148/radiol.2020201187, 201187.
[38] Moriguchi T, Harii N, Goto J, Haradaa D, Sugawaraa H, Takamino J, et al. A first case of meningitis/encephalitis associated with SARS-Coronavirus-2. Int J Infect Dis 2020;94:55–8. https://doi.org/10.1016/j.ijid.2020.03.062.
[39] Mao L, Jin H, Wang M, Hu Y, Chen S, He Q, et al. Neurologic manifestations of hospitalized patients with coronavirus disease 2019 in Wuhan, China. JAMA Neurol 2020;e201127.
[40] Lely A, Hamming I, van Goor H, Navis G. Renal ACE2 expression in human kidney disease. J Pathol 2004;204:587–93. https://doi.org/10.1002/path.1670.
[41] Bader M. ACE2, angiotensin-(1–7), and Mas: the other side of the coin. Pflügers Archiv Eur J Physiol 2013;465:79–85.
[42] Diao B, Wang C, Wang R, Feng Z, Tan Y, Wang H, Wang C, Liu L, Liu Y, Liu Y, et al. Human kidney is a target for novel severe acute respiratory syndrome coronavirus 2 (SARS-CoV-2). Infection medRxiv 2003.
[43] Pan L, Mu M, Yang P, et al. Clinical characteristics of COVID-19 patients with digestive symptoms in Hubei, China: a descriptive cross-sectional multicenter study. Am J Gastroenterol 2020;115(5):766–73.
[44] Parasa S, et al. Prevalence of gastrointestinal symptoms and fecal viral shedding in patients with coronavirus disease 2019: a systematic review and meta-analysis. JAMA Netw Open 2020;3, e2011335.
[45] Villapol S. Gastrointestinal symptoms associated with COVID-19: impact on the gut microbiome. Transl Res 2020;226:57–69. https://doi.org/10.1016/j.trsl.2020.08.004. Epub 2020 August 20. PMID: 32827705; PMCID: PMC7438210.
[46] Zhang H, Kang Z, Gong H, Xu D, Wang J, Li Z, et al. Digestive system is a potential route of COVID-19: an analysis of single-cell co-expression pattern of key proteins in viral entry process. Gut 2020;69(6):1010–8.
[47] Muus C, Luecken MD, Eraslan G, et al. Integrated analyses of single-cell atlases reveal age, gender, and smoking status associations with cell type-specific expression of mediators of SARS-CoV-2 viral entry and highlights inflammatory programs in putative target cells. bioRxiv 2020. https://doi.org/10.1101/2020.04.19.049254. http://biorxiv.org/content/early/2020/04/21/2020.04.19.049254.

5

Cytokine storm and stem cell activation in unveiling potential targets for diagnosis and therapy

Daniel Miranda, Jr.* and David Jesse Sanchez*

Pharmaceutical Sciences Department, College of Pharmacy, Western University of Health Sciences, Pomona, CA, United States

Introduction

As the COVID-19 pandemic unfolds, a cytokine storm, also known as cytokine release syndrome [1], has become a critical pathology of Severe Acute Respiratory Syndrome-Coronavirus-2 (SARS-CoV-2) infection [2]. Cytokine storms generally are induced and characterized by a heightened inflammatory or immune activated state that leads to dramatic changes in a person's physiology. The development of a cytokine storm during a COVID-19 infection often requires immediate, critical care to combat the extreme pathophysiology that can arise [3]. Due to the multiple possible triggering events of a cytokine storm, as well as the multifaceted outcomes that could impact a person's health, few therapeutic approaches are consistently used. Long-term COVID-19-induced impact or remodeling of the lungs and cardiovascular system has necessitated looking at treatment of a COVID-19 patient from additional perspectives. In this chapter, we will review cytokine storms and how they impact COVID-19 with a focus on the outcomes and possible utility of stems cells in combating the consequences of cytokine storms.

Cytokine storms

Cytokines are protein messengers that allow communication between cells, as well as inducing effector functions of the immune system. As such, they are key components of an effective immune response. From the discovery of Type I Interferons in the 1950s to the

* Authors contributed equally to this work

Stem Cells and COVID-19. https://doi.org/10.1016/B978-0-323-89972-7.00006-4

substantial present-day catalog of cytokines that have distinct and overlapping functions, the list of cytokines and their importance in immunity cannot be overstated. However, over time, we have come to appreciate that while cytokines are good for the immune system, they are only good in moderation. Sustained cytokine expression can lead to chronic inflammatory conditions, while hyperacute release of cytokines can induce a variety of pathologies including cytokine storms. Excessive release of cytokines during a cytokine storm can have life-threatening consequences.

For example, the development of Acute Respiratory Distress Syndrome (ARDS) is common when excessive inflammation or severe cytokine storms impact the lungs. Patients with ARDS have difficulty breathing and unless the precursors and inducers of ARDS are rapidly controlled then the patient has an increased risk of succumbing to the disease [4]. In COVID-19 infections, control of viral replication and control of the systemic inflammatory components are most important in controlling ARDS. This state of uncontrolled inflammation is exacerbated in the elderly and can thus lead to profound disease in those individuals who are infected.

Cytokine storms have now been seen in a variety of scenarios, many of which involve severe infection, or can be seen accompanying cancer treatments based on immunotherapies. Chimeric Antigen Receptor (CAR)-T cell-based therapies are now well known to have the capacity to induce side effect cytokine storms [5]. This led to the use of tocilizumab, an antiinterleukin-6 receptor monoclonal antibody, to help abate the symptoms associated with the excessive cytokine release during CAR-T cell therapy, even in FDA-approved CAR-T cell therapies [6,7]. With the expanded use of CAR-T cells, and the notoriety to have concomitant tocilizumab, an unexpected consequence of this CAR-T cell and tocilizumab combination was tocilizumab, being one of the first therapies used for severe cases of COVID-19 [8–10].

Cytokine storms in viral respiratory infections

Influenza shapes the idea of a cytokine storm

For viral respiratory infections, severe pathogenic influenza virus infection has become a standard to study the components of a cytokine storm, as well as its impact on disease and pathology [11]. In severe infections with avian influenza H5N1 [12], pandemic swine flu H1N1 [13], and influenza H7N9 [14], the associated enhanced production of cytokines is believed to be connected to outcomes such as ARDS and death. In these influenza infections, heightened levels of proinflammatory cytokines such as TNF-alpha are commonly seen [15,16]. However, as with all viral-induced cytokine storm there is no clear-cut pattern of cytokine release, even for diseases of the same general viral etiology.

On the other hand, Type I Interferons, potent antiviral cytokines, are often induced during influenza infections leading to a speed bump in the replication of virus [17,18]. Detailed analysis of gene expression in severe versus mild cases of H5N1 influenza noted that interferon expression was directly linked to a more controlled disease outcome [19]. This strongly implies that Type I Interferon-mediated control of viral replication is key in controlling the release of cytokines and the overall pathology of a severe viral infection.

In general, cytokine storms reported for severe influenza infections have focused on several key cytokines [20]. As mentioned, TNF-alpha is consistently elevated in influenza infections. Additionally, interleukein-6 (IL-6) and Interferon-gamma were seen to be elevated in serums of patients who had severe influenza infections with associated cytokine storm pathology [21].

While a variety of approaches to target influenza-induced cytokine storms have been used, there seems to be a focus on therapies that reduce the hyperactivation of the immune system [11]. While influenza set the foundation for understanding viral-induced cytokine storms, the emergence of human pathogenic coronaviruses began to show the importance of these etiologic agents triggering cytokine storms with different hallmarks.

Lessons from SARS

Previous coronavirus outbreaks have informed our understanding of the general course of both mild and severe coronavirus infections. In particular, lessons learned from the outbreaks of Severe Acute Respiratory Syndrome (SARS) and Middle East Respiratory Syndrome (MERS) over the past years have shown similar signs of this coronavirus-type infection. Several comprehensive reviews of the immune response to SARS-Coronavirus-1 (SARS-CoV-1) are available [22,23].

Some of the earliest milestones in a coronavirus infection were seen in SARS patients in 2003, where flu-like symptoms with fever were associated with infection progression [24]. However, closer immunological analysis uncovered that the classic antiviral immune profiles seen in influenza infections, where Interferon-gamma and TNF-alpha were in the forefront, were skewed in SARS patients. Compared to influenza infections, SARS patients had sharply decreased or undetectable Interferon-gamma levels [25,26] as well as low levels of the classic proinflammatory cytokine TNF-alpha. However, IL-6 continued to remain elevated in patients with severe SARS [27].

The lack of proper Type I Interferon control of SARS, due to viral antagonism of the innate immune system, has been proposed as a mechanism leading to cytokine storm [28]. Type I Interferons are a key component in the control of viral replication and, as seen in influenza infection, their impact on viral replication early on in infection may be important in tipping the scales toward health or disease.

This theme of coronaviruses changing and misdirecting the course of cytokine release in standard antiviral models is key in the induction of cytokine storms in these viral infections. However, as has been highlighted by the heterogeneity of COVID-19 patients, the exact cascade leading to a cytokine storm is not consistent among patients with severe coronavirus infection. Of note, a study by Huang *et al.* reports that while the signature profile of pathogenic cytokines is consistent in their patient cohort, they see a sharp elevation of Interferon-gamma, which is different than other studies that showed Interferon-gamma was low in this state of infection [29]. This study still showed a marked lack of TNF-alpha.

In terms of lymphocytes, SARS patients have a noted lymphopenia during acute SARS infection as opposed to other acute viral infections [30]. Interestingly, resolution of SARS infection leads to rapid return of lymphocyte levels leading investigators to speculate that the tissue infiltration of lymphocytes into target tissue could be a significant component of pathogenic SARS infection. Histopathology has not supported steady findings on this idea, as a lack of consistent tissue infiltration by lymphocytes has been observed in pathology [31].

The COVID-19 cytokine storm

The widespread COVID-19 pandemic led to the identification of a large number of severe cases which included pathology consistent with cytokine storm. Indeed, the diverse clinical outcomes allowed for a significant number of studies that sought to define and predict the course of COVID-19 disease. Importantly, past studies into other viruses also allowed clinicians and researchers to tease out nuanced details and begin to elucidate the COVID-19 cytokine storm.

The impact of Type I Interferon on SARS-CoV-2 infection has been directly compared to infection by SARS-CoV-1 [32]. In these in vitro studies, the authors concluded that SARS-CoV-2 is more sensitive to the antiviral effects of Type I Interferon. This strongly implies that antiviral gene programs that led to Type I Interferon release will control a SARS-CoV-2 infection. This idea has been supported by findings that severe SARS-CoV-2 infections became pronounced in a subgroup of individuals with genetic errors in the Type I Interferon immune pathway [33]. An additional finding also showed that a portion of individuals with severe SARS-CoV-2 infections had autoreactive antibodies specifically targeting Type I Interferons [34]. Taken together, while levels of Type I Interferon may increase, they are functionally lessened in many patients with severe COVID-19. Type I Interferons are integral in controlling SARS-CoV-2 infection and, without their proper function, SARS-CoV-2 virus may be able to replicate to higher levels than the immune system can properly deal with. Similar to SARS, Type II Interferon, Interferon-gamma, was found to be inconsistent in its response to SARS-CoV-2 infection, and TNF-alpha was lowered in patients with severe COVID-19 infection [35].

In the early months of the COVID-19 pandemic, the identification of the COVID-19-induced cytokine storm, together with the extensive experience of clinicians combating cytokine storms in other conditions, led to the use of different therapies. These therapies targeted the individual cytokines involved in the cytokine storm. Upregulated IL-6 levels during COVID-19 became a central focus in treating severe COVID-19 infection [36]. In fact, some of the earliest studies of people that succumbed to COVID-19 noticed the upregulation of IL-6 as a key diagnostic marker [37]. However, coming to a definitive conclusion to use IL-6 blocking antibodies, such as tocilizumab, as an efficacious treatment, has been complex [38,39]. Initially, several smaller studies suggested benefit could be achieved if tocilizumab was administered at the right point or timing of infection [40,41]. Additional clinical trials found that targeting IL-6 provided variable, often moderate [42] to no benefit over standard of care at the time [43,44].

Blocking one cytokine, though a standout component of the COVID-19 induced cytokine storm as IL-6, was not sufficient for a generalized standard of care for acute stage COVID-19 disease. Other studies found that using the IL-1 receptor antagonist, anakinra, compared to IL-6 blocking monoclonal antibodies, could provide some benefit to patients by showing a lowered mortality risk [45]. The complex guidelines to administer cytokine-targeting drugs in COVID-19 patients were overshadowed by the discovered benefit of using glucocorticoids, such as dexamethasone, that provided a more general control of the hyperimmune activation, leading to consistently better outcomes in the more advanced COVID-19 patients who had cytokine storms during their infections [46]. Overall, it seemed that the multiple facets of a cytokine storm were not easily targeted one by one, but instead required a more generalized, systematic change to the immune landscape to provide consistent therapeutic outcomes.

Cytokine storms and stem cells

Impact of cytokines on stem cell activation

Hematopoietic stem cells are the most well-known targets of cytokines [47]. The classic view of the interplay between cytokines and hematopoietic stem cells is based on the need for acute inflammation and cytokine release, calling forth and mobilizing hematopoietic stem cells from the bone marrow for advanced expansion of particular immune components. The use of filgrastim, a drug that is the cytokine Granulocyte Colony-Stimulating Factor (GCSF), extended the canonical role of cytokines as not only proliferation inducers, leading to increased terminally differentiated cells like a neutrophil, but also mobilization enhancers of the hematopoietic stem cells themselves [48].

However, recently, a more complex picture of the impact of cytokines on the hematopoietic stem cell compartment has started to emerge [49]. In particular, disease pathogenesis has led to elucidating the contribution of stem cells on the development of disease. For example, atherosclerosis has a multistep pathology with strong inflammatory instruction that involves a large collection of cytokines [50]. The accumulation of smooth muscle cells in the development of an atherosclerotic plaque has been linked to hematopoietic stem cell recruitment, induced by cytokine overproduction, and their differentiation into vascular cells.

When considering the key cytokines in a cytokine storm, the contribution of these cytokines and their impacts on stem cells is difficult to define. The role of Interferon-gamma on hematopoietic stem cells is often context dependent and distinct from its classic role in an active immune response. For instance, in mice, Interferon-gamma drives the differentiation of the myeloid-biased hematopoietic stem cells during viral infection [51]. The consequence of favoring myeloid-biased hematopoietic stem cells to differentiate is that the bone marrow appears to be acutely depleted of myeloid-biased hematopoietic stem cells. While transient, this does account for a loss of this component during viral infections and may be consequential to the entire homeostasis of the immune response, similar to the impact chronic infection has on hematopoietic stem cell differentiation [52].

Type I Interferons, which would normally control a viral infection, have had different impacts on hematopoietic stem cells. Hematopoietic stem cells can start actively cycling when exposed to acute levels of Interferon-alpha; however, chronic stimulation actually limits the ability of the stem cells to grow and expand [53]. This strongly implies that the impact of a viral infection on the homeostasis of hematopoietic stem cells is being regulated by cytokine levels.

Using mesenchymal stem cells as COVID-19 therapies

Hematopoietic stem cells, as a natural foundation of the different components of the immune response, have mostly been the focus of research into how stem cells and cytokines interact. Mobilization and differentiation of the hematopoietic stem cell component are indeed one of the main aspects of how hematopoietic stem cells contribute to immune responses and that these responses are dependent on cytokine instruction. Mesenchymal stem cells, on the other hand, are stem cells from a stromal compartment that are well known for their ability to induce control of the immune system, as well as, repair of areas [54].

The ability of mesenchymal stem cells to modulate the immune system has been extensively studied and seems to be pleiotropic. For example, mesenchymal stem cells derived from human umbilical cord can control acute lung injury in mice [55]. Other models include lipopolysaccharide-induced ARDS in mice, where less neutrophil infiltration and fibrotic outcomes are seen in the mice after mesenchymal stem cell administration [56]. The ability of mesenchymal stem cells to push toward a favorable balance between different subsets of T cells, for example regulatory and TH17 T cells, was found to be an important outcome in lessening disease in other lipopolysaccharide-induced disease. Initially, the main function of mesenchymal stem cells seemed to focus on the ability to dampen T cell responses based on the activity of indoleamine 2, 3-dioxygenase [57]. However, others have shown that mesenchymal stem cells have many antiinflammatory mechanisms. For example, Inducible T Cell Costimulator Ligand (ICOSL) expression in human bone marrow-derived mesenchymal stem cells has been found to promote the activation of T regulatory cells [58].

Conclusion

As our ability to track individual immune components, such as cytokines, has developed over the past few decades, we have begun to define hallmarks in disease manifestation. In particular, overproduction of cytokines, or unbalanced cytokine production, in the midst of a strong viral infection, can become detrimental to a host to the point where this cytokine storm can lead to increased mortality. Influenza infections have set the groundwork for understanding how viral infection can lead to cytokine storms and the consequences that arise from such events. Importantly, cytokine storms are not a generalized overproduction of all cytokines but rather a skewing of cytokine production toward particular ones, to either be overexpressed or underexpressed during the infection. Knowing which cytokine to target in a cytokine storm is the purpose of much ongoing work.

The development of targeted immunotherapies to dampen cytokine function has become a key part of mitigating disease treatments that have a substantial immunologic focus. For example, targeted immunotherapies, such as blocking IL-6 with monoclonal antibodies or using receptor antagonist against IL-1, such as anakinra, represent strategies that have had considerable success outside of viral infections. While direct-acting antivirals represent a bull's eye to block viral infection, the use of immune modulating therapies has become integral in standards of care for viral infection, to reduce the damage caused by excessive inflammation or systemic immune responses.

Since the early 2000s, pathogenic coronaviruses have slowly revealed themselves, through the outbreaks of SARS, MERS, and the pandemic spread of COVID-19, to be major factors in the health and well-being of our global society. The spread and pathology of these viruses leads to a variety of diseases, especially in the respiratory tract, with outcomes such as ARDS. The increased risk of mortality after a pathogenic coronavirus infection gave impetus to understand possible targets to ameliorate the disease outcomes. Cytokine storms became a key focus as to how acute disease manifested in COVID-19 patients, and the need to dampen this immunologic event, to lower the risk of mortality or long-term consequences, became clear.

The number of cytokines upregulated in COVID-19 infection makes targeting any one cytokine difficult. For example, targeting IL-6 was a standard approach early on in the pandemic for treatment of severe COVID-19 cases. However, the eventual use of glucocorticoids such as dexamethasone which more generally dampened the immune system showed that honing in on the right cytokine or signaling component is not clear-cut and would take time. Use of general immune dampening therapies such as dexamethasone as well as clinical trials to determine the possible use of mesenchymal stem cells, that have more broad immune modulation properties, have been placed at the forefront of new potential therapies. As COVID-19 is a heterogenous disease, with a range of disease manifestations from asymptomatic to aggressive ARDS, the use of diagnostics tests to determine cytokine levels is critical to target and stop the progression of a cytokine storm. However, the eventual receding of the pandemic may slowly lessen the momentum to determine a targeted approach to control the cytokine storm. Hopefully, the information and findings for treating COVID-19 can lead to new therapies for cytokine storms and harness the destructive force of the immune system into a stronger protection against these severe viral infections (Fig. 1).

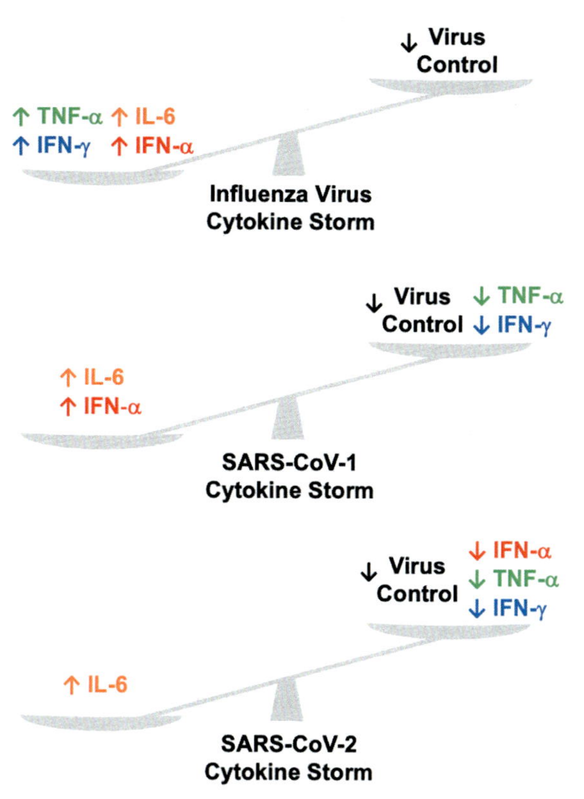

Fig. 1 Balance of cytokines during different viral infections and cytokine storms. Depiction of the balance and imbalance of cytokines and virus control during different viral infection-induced Cytokine Storms. Represented are modulation of cytokines and control of viral replication in influenza virus, SARS-CoV-1 and -2, cytokine storms.

Acknowledgments

DJS was partially supported by NIH R15AI138847 and NIH R21CA261297.

References

[1] Liu B, et al. Can we use interleukin-6 (IL-6) blockade for coronavirus disease 2019 (COVID-19)-induced cytokine release syndrome (CRS)? J Autoimmun 2020;111:102452.

[2] Olbei M, et al. SARS-CoV-2 causes a different cytokine response compared to other cytokine storm-causing respiratory viruses in severely ill patients. Front Immunol 2021;12:629193.

[3] Webb BJ, et al. Clinical criteria for COVID-19-associated hyperinflammatory syndrome: a cohort study. Lancet Rheumatol 2020;2(12):e754–63.

[4] Torres Acosta MA, Singer BD. Pathogenesis of COVID-19-induced ARDS: implications for an ageing population. Eur Respir J 2020;56(3).
[5] Morgan RA, et al. Case report of a serious adverse event following the administration of T cells transduced with a chimeric antigen receptor recognizing ERBB2. Mol Ther 2010;18(4):843–51.
[6] Grupp SA, et al. Chimeric antigen receptor-modified T cells for acute lymphoid leukemia. N Engl J Med 2013;368(16):1509–18.
[7] Le RQ, et al. FDA approval summary: tocilizumab for treatment of chimeric antigen receptor T cell-induced severe or life-threatening cytokine release syndrome. Oncologist 2018;23(8):943–7.
[8] Rosas IO, et al. Tocilizumab in hospitalized patients with severe Covid-19 pneumonia. N Engl J Med 2021;384(16):1503–16.
[9] Rossi B, et al. Effect of tocilizumab in hospitalized patients with severe COVID-19 pneumonia: a case-control cohort study. Pharmaceuticals (Basel) 2020;13(10).
[10] Valenzuela O, et al. First report of tocilizumab use in a cohort of Latin American patients hospitalized for severe COVID-19 pneumonia. Front Med (Lausanne) 2020;7:596916.
[11] Liu Q, Zhou YH, Yang ZQ. The cytokine storm of severe influenza and development of immunomodulatory therapy. Cell Mol Immunol 2016;13(1):3–10.
[12] de Jong MD, et al. Fatal outcome of human influenza a (H5N1) is associated with high viral load and hypercytokinemia. Nat Med 2006;12(10):1203–7.
[13] Gao R, et al. Cytokine and chemokine profiles in lung tissues from fatal cases of 2009 pandemic influenza A (H1N1): role of the host immune response in pathogenesis. Am J Pathol 2013;183(4):1258–68.
[14] Chi Y, et al. Cytokine and chemokine levels in patients infected with the novel avian influenza a (H7N9) virus in China. J Infect Dis 2013;208(12):1962–7.
[15] Cheung CY, et al. Induction of proinflammatory cytokines in human macrophages by influenza a (H5N1) viruses: a mechanism for the unusual severity of human disease? Lancet 2002;360(9348):1831–7.
[16] de Wit E, et al. 1918 H1N1 influenza virus replicates and induces proinflammatory cytokine responses in extrarespiratory tissues of ferrets. J Infect Dis 2018;217(8):1237–46.
[17] Mi Z, Ma Y, Tong Y. Avian influenza virus H5N1 induces rapid interferon-beta production but shows more potent inhibition to retinoic acid-inducible gene I expression than H1N1 in vitro. Virol J 2012;9:145.
[18] Zhou J, et al. Biological features of novel avian influenza A (H7N9) virus. Nature 2013;499(7459):500–3.
[19] Muramoto Y, et al. Disease severity is associated with differential gene expression at the early and late phases of infection in nonhuman primates infected with different H5N1 highly pathogenic avian influenza viruses. J Virol 2014;88(16):8981–97.
[20] Beigel JH, et al. Avian influenza A (H5N1) infection in humans. N Engl J Med 2005;353(13):1374–85.
[21] To KF, et al. Pathology of fatal human infection associated with avian influenza A H5N1 virus. J Med Virol 2001;63(3):242–6.
[22] Chen J, Subbarao K. The Immunobiology of SARS*. Annu Rev Immunol 2007;25:443–72.
[23] Totura AL, Baric RS. SARS coronavirus pathogenesis: host innate immune responses and viral antagonism of interferon. Curr Opin Virol 2012;2(3):264–75.
[24] Peiris JS, et al. Clinical progression and viral load in a community outbreak of coronavirus-associated SARS pneumonia: a prospective study. Lancet 2003;361(9371):1767–72.
[25] Jiang Y, et al. Characterization of cytokine/chemokine profiles of severe acute respiratory syndrome. Am J Respir Crit Care Med 2005;171(8):850–7.

[26] Tang NL, et al. Early enhanced expression of interferon-inducible protein-10 (CXCL-10) and other chemokines predicts adverse outcome in severe acute respiratory syndrome. Clin Chem 2005;51(12):2333–40.
[27] Zhang Y, et al. Analysis of serum cytokines in patients with severe acute respiratory syndrome. Infect Immun 2004;72(8):4410–5.
[28] Channappanavar R, Perlman S. Pathogenic human coronavirus infections: causes and consequences of cytokine storm and immunopathology. Semin Immunopathol 2017;39(5):529–39.
[29] Huang KJ, et al. An interferon-gamma-related cytokine storm in SARS patients. J Med Virol 2005;75(2):185–94.
[30] Li T, et al. Significant changes of peripheral T lymphocyte subsets in patients with severe acute respiratory syndrome. J Infect Dis 2004;189(4):648–51.
[31] Gu J, et al. Multiple organ infection and the pathogenesis of SARS. J Exp Med 2005;202(3):415–24.
[32] Schroeder S, et al. Interferon antagonism by SARS-CoV-2: a functional study using reverse genetics. Lancet Microbe 2021;2(5):e210–8.
[33] Zhang Q, et al. Inborn errors of type I IFN immunity in patients with life-threatening COVID-19. Science 2020;370(6515).
[34] Bastard P, et al. Autoantibodies against type I IFNs in patients with life-threatening COVID-19. Science 2020;370(6515).
[35] Tang Y, et al. Cytokine storm in COVID-19: the current evidence and treatment strategies. Front Immunol 2020;11:1708.
[36] Chen X, et al. Detectable serum severe acute respiratory syndrome coronavirus 2 viral load (RNAemia) is closely correlated with drastically elevated interleukin 6 level in critically ill patients with coronavirus disease 2019. Clin Infect Dis 2020;71(8):1937–42.
[37] Ruan Q, et al. Clinical predictors of mortality due to COVID-19 based on an analysis of data of 150 patients from Wuhan, China. Intensive Care Med 2020;46(5):846–8.
[38] Attaway AH, et al. Severe covid-19 pneumonia: pathogenesis and clinical management. BMJ 2021;372, n436.
[39] Huang E, Jordan SC. Tocilizumab for Covid-19 - the ongoing search for effective therapies. N Engl J Med 2020;383(24):2387–8.
[40] Luo P, et al. Tocilizumab treatment in COVID-19: a single center experience. J Med Virol 2020;92(7):814–8.
[41] Guaraldi G, et al. Tocilizumab in patients with severe COVID-19: a retrospective cohort study. Lancet Rheumatol 2020;2(8):e474–84.
[42] Hermine O, et al. Effect of tocilizumab vs usual care in adults hospitalized with COVID-19 and moderate or severe pneumonia: a randomized clinical trial. JAMA Intern Med 2021;181(1):32–40.
[43] Stone JH, et al. Efficacy of tocilizumab in patients hospitalized with Covid-19. N Engl J Med 2020;383(24):2333–44.
[44] Salvarani C, et al. Effect of tocilizumab vs standard care on clinical worsening in patients hospitalized with COVID-19 pneumonia: a randomized clinical trial. JAMA Intern Med 2021;181(1):24–31.
[45] Cavalli G, et al. Interleukin-1 and interleukin-6 inhibition compared with standard management in patients with COVID-19 and hyperinflammation: a cohort study. Lancet Rheumatol 2021;3(4):e253–61.
[46] RECOVERY Collaborative Group, Horby P, et al. Dexamethasone in hospitalized patients with Covid-19. N Engl J Med 2021;384(8):693–704.
[47] Schuettpelz LG, Link DC. Regulation of hematopoietic stem cell activity by inflammation. Front Immunol 2013;4:204.
[48] Bendall LJ, Bradstock KF. G-CSF: from granulopoietic stimulant to bone marrow stem cell mobilizing agent. Cytokine Growth Factor Rev 2014;25(4):355–67.

[49] Pietras EM. Inflammation: a key regulator of hematopoietic stem cell fate in health and disease. Blood 2017;130(15):1693–8.

[50] Fatkhullina AR, Peshkova IO, Koltsova EK. The role of cytokines in the development of atherosclerosis. Biochemistry (Mosc) 2016;81(11):1358–70.

[51] Matatall KA, et al. Type II interferon promotes differentiation of myeloid-biased hematopoietic stem cells. Stem Cells 2014;32(11):3023–30.

[52] Matatall KA, et al. Chronic infection depletes hematopoietic stem cells through stress-induced terminal differentiation. Cell Rep 2016;17(10):2584–95.

[53] Essers MA, et al. IFNalpha activates dormant haematopoietic stem cells in vivo. Nature 2009;458(7240):904–8.

[54] Dominici M, et al. Minimal criteria for defining multipotent mesenchymal stromal cells. The international society for cellular therapy position statement. Cytotherapy 2006;8(4):315–7.

[55] Zhu H, et al. Therapeutic effects of human umbilical cord-derived mesenchymal stem cells in acute lung injury mice. Sci Rep 2017;7:39889.

[56] Jung YJ, et al. The effect of human adipose-derived stem cells on lipopolysaccharide-induced acute respiratory distress syndrome in mice. Ann Transl Med 2019;7(22):674.

[57] Meisel R, et al. Human bone marrow stromal cells inhibit allogeneic T-cell responses by indoleamine 2,3-dioxygenase-mediated tryptophan degradation. Blood 2004;103(12):4619–21.

[58] Lee HJ, et al. ICOSL expression in human bone marrow-derived mesenchymal stem cells promotes induction of regulatory T cells. Sci Rep 2017;7:44486.

6

Mesenchymal stem cells: Novel avenues in combating COVID-19

Anwesha Mukherjee and Bodhisatwa Das
Department of Biomedical Engineering, Indian Institute of Technology Ropar, Punjab, India

Introduction

On December 31, 2019, a novel severe acute respiratory syndrome coronavirus 2 (SARS-CoV-2) caused an outbreak of coronavirus disease (COVID-19) in Wuhan city, Hubei province, China, and subsequently spread rapidly to the whole world. The World Health Organization (WHO) declared the SARS-CoV-2 outbreak as a Public Health Emergency of International Concern (PHEIC) on January 30, 2020, and announced the outbreak of COVID-19 as a pandemic on 11 March 2020. COVID-19 has spread worldwide infecting several countries with extreme severity including the USA, India, Brazil, Russia, France, the United Kingdom, Italy, Spain, etc. As of August 7th, 2021, the following are the cases and deaths statistics for some countries [1].

Coronaviruses are enveloped positive-sense, single-stranded RNA viruses anchored with large genomes ranging from 27 to 30 kb and possess the capability of infecting a wide variety of mammalian and avian species [2–5]. Coronaviruses consist of large spherical particles of average diameters of 80–120 nm with special surface projections and their average total molecular mass is of the order of 40,000 KD. The coronavirus core particle is enveloped by a protective capsid implanted with several protein molecules. The lipid bilayer which forms the viral envelope is anchored with the membrane (M), envelope (E), and spike (S) structural proteins with the molar ratio of E:S:M as approximately 1:20:300. The E and M proteins are structural proteins whereas S proteins interact with the host cells and bind virion on the cell surface. In the outer membrane, the set of proteins projected out from the particle are known as spike proteins. The spikes give the distinguishing feature of the corona- or halo-like surface which appears like a crown under the electron microscope, from which the name coronavirus was derived. Nucleocapsid made from multiple types of nucleocapsid (N) protein is enclosed inside the envelope and is attached to the

Stem Cells and COVID-19. https://doi.org/10.1016/B978-0-323-89972-7.00009-X

positive-sense single-stranded RNA genome in a beads-on-a-string type configuration. Nucleocapsid (N) protein is composed of two domains that can combine with the RNA to help replication and performs as a suppressor of the RNAi system of the host cell and assists the viral replication. The S protein is made up of two subunits S_1 (specific receptor binding domain known as RBD) and S_2 (CoV S2 glycoprotein). The head of the spike is formed by the S_1 subunit with the receptor-binding domain (RBD). The stem that connects the spike with the envelope is composed of the S_2 subunit and is fused by activation of protease. Three S1 units are connected to two S_2 subunits in the active state. S_1 proteins possess two infection-specific receptor-binding domains such as the N-terminal domain (S1-NTD) and C-terminal domain (S1-CTD). S1-CTDs recognize the protein receptor angiotensin-converting enzyme 2 (ACE2) and transmembrane protease serine 2 (TMPRSS2) of the host cell. Coronaviruses belong to the family Corona viridae in the order Nidovirales, suborder Cornidovirineae and are genetically classified into four major genera: alpha-coronavirus, beta-coronavirus, gamma-corona virus, and delta-coronavirus. Among them, alpha- and beta-coronaviruses are pathogenic to mammals, gamma-coronaviruses infect avian species, and delta coronaviruses infect both mammalian and avian species (Table 1).

The SARS-CoV-2 virus attacks the pulmonary epithelial cells as the initial site of infection by recognizing the host ACE2 receptor and TMPRSS2 by its spike glycoprotein. In the initial stage, the patients don't develop any severe clinical manifestations but show mild symptoms such as fever, cough, muscle soreness, or body ache. In the later

Table 1 Total number of infection and fatality cases in some of the countries On 07.08.2021 at 10–45 AM.

Place	No. of cases	Deaths
World	20.1 Cr	42.27 L
United States	3.57 Cr + 1.68 L	6.16 L + 849
India	3.19 Cr	4.27 L
Brazil	2.01 Cr	5.61 L
Russia	63 L	1.6 L
France	62.6 L	1.12 L
United Kingdom	59.8 L	1.3 L
Spain	45.7 L	81,931
Italy	43.82 L	1.28 L
Germany	37.9 L	91,761

stages of disease progression, the health condition of the patients deteriorates suddenly. An exaggerated immune response develops from the internalization and gradual replication of the virus, stimulating the release of many proinflammatory cytokines and chemokines. The immune system appears to be unable to turn itself off and the release of excessive cytokines giving rise to a cytokine storm followed by inefficient gas exchange, ARDS (Acute Respiratory Distress Syndrome). The unchecked inflammation causes multiple health hazards in the patient leading to multiple organ failure and severe difficulty in breathing and ultimately results in death within a very short period. Till now no special treatment or effective antiviral drug has been developed for the treatment of COVID-19. To prevent or control infection, supplemental oxygen and mechanical ventilation support are provided to the patients when needed. Corticosteroid-mediated inflammation reduction, convalescent plasma therapy, antibiotics for treatment of secondary bacterial sepsis and nonspecific antivirals, etc., are employed for the clinical management of the patients currently.

The cytokine storm phase should be avoided for the treatment of COVID-19 and the reduction of the death rate. According to the FDA regulations, stem cell therapy can be used for the treatment of diseases with high morbidity rates. Stem cell therapy is one of the potential candidates for novel cellular regenerative technologies with promises to treat incurable diseases. Mesenchymal stromal/stem cell (MSC) therapy has attracted much attention because of its high proliferation rate, low invasive nature, and source potential free of ethical and social issues [6,7]. MSC therapy is now used for the treatment of several inflammatory and autoimmune disorders, including multiple sclerosis, Crohn's disease, diabetes, graft-versus-host disease(GvHD) [8–10]. Application of MSCs in animal models has resulted in the promising treatment of lung diseases, such as pulmonary fibrosis, chronic obstructive bronchiolitis, and bronchopulmonary dysplasia [11]. MSCs are characterized by powerful antimicrobial, antiinflammatory, and immunomodulatory properties and possess unique potential to regenerate damaged cells [12,13]. The function of various immune cells can be modulated by MSCs due to their special immunoregulatory properties. The innate and adaptive immune responses can also be altered by MSCs. These unique characteristics make MSC therapy a promising candidate for the treatment of COVID-19.

Alveolar cell and SARS-COVID-19

The architecture of the human lung is built with a tree-like air pathway and a honeycomb-like gas exchange compartment. In humans, the lungs function as the major respiratory system with two parts situated near the backbone on either side of the heart [14]. They draw oxygen from the atmosphere to transfer to the bloodstream and release carbon dioxide in the atmosphere from the bloodstream. There are two lungs (right and left) in the thoracic cavity of the chest. The lower respiratory tract starts from the trachea and bifurcates into the bronchi and bronchioles and collects air through the conducting zone which terminates to terminal bronchioles. The terminal bronchiole splits into the respiratory bronchioles and this splits into alveolar ducts and hence alveolar sacs which accommodate alveoli (300–500 million alveoli in the lungs) sparsely situated on the walls of the respiratory bronchioles and alveolar ducts [15]. Gas exchange occurs in alveoli, the structure of which is a hollow cup-shaped cavity. The basic functional units of respiration are termed the acini containing clusters of alveoli. An epithelium layer containing very thin, flattened, squamous epithelium cells exists in the alveoli and a part of the alveolar respiratory membrane is formed by the epithelial lining to facilitate the gas exchange. Besides the epithelial layer, the membrane consists of a layer fluid lining containing surfactant, the basement membrane and capillary membrane, a thin interstitial space between epithelial lining. The alveoli are interconnected through passages known as the pores of Kohn. A network of capillaries surrounds the alveolar membrane which functions as the gas exchange surface. The diffusion of oxygen takes place into the capillaries through the membrane and carbon dioxide comes out from the capillaries into alveoli to be exhaled out. Pleural fluid in the pleural sac surrounding the lungs causes smooth sliding of the inner and outer walls during breathing without friction. The pulmonary alveolar epithelium is made of three types of epithelial cells (pneumocytes): alveolar type I and alveolar type II cells and a large phagocytic cell known as an alveolar macrophage with distinct functional and structural specialization. The structure of the alveoli is formed by the thin (~ 25 nm), squamous and flat type I cells and these cells cover more than 95% of the gas exchange surface whereas the rest 5% surface is occupied by the type II cells [16–18]. Type I cells take part in the gas exchange between the air in the alveoli and blood in the capillaries. The cellular division and repair are not observed in the case of type I alveolar cells, but they are susceptible to external stimuli. Type II cells are smaller, cuboidal cells with the property of synthesizing and secreting surfactant to lower surface tension, thereby preventing the alveoli from collapsing [19,20]. An important role is played by the type II cell in fluid balance and ion transport in the alveolar epithelium. Due to proliferation and differentiation properties, the type II cells can self-renew and divide into type I cell [21–24]. Moreover, in the case of toxic responses, they become very active in the immunological defense of the alveoli through the secretion of a variety of cytokines and chemokines [25,26]. Third pneumocytes are alveolar macrophage cells (or dust cells) which are primary phagocytes of the inherent immune system near the pneumocytes at the major boundaries between the body and the outer world. The infectious, toxic, or allergic particles or dust are removed from the respiratory tract by these pneumocytes.

The major but not exclusive target of SARS-COVID-19 is the lung which is affected by Acute Respiratory Distress Syndrome (ARDS) [27–31]. Therefore it tends to develop a hyperreactive immune response that is considered the main cause of morbidity and mortality. In the lung, the alveolar epithelial type II cell is the main target of the virus. Once the alveolar epithelium is damaged, the progress of pathogenesis leads to ARDS. ARDS is not a disease itself, but a manifestation of lung damage caused by excessive inflammation and tissue destruction. The greater affinity of the corona virus-19 for the angiotensin-converting enzyme receptor type-2, ACE2, enables it to attach more effectively to the main target, the alveolar epithelial cell II (AEC-II cells) which play an important role in causing inflammatory responses [32,33]. Lung injury caused by infection of AEC II inhibits the production of the surfactant, stabilization of the air pathway barrier, immune defense, and proliferation and differentiation [34,35]. In ARDS there are three identified phases: the exudative phase, the proliferative phase, and the fibrotic phase. In the first phase, the cytokine storm is observed with an enhancement of inflammation along with an accumulation of neutrophils, monocytes, and macrophages [36,37]. In the second phase, the recovery of normal lung function and the restoration of normal alveolar-endothelial structure may occur. In the third phase, mortality rate increases.

The respiratory system fails due to cytokine storm, damage of parenchyma, fluid-filled alveoli, and immune infiltration. Maurizio Carcaterra and Cristina Caruso proposed a hypothesis that a dysregulation of the Nuclear Factor kappa-light-chain enhancer of activated B cells pathway (NF-κB pathway) gives rise to ARDS causing multiple organ failure and death [38].

Cytokine storm

COVID-19 affects different people in different ways. In the early stages of infection, major clinical symptoms are not observed in most COVID-19 patients infected from severe acute respiratory syndrome coronavirus 2 (SARS COVID 2). Most infected people develop mild to moderate illness with common symptoms like high temperature, dry cough, soreness of throat, muscle and body pain, tiredness [39]. Severe COVID-19 causes viral pneumonia leading to acute respiratory distress syndrome (ARDS) [40,41]. With the progress of COVID-19 pneumonia, an increasing number of the air sacs become filled with fluid leaking from the tiny blood vessels in the lungs resulting in shortness of breath and can lead to acute respiratory syndrome (ARDS) and acute lung injury (ALI). Consequently, multiorgan failure takes place to quicken the death of the patients [42]. Cytokine storm induced by the virus is detected from the clinical studies of COVID-19 critical patients and may be linked to the cause of ARDS and multiple organ failure [43,44]. A cytokine storm is a physiological reaction in the immune system of the body that responds to the infection by external stimuli leading to an uncontrolled and excessive release of proinflammatory molecules called cytokines [45]. This occurs due to the overactivity of the immune system. This activates more immune cells causing hyperinflammation and gives rise to multiorgan failure and death of the patient [45]. Therefore the deterioration of patients can be prevented by the effective suppression of cytokine storms. Complication due to cytokine storm is also observed in the case of other respiratory diseases infected by coronaviruses such as SARS, MERS and in noninfectious diseases such as multiple sclerosis and pancreatitis [43,46–48].

Mechanism of cytokine storm

Cells secrete a diverse group of small proteins known as cytokines for intercellular signaling and communication purpose. At the heart of cytokine storm, there are various kinds of cytokines such as interferons, interleukins, chemokines, colony-stimulating factors, tumor necrosis factors. Interferons (IFNS) regulate innate immunity, activate antiviral properties, and control antiproliferative effects. Interleukins (ILs) control the growth and differentiation of leukocytes and many of them are proinflammatory. The function of chemokines is control of chemotaxis, recruitment of leucocytes, and many of those are proinflammatory. Colony-stimulating factors (CSFs) stimulate hematopoietic progenitor cell proliferation whereas tumor necrosis factor (TNFs) controls the activity of cytotoxic T lymphocytes and are proinflammatory [49,50].

The release of many proinflammatory cytokines is stimulated during the deadly unregulated inflammatory response. Coronavirus infection leads to the secretion of the following cytokines.

(IL)-1β, IL-2, IL-6, IL-7, IL-12, IL-18, IL-33, (IFN)-α, IFN-γ, (TNFα), granulocyte colony-stimulating factor (GSCF), interferon-γ inducible protein 10 (IP10), monocyte chemoattractant protein 1 (MCP1), macrophage inflammatory protein 1-α (MIP1A), and transforming growth factor-beta (TGF-β) such as chemokines [51–53].

The level of inflammatory factors among patients with COVID-19 was reported by measuring the cytokines of patients with COVID-19 (Fig. 1) [54]. They observed enhanced levels of IL-1B, IL-1 RA, IL-7, IL-8, IL-9, IL-10, fibroblast growth factor (FGF), granulocyte-macrophage colony-stimulating factor (GM-CSF), IFN-γ, G-CSF, IP10, MCP1, MIP1A, PDGF, TNFα, and vascular endothelial growth factor (VEGF)

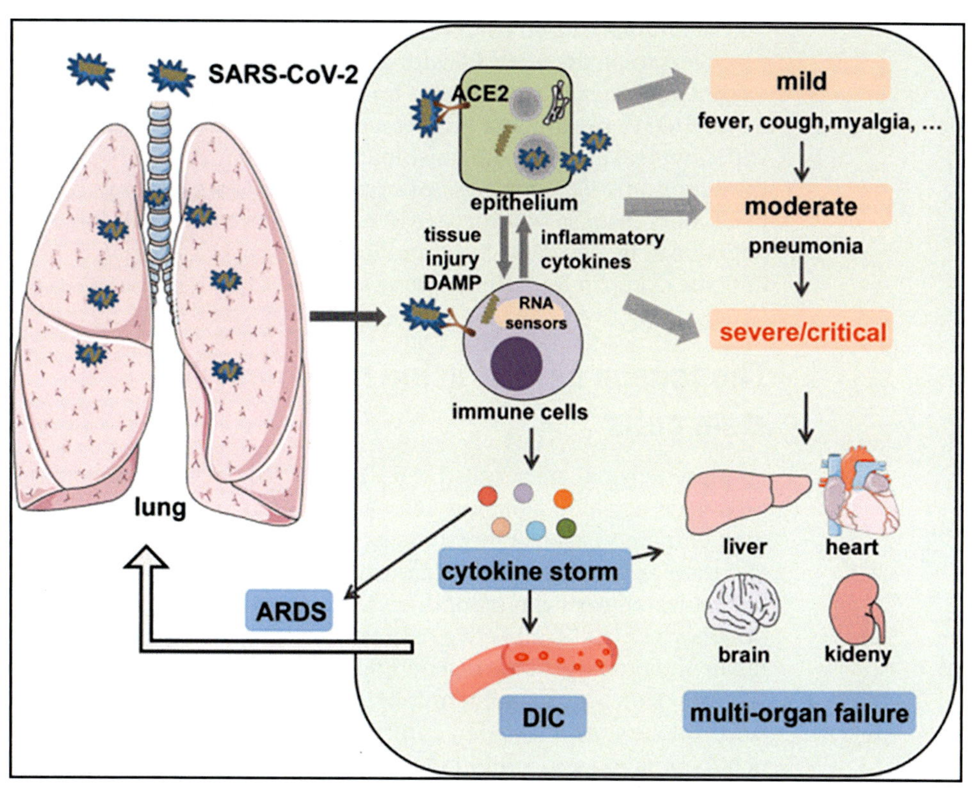

Fig. 1 The effect of COVID-19 infection on the respiratory tract and secondary organ failure: direct infection creates local inflammation on the alveolar epithelial membrane, wall thickening, and pneumonia; indirect effect involves hyperactivation of immune cells, chronic burst release of inflammatory cytokines leading to ARDS and multiorgan failure which is a major reason for patient mortality. Reused with the permission of the publisher. Yang L, Xie X, Tu Z, Fu J, Xu D, Zhou Y. The signal pathways and treatment of cytokine storm in COVID-19. Signal Transduct Target Ther 2021;6(1):255.

in their specimens. In the case of seriously affected patients, a higher level of TNFα was estimated. A significant enhancement of IL-6 levels was detected in the serum of nonsurvivor COVID-19 patients as compared with survivors [55].

At present no successful antiviral treatment is available for COVID-19 and the patients are treated only with symptomatic management of the disease because there is no successful result from the worldwide frantic search for COVID-19 treatment. COVID-19 infected cytokine storm can cause organ failure, edema, cardiac attack, and ARDS (which is an inflammatory condition). Uncontrolled inflammation leads to dysfunction of lungs(pneumonia). In pneumonia, the lungs become filled with fluid and inflamed, leading to breathing difficulties.

Pneumonia caused by COVID-19 tends to take hold in both lungs. Alveolar sacs in the lungs fill with fluid, limiting their ability to take in oxygen and causing shortness of breath, cough, and other symptoms. As COVID-19 pneumonia progresses, more of the sacs become filled with fluid leaking from the tiny blood vessels in the lungs. Over time, the symptoms like shortness of breathing appears leading to acute respiratory distress syndrome (ARDS). Therefore the condition of the patient can be improved by controlling the cytokine storm and inhibiting the overactivity of the immune system.

The reason behind using mesenchymal stem cells

Mesenchymal stem cells (MSCs) are a type of multipotent stem cells or medicinal signaling cells which undergo self-renewal and differentiation because they possess the dual capability of high proliferation and differentiation into multiple tissue types such as bone (osteoblasts), cartilage (chondrocytes), muscle (myocytes), fat cells (adipocytes), and connective tissues. They play an important role in repair and promote tissue regeneration. Stem cell therapy with administration of mesenchymal stem cells into the immune system may control cytokine storm caused by coronavirus infection through reduction of inflammation and inhibition of hyperactivation of the immune system. Mesenchymal cells can be stored so that they can be repetitively used for various therapeutic purposes. In the current chapter, we will divide the application of MSCs in COVID-19 into three different areas.

- MSCs for regeneration of lung tissue (tissue engineering and bioprinting)
- MSCs for development of drug screening models (static models and organ on chip)
- MSCs for alleviating the cytokine storm

The tissue engineering and regenerative medicine approaches have allowed us to functionally restore either damaged alveoli or complete lungs using biomaterial-based scaffolds and microfabrication techniques [56]. The major role of the lungs is gas transport. It is also very rich in blood capillaries. It is an inflatable structure that should not collapse due to fluctuation of air pressure. As the biological gas transport across the lung alveoli is purely based on Fick's diffusion boundary, the gas transport distance limit is around 250–300 μm [57]. Thus the tissue engineering scaffold material should mimic the diffusivity of the natural alveolar epithelial membrane and elasticity to survive the pressure fluctuation. The major artificial polymers used to develop lung tissue equivalent scaffolds were mainly from elastomers like silicone rubber and poly-dimethyl-siloxane (PDMS). MEMS fabrication and micromilling were explored to develop these devices. However, most of these were cell free and couldn't mimic the diffusivity [58]. So, over a period, cell-seeded tissue-engineered lung tissue equivalent was prepared. Also, biological ECM-based materials like collagen, fibrin, biopolymers (chitosan, alginate) and decellularized ECM are considered for scaffolding material for lung tissue engineering. While initially, primary cells like lung epithelial cells and endothelial cells were explored for this purpose, stem cells were observed to be better alternatives for the same purpose. Three major subpopulations are currently observed to be explored for tissue engineering. Induced pluripotent stem cells (iPSCs) are originally fibroblast-like cells, genetically engineered to restore their plasticity. iPSCs are not yet used in clinical practice. However, it is already shown to be useful in the reconstruction of airways and alveolar epithelium (Fig. 2) [59,60]. Mesenchymal stem cells (MSCs) are another major group of stem cells that are utilized for lung tissue engineering. It is observed

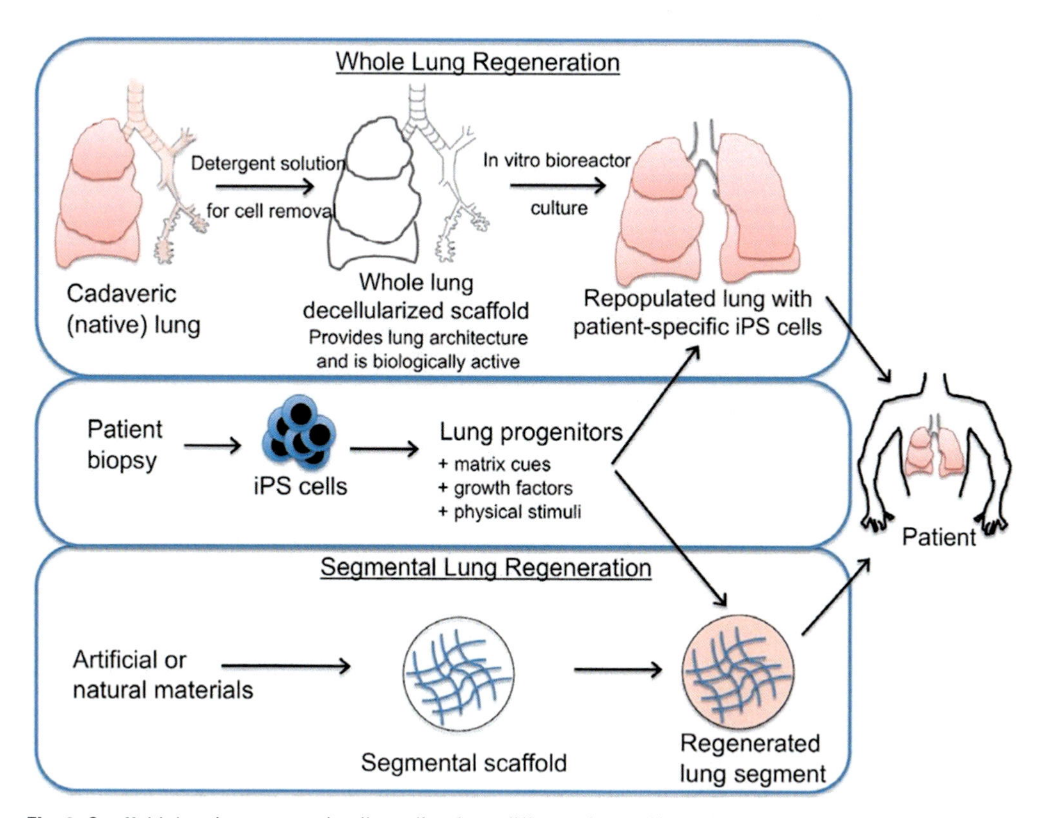

Fig. 2 Scaffold development and cell seeding for a different form of lung tissue constructs using iPSCs. Reused with the permission of the publisher. Petersen TH, Calle EA, Niklason LE. Strategies for lung regeneration. Mater Today 2011;14(5):196–201.

that, there is a group of MSCs already resident to lung tissue, which can give rise to alveolar epithelial cells type-2 (AT-II) cells and epithelial progenitors. Umbilical cord and Wharton's jelly derived MSCs also have been observed to develop pulmonary epithelial (both type I and II) like cells [61,62]. The other group of cells used to seed the tissue engineering grafts is endothelial cells (HUVECs) or their progenitors. While the AT-I and AT-II mimic the airway side of alveoli, the endothelial cells are distributed to the other side of the alveolar membrane mimicking the capillary network of the lungs.

However, it is always observed that in vitro culture might lead to differentiation and appropriate marker expression of lung tissues. But, coculture models, 3D scaffolds, and most importantly dynamic cell culture (bioreactors, microfluidic devices) are more effective in generating functional tissue construct [63]. MSCs seeded in 3D decellularized ECM (from the cadaveric lung) could be seeded with vascular endothelial and pulmonary epithelial cells to generate alveoli-like structures and organoids which can be useful for the recovery of lung tissue (Fig. 3).

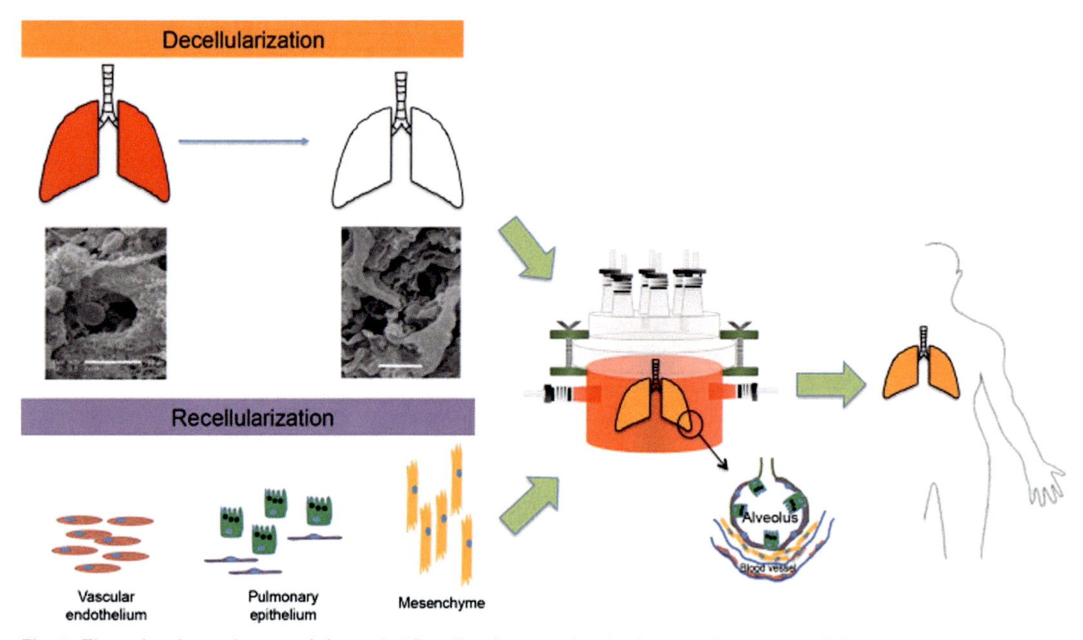

Fig. 3 The role of coculture and dynamic 3D cell culture technologies to enhance the differentiation potential of MSCs into lung tissue constructs. Adapted from Raredon MSB, Yuan Y, Niklason LE. Chapter 68—Lung tissue engineering. In: Lanza R, Langer R, Vacanti JP, Atala A, editors. Principles of tissue engineering. 5th ed. Academic Press; 2020, pp. 1273–1285 with publisher's permission.

Bioprinting and COVID-19

Three-dimensional bioprinting (3D bioprinting) is the application of the principles of conventional 3D printing techniques by utilizing living cells and biocompatible materials known as bio-inks to assemble living structures that mimic the behavior of the natural tissues [64,65].

After printing, they also imitate the extracellular matrix environment with the properties of adhesion, proliferation, and differentiation. In 3D bioprinting, bio-ink materials are deposited by layer-by-layer process and the tissue-like bioprinted structures are mainly utilized in the fields of tissue engineering, bioengineering, material science, pharmaceutical development, and drug validation [66]. The 3D bioprinting is being used to engineer scaffolds for skin grafts, bone grafts, implants, and biomedical devices. Bioprinting takes place in three steps.

Prebioprinting

In this stage, a digital model is produced obtaining a biopsy of the organ to be imitated by computerized tomography or magnetic resonance imaging. The scan data are fed in a computer-generated design (CAD) program. This 3D model file generates a series of thin layers to construct the original structure when arranged vertically. The model is converted into path data compiled into a g-code file which is fed into the 3D printer. The bioprinter follows the instructions of the file during printing. In the process, some cells are chosen, and they grow in number. A liquefied material is mixed with the cells to supply oxygen and nutrients for survival of the cells.

Bioprinting

The bio-inks thus produced are placed into a cartridge and to create the target structure the required print heads are chosen. The bioprinted pretissue with three-dimensional structures is placed in the incubator to grow into matured tissue.

Postbioprinting

For the stability of the printed structure, this postbioprinting stage is very important. The printed structure should be stimulated mechanically and chemically for remodeling and growth of cells.

There are multiple bioprinting technologies currently in use (Fig. 4). All of the bioprinting techniques have their advantages and disadvantages [67,68].

Stereolithography

Stereolithography (SLA) is a very competitive technique for bioprinting due to its high resolution, high speed, and high cell viability. The SLA systems consist of photosensitive resin, a laser, and a printing stage.

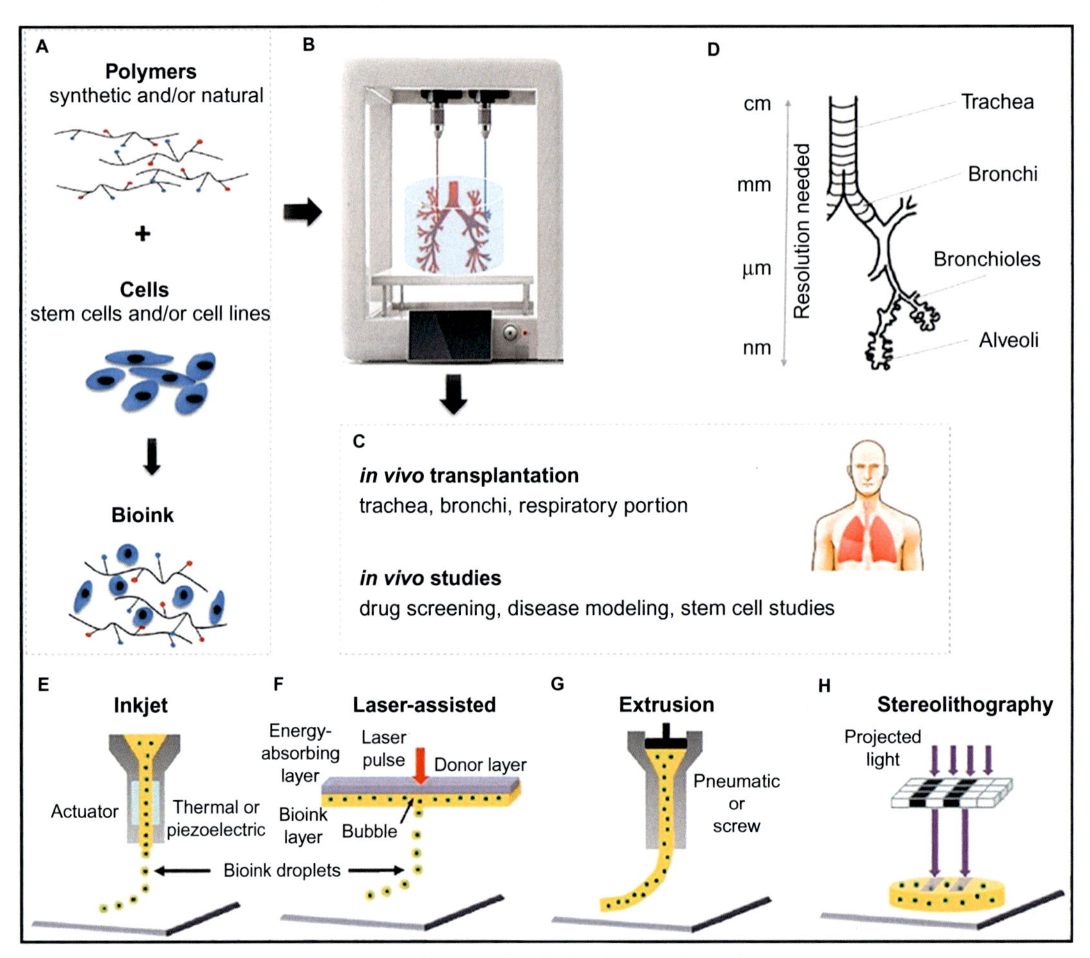

Fig. 4 The major components and methods involved in 3D bioprinting of lung tissues. Reused with the permission of the publisher. Mahfouzi SH, Safiabadi Tali SH, Amoabediny G. 3D bioprinting for lung and tracheal tissue engineering: criteria, advances, challenges, and future directions. Bioprinting 2021;21:e00124.

The movement of the laser in the *x*-*y* axis directions induces polymerization and solidification of the top layer of the resin in the reservoir. For the next layer of printing, the stage is lowered in the z-axis direction in the bottom-up printing process whereas in the top-down printing light is projected from under the vat. For both processes, the support material is needed. For bioprinting using SLA, the reservoir is filled with cells and photocurable hydrogel, and a laser with a specific wavelength is employed to save the cellular materials from any damage. In this technique, the solution used should be photoreactive and photocurable.

Inkjet bioprinting

In this bioprinting stage, the material jetting method is employed like the commercial inkjet printers. With the help of thermal, electromagnetic, piezoelectric, or acoustic mechanisms, droplets of the biomaterials are projected on the printing stage substrates. Due to high cell viability in bio-inks, thermal deposition technique is extensively used. The temperature is raised to 300°C for few microseconds to produce an inflated bubble that pushes the ink out of the nozzle on the substrate. This technique produces very fast and precise deposition of very small droplets of a diameter of the order of 20 μm. In thermal technique, the droplets are unequal and smooth printing is not possible due to frequent nozzle blockages. The viability of the bio-inks is effected by shear and thermal stress. In the piezoelectric method, heat is not used, and drops are produced by the transient pressure from the piezoelectric actuator. No orifice clogging occurs, and the generated droplets remain directional with regular and equal size. Greater than 90% viability is observed for piezoelectrically deposited cells.

Laser-assisted printing

Laser-assisted printing setup is composed of a pulsed laser source, a material to be printed (a multilayered ribbon), and a collecting substrate. The ribbon is transparent to the laser radiation wavelength and is coated with bio-inks. A focused laser beam generates high pressure due to the absorption of energy by the substrate and the droplets of biomaterials evaporate to be propelled on the receiving substrate. High resolution and high viability are observed in laser-assisted printing. A biopolymer or cell culture material is coated on the collecting substrate for maintenance of cellular adhesion and sustained growth after the transfer of cells.

Extrusion printing

Extrusion printing employs fused deposition modeling, pneumatic and mechanical extrusion technologies consisting here of three units: a reservoir (e.g., syringe) that contains the bio-ink; a printing head through which the bio-ink is ejected from the reservoir; a printing stage where the deposited bio-ink is collected along with a control system responsible for controlling printing speed, temperature, and location. Axis control varies depending on the system and allows for stage movement in three directions. A syringe and needle thrust out a continuous string of biomaterials by gravitational and/or mechanical forces. The resolution of bioprinting depends on the diameter of the needle and the amount of mechanical force exerted. The integrity of the cell membrane is lost due to the pressure and shear forces, thereby limiting the resolution to 100 μm. Extrusion printing has fast deposition speed, poor resolution, and clogging problems. Multihead and core-shell nozzle systems can print multiple biomaterials simultaneously.

The 3D printing together with tissue engineering technology is often implemented to get the best possible picture of the disease. The disease progression is modeled in vitro, the novel disease-driver gene target is identified, and the medicinal chemistry that modulates these targets is advanced to develop drugs [69].

In vitro organ models and organ chips for lungs modeling

In vitro modeling is generally used before in vivo testing. The bioprinted disease models provide a more accurate disease model for the identification of drug targets than the animal models. Due to the difference in cellular microenvironments, there exists a limitation in clinical translatability in animal models. It is found that about 80% of the therapeutics assessed effective in animals fails during the human clinical trial. The 3D in vitro model produces a cellular architecture that bears close resemblance with the human physiological environment. The features such as the substrate topography and stiffness, mechanical forces, and density affect the cell behavior which can be controlled by 3D bioprinting. Previous research has already shown that 3D human lung tissue better mimics viral infectivity compared to regular cell culture. The lung tissue model is successfully developed by 3D bioprinting using chitosan-collagen scaffolds cultured with human epithelial cells. Changes in marker protein expression and release of proinflammatory cytokines are observed in case of infection of lung model by H1N1 and H3N2 influenza [70]. Worldwide research works are in progress to design a 3D lung tissue model for SARS-Cov-2 employing CAD software to design microcellular structure. A bio-ink is generated by selecting proper biomaterials and lung tissue cells [71,72]. With the completion of the culture, the bioprinted construct is infected with the virus and then assessed technically. A novel biological pathway is identified that is a major contributor to disease progression. Physiological conditions produced by the 3D model closely represent that in the in vivo studies (Fig. 5). This will give rise to a more successful clinical translation of therapeutics.

The application of microfluidics and tissue organoids also helped us to develop models analogous to whole lung tissue with proper functionality. The proper differentiation of MSCs and the development of alveoli-like structures are related to proper biophysical stimuli. Providing periodic strain, air-liquid interface, and natural ECM equivalent elastic moduli for the basement membrane are some specific situations that can make them in vitro system much closer to the natural system [70]. In Fig. 6, we get a complete overview of different

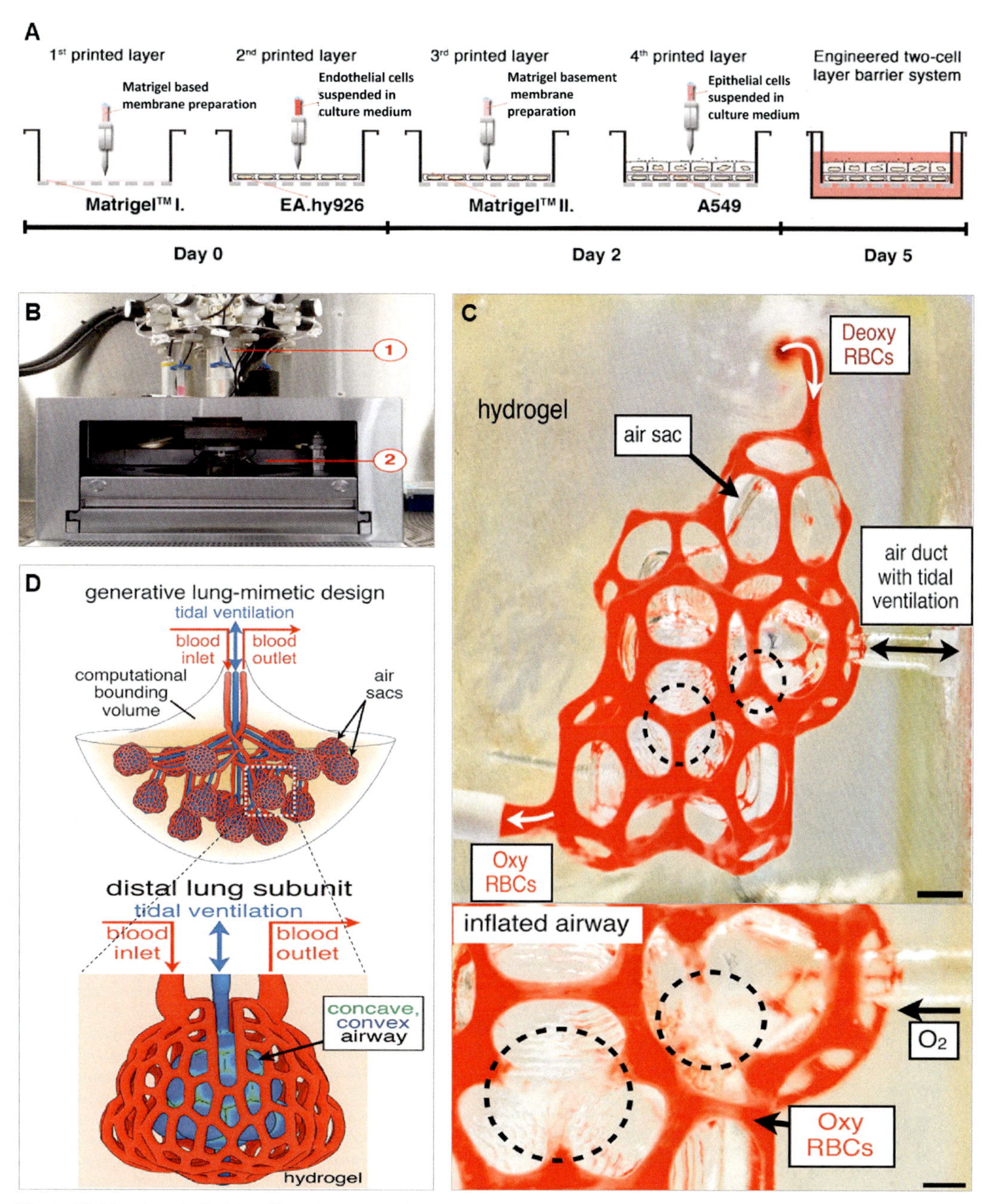

Fig. 5 3D bioprinted air-*blood* interface model to understand gas transport in lung microtissue. Reused with permission from the publisher. Mahfouzi SH, Safiabadi Tali SH, Amoabediny G. 3D bioprinting for lung and tracheal tissue engineering: criteria, advances, challenges, and future directions. Bioprinting 2021;21:e00124.

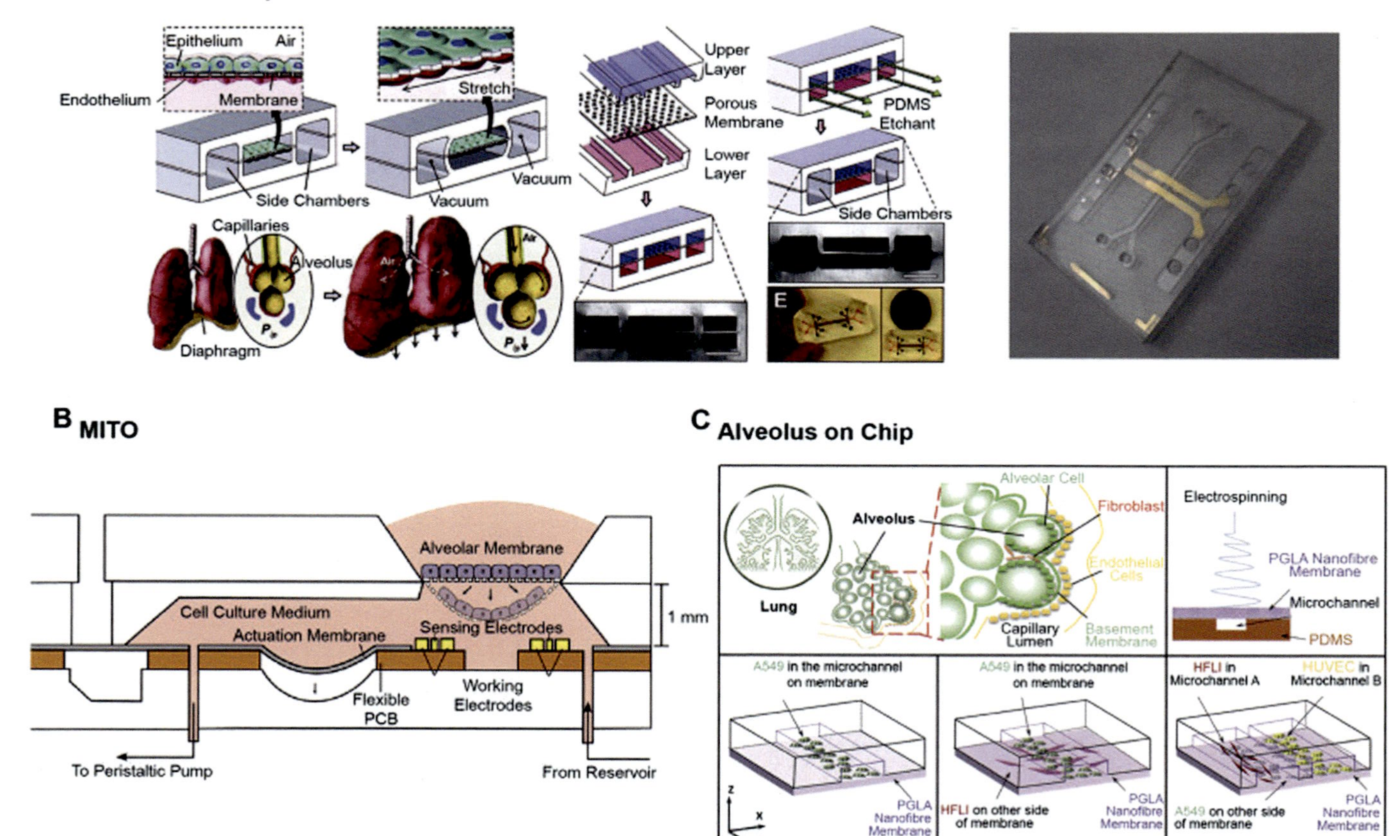

Fig. 6 (A) Development of the alveolus on a chip model via microfabrication, (B) The cross-sectional view of how chambers are designed to mimic the alveolar microenvironment, (C) The engineering strategies to generate the coculture and air-liquid interface for the chip. Figure reused with permission of the publisher. Mahfouzi SH, Safiabadi Tali SH, Amoabediny G. 3D bioprinting for lung and tracheal tissue engineering: criteria, advances, challenges, and future directions. Bioprinting 2021;21:e00124.

microfabrication methods together that can be useful for the development of an alveolar gas exchange model on a chip. However, the COVID-19 infections are different in their own way. While it mostly enters the body via the nasopharyngeal system as an airborne disease, it is observed to have affected organs like the heart, liver, kidney, and GI tract. Thus the COVID-19 infection model needs a complex organ model involving different organs in a single chip. MSCs can be explored for precursors to the major cells involved in most of these tissues, including hepatocytes, goblet cells, cardiomyocytes, etc. In recent years, multiorgan modeling chips have been developed to mimic physiological crosstalk between multiple organs. Even up to four organs have been observed to be modeled in a single chip [73–75]. However, COVID-19 affects the immune system and there is evidence of creating neuropathological conditions. Mimicking these conditions is challenging and yet to be discovered [76].

Paracrine and immunomodulatory effects of MSCs

Currently, MSCs get attention in clinical trials due to their immunomodulatory and regenerative properties. After the intravenous transplantation of MSCs, a significant number of cells build up in the lung, along with the immunomodulatory effect that could protect alveolar epithelial cells, recover the pulmonary microenvironment, inhibit pulmonary fibrosis, and cure lung dysfunction [77]. Immune design of COVID-19 including lymphopenia causes dysfunction and activation of lymphocytes, abnormalities of granulocytes and monocytes, and increases the production of cytokines and antibodies. Lymphopenia is a significant feature of patients with COVID-19, especially in severe cases. Patients with severe COVID-19 are more likely to exhibit lymphopenia on admission, indicating a significant predictor for severe patients [78–80].

With the intravenous introduction of mesenchymal stem cells, a part of it is entrapped in the lungs, resulting in the secretion of some soluble mediators such as antimicrobial peptides, antiinflammatory cytokines, extracellular vesicles, and angiogenic growth factors [81–83]. These mediators are controlled by the differentiation of the damage and pathogen-associated molecular pathogen receptors on the surface of the MSCs with TLRs (toll-like receptors) [84]. The disrupted alveolar-capillary barriers caused by ARDS are restored by Keratinocyte Growth Factor (KGF) and angiopoietin-1 (Ang-1) released by MSCs [85]. It is reported that extracellular vesicles with specific miRNA and inhibitory mRNAs control the protective effects of MSCs in noninfectious and acute lung injuries or bacterial sepsis in the lungs [85,86]. MSCs release different paracrine factors. The interaction of these factors with immune cells gives rise to immunomodulation. Improvement of the condition of COVID-19 patients after the injection is caused by the vigorous antiinflammatory activities of MSCs. The number of peripheral lymphocytes is found to increase while the level of C-reactive protein (CRP) decreases, and the number of regulatory dendritic cells ($CD14^+$ $CD11b^{mid}$) increases after infusion [87]. MSCs via paracrine and juxtacrine signaling crosstalk with most of the immune cells. It can generate an immunomodulatory effect via direct up or downregulation of specific genes or its secretome can also affect hyperimmune system activation and alleviate the cytokine storm (Fig. 7) [88,89].

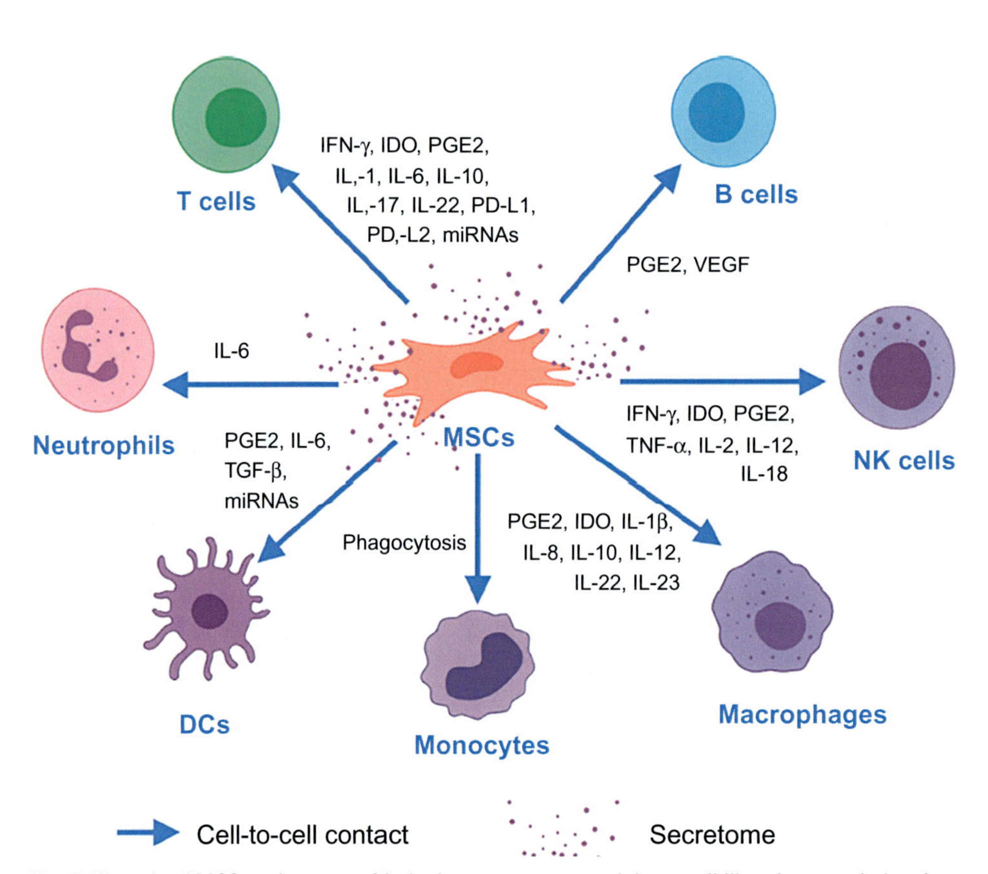

Fig. 7 The role of MSCs to interact with the immune system and the possibility of counterbalancing ARDS-related complications. Reused with permission from the publisher. Song N, Scholtemeijer M, Shah K. Mesenchymal stem cell immunomodulation: mechanisms and therapeutic potential. Trends Pharmacol Sci 2020;41(9):653–664.

Conclusion

COVID-19 research is suggesting that this pandemic will have long-term repercussions on the health of the infected person. While lung injury, pneumonia, and breathing difficulty might be the tip of the iceberg, the disease might have long-term effects related to different immunological disorders. Currently, our knowledge is limited only up to cytokine storm which is an effect after the hyperactivation of the immune system. But inflammation is a necessary evil, and it can lead to multiple diseases like autoimmune diseases, degenerative disorders, even cancer. The current manifestation of the disease might be the beginning of just a long-lasting effect that is yet to be witnessed. MSCs are extremely important in terms of therapy. Cell transplantation therapy can be a major answer to the degeneration occurring due to primary and secondary effects of the COVID-19 infection. It is also observed that implantation of MSCs can downregulate inflammatory cytokine secretion and hyperactivation of proinflammatory macrophages. Therefore cell delivery could be beneficial for saving the tissues from inflammatory damages also. COVID-19 therapeutic evaluation process has a major limitation. We don't have a proper animal model to test these therapeutics. Organ on chip models can fill this void and we can evaluate the efficacy of therapeutics both at single and multiorgan levels in a very controlled environment level. MSCs allow us to generate pure cellular phenotypes via in vitro differentiation and make organ chip models robust and repeatable. Thus MSCs can be explored in near future as biological theranostics.

References

[1] World Health Organization, (2021). https://covid19.who.int/.

[2] Masters PS. The molecular biology of coronaviruses, advances in virus research. Academic Press; 2006. p. 193–292.

[3] Mittal A, Manjunath K, Ranjan RK, Kaushik S, Kumar S, Verma V. COVID-19 pandemic: insights into structure, function, and hACE2 receptor recognition by SARS-CoV-2. PLoS Pathog 2020;16(8), e1008762.

[4] Alipoor SD, Mortaz E, Jamaati H, Tabarsi P, Bayram H, Varahram M, et al. COVID-19: molecular and cellular response. Front Cell Infect Microbiol 2021;11(27):1–16. https://doi.org/10.3389/fcimb.2021.563085.

[5] Li F. Structure, function, and evolution of coronavirus spike proteins. Annu Rev Virol 2016;3(1):237–61.

[6] Leng Z, Zhu R, Hou W, Feng Y, Yang Y, Han Q, Shan G, Meng F, Du D, Wang S, Fan J, Wang W, Deng L, Shi H, Li H, Hu Z, Zhang F, Gao J, Liu H, Li X, Zhao Y, Yin K, He X, Gao Z, Wang Y, Yang B, Jin R, Stambler I, Lim LW, Su H, Moskalev A, Cano A, Chakrabarti S, Min KJ, Ellison-Hughes G, Caruso C, Jin K, Zhao RC. Transplantation of ACE2(−) mesenchymal stem cells improves the outcome of patients with COVID-19 pneumonia. Aging Dis 2020;11(2):216–28.

[7] Harrell CR, Jovicic N, Djonov V, Arsenijevic N, Volarevic V. Mesenchymal stem cell-derived exosomes, and other extracellular vesicles as new remedies in the therapy of inflammatory diseases. Cell 2019;8(12):1605.

[8] Thanunchai M, Hongeng S, Thitithanyanont A. Mesenchymal stromal cells and viral infection. Stem Cells Int 2015;2015, 860950.

[9] Baron F, Storb R. Mesenchymal stromal cells: a new tool against graft-versus-host disease? Biol Blood Marrow Transplant 2012;18(6):822–40.

[10] Konari N, Nagaishi K, Kikuchi S, Fujimiya M. Mitochondria transfer from mesenchymal stem cells structurally and functionally repairs renal proximal tubular epithelial cells in diabetic nephropathy in vivo. Sci Rep 2019;9(1):5184.

[11] Wei X, Yang X, Han ZP, Qu FF, Shao L, Shi YF. Mesenchymal stem cells: a new trend for cell therapy. Acta Pharmacol Sin 2013;34(6):747–54.

[12] Khoury M, Cuenca J, Cruz FF, Figueroa FE. Current status of cell-based therapies for respiratory virus infections: applicability to COVID-19. Eur Respir J 2020;55(6):1–18. https://doi.org/10.1183/13993003.00858-2020. 2000858.

[13] Paliwal S, Chaudhuri R, Agrawal A, Mohanty S. Regenerative abilities of mesenchymal stem cells through mitochondrial transfer. J Biomed Sci 2018;25(1):31.

[14] Hansen JE, Ampaya EP, Bryant GH, Navin JJ. Branching pattern of airways and air spaces of a single human terminal bronchiole. J Appl Physiol 1975;38(6):983–9.

[15] Knudsen L, Ochs M. The micromechanics of lung alveoli: structure and function of surfactant and tissue components. Histochem Cell Biol 2018;150(6):661–76.

[16] Berthiaume Y, Voisin G, Dagenais A. The alveolar type I cells: the new knight of the alveolus? J Physiol 2006;572(Pt 3):609–10.

[17] Weibel ER. The mystery of "non-nucleated plates" in the alveolar epithelium of the lung explained. Acta Anat 1971;78(3):425–43.

[18] Ochs M, Nyengaard JR, Jung A, Knudsen L, Voigt M, Wahlers T, Richter J, Gundersen HJG. The number of alveoli in the human lung. Am J Respir Crit Care Med 2004;169(1):120–4.

[19] Crapo JD, Barry BE, Gehr P, Bachofen M, Weibel ER. Cell number and cell characteristics of the normal human lung. Am Rev Respir Dis 1982;126(2):332–7.

[20] Wright JR. Immunoregulatory functions of surfactant proteins. Nat Rev Immunol 2005;5(1):58–68.

[21] Barkauskas CE, Cronce MJ, Rackley CR, Bowie EJ, Keene DR, Stripp BR, Randell SH, Noble PW, Hogan BLM. Type 2 alveolar cells are stem cells in adult lung. J Clin Invest 2013;123(7):3025–36.

[22] Desai TJ, Brownfield DG, Krasnow MA. Alveolar progenitor and stem cells in lung development, renewal and cancer. Nature 2014;507(7491):190–4.

[23] Fehrenbach H. Alveolar epithelial type II cell: defender of the alveolus revisited. Respir Res 2001;2(1):33.

[24] Jain R, Barkauskas CE, Takeda N, Bowie EJ, Aghajanian H, Wang Q, Padmanabhan A, Manderfield LJ, Gupta M, Li D, Li L, Trivedi CM, Hogan BLM, Epstein JA. Plasticity of Hopx+ type I alveolar cells to regenerate type II cells in the lung. Nat Commun 2015;6(1):6727.

[25] Mason RJ. Biology of alveolar type II cells. Respirology 2006;11 Suppl:S12–5.

[26] Vanderbilt JN, Mager EM, Allen L, Sawa T, Wiener-Kronish J, Gonzalez R, Dobbs LG. CXC chemokines and their receptors are expressed in type II cells and upregulated following lung injury. Am J Respir Cell Mol Biol 2003;29(6):661–8.

[27] Katsura H, Sontake V, Tata A, Kobayashi Y, Edwards CE, Heaton BE, Konkimalla A, Asakura T, Mikami Y, Fritch EJ, Lee PJ, Heaton NS, Boucher RC, Randell SH, Baric RS, Tata PR. Human lung stem cell-based alveolospheres provide insights into SARS-CoV-2-mediated interferon responses and pneumocyte dysfunction. Cell Stem Cell 2020;27(6):890–904.e8.

[28] Mason RJ. Thoughts on the alveolar phase of COVID-19. Am J Physiol Lung Cell Mol Physiol 2020;319(1):L115–l120.

[29] Denney L, Ho L-P. The role of respiratory epithelium in host defence against influenza virus infection. Biom J 2018;41(4):218–33.
[30] Thompson BT, Chambers RC, Liu KD. Acute respiratory distress syndrome. N Engl J Med 2017;377(6):562–72.
[31] Ware LB, Matthay MA. The acute respiratory distress syndrome. N Engl J Med 2000;342(18):1334–49.
[32] Ziegler CGK, Allon SJ, Nyquist SK, Mbano IM, Miao VN, Tzouanas CN, Cao Y, Yousif AS, Bals J, Hauser BM, Feldman J, Muus C, Wadsworth 2nd MH, Kazer SW, Hughes TK, Doran B, Gatter GJ, Vukovic M, Taliaferro F, Mead BE, Guo Z, Wang JP, Gras D, Plaisant M, Ansari M, Angelidis I, Adler H, Sucre JMS, Taylor CJ, Lin B, Waghray A, Mitsialis V, Dwyer DF, Buchheit KM, Boyce JA, Barrett NA, Laidlaw TM, Carroll SL, Colonna L, Tkachev V, Peterson CW, Yu A, Zheng HB, Gideon HP, Winchell CG, Lin PL, Bingle CD, Snapper SB, Kropski JA, Theis FJ, Schiller HB, Zaragosi LE, Barbry P, Leslie A, Kiem HP, Flynn JL, Fortune SM, Berger B, Finberg RW, Kean LS, Garber M, Schmidt AG, Lingwood D, Shalek AK, Ordovas-Montanes J. SARS-CoV-2 receptor ACE2 is an interferon-stimulated gene in human airway epithelial cells and is detected in specific cell subsets across tissues. Cell 2020;181(5):1016–1035.e19.
[33] Lukassen S, Chua RL, Trefzer T, Kahn NC, Schneider MA, Muley T, Winter H, Meister M, Veith C, Boots AW, Hennig BP, Kreuter M, Conrad C, Eils R. SARS-CoV-2 receptor ACE2 and TMPRSS2 are primarily expressed in bronchial transient secretory cells. EMBO J 2020;39(10), e105114.
[34] Zuo YY, Veldhuizen RA, Neumann AW, Petersen NO, Possmayer F. Current perspectives in pulmonary surfactant—inhibition, enhancement and evaluation. Biochim Biophys Acta 2008;1778(10):1947–77.
[35] Gong MN, Wei Z, Xu LL, Miller DP, Thompson BT, Christiani DC. Polymorphism in the surfactant protein-B gene, gender, and the risk of direct pulmonary injury and ARDS. Chest 2004;125(1):203–11.
[36] Pierrakos C, Karanikolas M, Scolletta S, Karamouzos V, Velissaris D. Acute respiratory distress syndrome: pathophysiology and therapeutic options. J Clin Med Res 2012;4(1):7–16.
[37] Polak SB, Van Gool IC, Cohen D, von der Thüsen JH, van Paassen J. A systematic review of pathological findings in COVID-19: a pathophysiological timeline and possible mechanisms of disease progression. Mod Pathol 2020;33(11):2128–38.
[38] Carcaterra M, Caruso C. Alveolar epithelial cell type II as main target of SARS-CoV-2 virus and COVID-19 development via NF-Kb pathway deregulation: a physio-pathological theory. Med Hypotheses 2021;146, 110412.
[39] Lai C-C, Shih T-P, Ko W-C, Tang H-J, Hsueh P-R. Severe acute respiratory syndrome coronavirus 2 (SARS-CoV-2) and coronavirus disease-2019 (COVID-19): the epidemic and the challenges. Int J Antimicrob Agents 2020;55(3):105924.
[40] Anon. An update on the epidemiological characteristics of novel coronavirus pneumonia (COVID-19). Zhonghua Liu Xing Bing Xue Za Zhi 2020;41(2):139–44.
[41] Grasselli G, Tonetti T, Filippini C, Slutsky AS, Pesenti A, Ranieri VM. Pathophysiology of COVID-19-associated acute respiratory distress syndrome—authors' reply. Lancet Respir Med 2021;9(1):e5–6.
[42] Chousterman BG, Swirski FK, Weber GF. Cytokine storm and sepsis disease pathogenesis. Semin Immunopathol 2017;39(5):517–28.
[43] Shimabukuro-Vornhagen A, Gödel P, Subklewe M, Stemmler HJ, Schlößer HA, Schlaak M, Kochanek M, Böll B, von Bergwelt-Baildon MS. Cytokine release syndrome. J Immunother Cancer 2018;6(1):56.
[44] Tisoncik JR, Korth MJ, Simmons CP, Farrar J, Martin TR, Katze MG. Into the eye of the cytokine storm. Microbiol Mol Biol Rev 2012;76(1):16–32.
[45] Ye Q, Wang B, Mao J. The pathogenesis and treatment of the 'Cytokine Storm' in COVID-19. J Infect 2020;80(6):607–13.

[46] Kuiken T, Fouchier RA, Schutten M, Rimmelzwaan GF, van Amerongen G, van Riel D, Laman JD, de Jong T, van Doornum G, Lim W, Ling AE, Chan PK, Tam JS, Zambon MC, Gopal R, Drosten C, van der Werf S, Escriou N, Manuguerra JC, Stöhr K, Peiris JS, Osterhaus AD. Newly discovered coronavirus as the primary cause of severe acute respiratory syndrome. Lancet 2003;362(9380):263–70.
[47] Peiris JS, Lai ST, Poon LL, Guan Y, Yam LY, Lim W, Nicholls J, Yee WK, Yan WW, Cheung MT, Cheng VC, Chan KH, Tsang DN, Yung RW, Ng TK, Yuen KY. Coronavirus as a possible cause of severe acute respiratory syndrome. Lancet 2003;361(9366):1319–25.
[48] Zaki AM, van Boheemen S, Bestebroer TM, Osterhaus AD, Fouchier RA. Isolation of a novel coronavirus from a man with pneumonia in Saudi Arabia. N Engl J Med 2012;367(19):1814–20.
[49] Kavianpour M, Saleh M, Verdi J. The role of mesenchymal stromal cells in immune modulation of COVID-19: focus on cytokine storm. Stem Cell Res Ther 2020;11(1):404.
[50] Metcalfe SM. Mesenchymal stem cells and management of COVID-19 pneumonia. Med Drug Discov 2020;5, 100019.
[51] Liu B, Li M, Zhou Z, Guan X, Xiang Y. Can we use interleukin-6 (IL-6) blockade for coronavirus disease 2019 (COVID-19)-induced cytokine release syndrome (CRS)? J Autoimmun 2020;111, 102452.
[52] Conti P, Ronconi G, Caraffa A, Gallenga CE, Ross R, Frydas I, Kritas SK. Induction of pro-inflammatory cytokines (IL-1 and IL-6) and lung inflammation by Coronavirus-19 (COVI-19 or SARS-CoV-2): anti-inflammatory strategies. J Biol Regul Homeost Agents 2020;34(2):327–31.
[53] Huang C, Wang Y, Li X, Ren L, Zhao J, Hu Y, Zhang L, Fan G, Xu J, Gu X, Cheng Z, Yu T, Xia J, Wei Y, Wu W, Xie X, Yin W, Li H, Liu M, Xiao Y, Gao H, Guo L, Xie J, Wang G, Jiang R, Gao Z, Jin Q, Wang J, Cao B. Clinical features of patients infected with 2019 novel coronavirus in Wuhan, China. Lancet 2020;395(10223):497–506.
[54] Yang L, Xie X, Tu Z, Fu J, Xu D, Zhou Y. The signal pathways and treatment of cytokine storm in COVID-19. Signal Transduct Target Ther 2021;6(1):255.
[55] Zhou F, Yu T, Du R, Fan G, Liu Y, Liu Z, Xiang J, Wang Y, Song B, Gu X, Guan L, Wei Y, Li H, Wu X, Xu J, Tu S, Zhang Y, Chen H, Cao B. Clinical course and risk factors for mortality of adult inpatients with COVID-19 in Wuhan, China: a retrospective cohort study. Lancet 2020;395(10229):1054–62.
[56] Hoganson DM, Bassett EK, Vacanti JP. Lung tissue engineering. Front Biosci (Landmark Ed) 2014;19(8):1227–39.
[57] Åberg C, Sparr E, Larsson M, Wennerström H. A theoretical study of diffusional transport over the alveolar surfactant layer. J R Soc Interface 2010;7(51):1403–10.
[58] Hoganson DM, Anderson JL, Weinberg EF, Swart E, Orrick BK, Borenstein JT, Vacanti JP. Branched vascular network architecture: a new approach to lung assist device technology. J Thorac Cardiovasc Surg 2010;140(5):990–5.
[59] Huang SX, Islam MN, O'Neill J, Hu Z, Yang YG, Chen YW, Mumau M, Green MD, Vunjak-Novakovic G, Bhattacharya J, Snoeck HW. Efficient generation of lung and airway epithelial cells from human pluripotent stem cells. Nat Biotechnol 2014;32(1):84–91.
[60] Huang SX, Green MD, de Carvalho AT, Mumau M, Chen YW, D'Souza SL, Snoeck HW. The in vitro generation of lung and airway progenitor cells from human pluripotent stem cells. Nat Protoc 2015;10(3):413–25.
[61] Carraro G, Perin L, Sedrakyan S, Giuliani S, Tiozzo C, Lee J, Turcatel G, De Langhe SP, Driscoll B, Bellusci S, Minoo P, Atala A, De Filippo RE, Warburton D. Human amniotic fluid stem cells can integrate and differentiate into epithelial lung lineages. Stem Cells 2008;26(11):2902–11.
[62] De Paepe ME, Mao Q, Ghanta S, Hovanesian V, Padbury JF. Alveolar epithelial cell therapy with human cord blood-derived hematopoietic progenitor cells. Am J Pathol 2011;178(3):1329–39.

[63] Raredon MSB, Yuan Y, Niklason LE. Chapter 68—Lung tissue engineering. In: Lanza R, Langer R, Vacanti JP, Atala A, editors. Principles of tissue engineering. 5th ed. Academic Press; 2020. p. 1273–85.

[64] Shafiee A, Atala A. Printing technologies for medical applications. Trends Mol Med 2016;22(3):254–65.

[65] Murphy SV, Atala A. 3D bioprinting of tissues and organs. Nat Biotechnol 2014;32(8):773–85.

[66] Hong N, Yang G-H, Lee J, Kim G. 3D bioprinting and its in vivo applications. J Biomed Mater Res B Appl Biomater 2018;106(1):444–59.

[67] Mahfouzi SH, Safiabadi Tali SH, Amoabediny G. 3D bioprinting for lung and tracheal tissue engineering: criteria, advances, challenges, and future directions. Bioprinting 2021;21, e00124.

[68] Ashammakhi N, Ahadian S, Xu C, Montazerian H, Ko H, Nasiri R, Barros N, Khademhosseini A. Bioinks and bioprinting technologies to make heterogeneous and biomimetic tissue constructs. Mater Today Bio 2019;1, 100008.

[69] de Melo BAG, Benincasa JC, Cruz EM, Maricato JT, Porcionatto MA. 3D culture models to study SARS-CoV-2 infectivity and antiviral candidates: from spheroids to bioprinting. Biom J 2021;44(1):31–42.

[70] Bhowmick R, Derakhshan T, Liang Y, Ritchey J, Liu L, Gappa-Fahlenkamp H. A three-dimensional human tissue-engineered lung model to study influenza A infection. Tissue Eng Part A 2018;24(19–20):1468–80.

[71] Shpichka A, Bikmulina P, Peshkova M, Kosheleva N, Zurina I, Zahmatkesh E, Khoshdel-Rad N, Lipina M, Golubeva E, Butnaru D, Svistunov A, Vosough M, Timashev P. Engineering a model to study viral infections: bioprinting, microfluidics, and organoids to defeat coronavirus disease 2019 (COVID-19). Int J Bioprint 2020;6(4):302.

[72] Gold K, Gaharwar AK, Jain A. Emerging trends in multiscale modeling of vascular pathophysiology: organ-on-a-chip and 3D printing. Biomaterials 2019;196:2–17.

[73] Maschmeyer I, Lorenz AK, Schimek K, Hasenberg T, Ramme AP, Hübner J, Lindner M, Drewell C, Bauer S, Thomas A, Sambo NS, Sonntag F, Lauster R, Marx U. A four-organ-chip for interconnected long-term co-culture of human intestine, liver, skin and kidney equivalents. Lab Chip 2015;15(12):2688–99.

[74] Vernetti L, Gough A, Baetz N, Blutt S, Broughman JR, Brown JA, Foulke-Abel J, Hasan N, In J, Kelly E, Kovbasnjuk O, Repper J, Senutovitch N, Stabb J, Yeung C, Zachos NC, Donowitz M, Estes M, Himmelfarb J, Truskey G, Wikswo JP, Taylor DL. Corrigendum: functional coupling of human microphysiology systems: intestine, liver, kidney proximal tubule, blood-brain barrier and skeletal muscle. Sci Rep 2017;7:44517.

[75] Skardal A, Murphy SV, Devarasetty M, Mead I, Kang HW, Seol YJ, Shrike Zhang Y, Shin SR, Zhao L, Aleman J, Hall AR, Shupe TD, Kleensang A, Dokmeci MR, Jin Lee S, Jackson JD, Yoo JJ, Hartung T, Khademhosseini A, Soker S, Bishop CE, Atala A. Multi-tissue interactions in an integrated three-tissue organ-on-a-chip platform. Sci Rep 2017;7(1):8837.

[76] Kremer S, Lersy F, de Sèze J, Ferré J-C, Maamar A, Carsin-Nicol B, Collange O, Bonneville F, Adam G, Martin-Blondel G, Rafiq M, Geeraerts T, Delamarre L, Grand S, Krainik A, Caillard S, Constans JM, Metanbou S, Heintz A, Helms J, Schenck M, Lefèbvre N, Boutet C, Fabre X, Forestier G, de Beaurepaire I, Bornet G, Lacalm A, Oesterlé H, Bolognini F, Messié J, Hmeydia G, Benzakoun J, Oppenheim C, Bapst B, Megdiche I, Henry Feugeas M-C, Khalil A, Gaudemer A, Jager L, Nesser P, Talla Mba Y, Hemmert C, Feuerstein P, Sebag N, Carré S, Alleg M, Lecocq C, Schmitt E, Anxionnat R, Zhu F, Comby P-O, Ricolfi F, Thouant P, Desal H, Boulouis G, Berge J, Kazémi A, Pyatigorskaya N, Lecler A, Saleme S, Edjlali-Goujon M, Kerleroux B, Zorn P-E, Matthieu M, Baloglu S, Ardellier F-D, Willaume T, Brisset JC, Boulay C, Mutschler V, Hansmann Y, Mertes P-M, Schneider F, Fafi-Kremer S, Ohana M, Meziani F, David J-S, Meyer N, Anheim M, Cotton F. Brain MRI findings in severe COVID-19: a retrospective observational study. Radiology 2020;297(2):E242–51.

[77] Golchin A, Seyedjafari E, Ardeshirylajimi A. Mesenchymal stem cell therapy for COVID-19: present or future. Stem Cell Rev Rep 2020;16(3):427–33.
[78] Yang L, Liu S, Liu J, Zhang Z, Wan X, Huang B, Chen Y, Zhang Y. COVID-19: immunopathogenesis and immunotherapeutics. Signal Transduct Target Ther 2020;5(1):128.
[79] Liu Y, Sun W, Li J, Chen L, Wang Y, Zhang L, Yu L. Clinical features and progression of acute respiratory distress syndrome in coronavirus disease 2019. medRxiv 2020. 2020.02.17.20024166.
[80] Lippi G, Plebani M. Laboratory abnormalities in patients with COVID-2019 infection. Clin Chem Lab Med 2020;58(7):1131–4.
[81] Lee RH, Pulin AA, Seo MJ, Kota DJ, Ylostalo J, Larson BL, Semprun-Prieto L, Delafontaine P, Prockop DJ. Intravenous hMSCs improve myocardial infarction in mice because cells embolized in lung are activated to secrete the anti-inflammatory protein TSG-6. Cell Stem Cell 2009;5(1):54–63.
[82] Krasnodembskaya A, Song Y, Fang X, Gupta N, Serikov V, Lee J-W, Matthay MA. Antibacterial effect of human mesenchymal stem cells is mediated in part from secretion of the antimicrobial peptide LL-37. Stem Cells 2010;28(12):2229–38.
[83] Hu S, Park J, Liu A, Lee J, Zhang X, Hao Q, Lee JW. Mesenchymal stem cell microvesicles restore protein permeability across primary cultures of injured human lung microvascular endothelial cells. Stem Cells Transl Med 2018;7(8):615–24.
[84] Liotta F, Angeli R, Cosmi L, Filì L, Manuelli C, Frosali F, Mazzinghi B, Maggi L, Pasini A, Lisi V, Santarlasci V, Consoloni L, Angelotti ML, Romagnani P, Parronchi P, Krampera M, Maggi E, Romagnani S, Annunziato F. Toll-like receptors 3 and 4 are expressed by human bone marrow-derived mesenchymal stem cells and can inhibit their T-cell modulatory activity by impairing Notch signaling. Stem Cells 2008;26(1):279–89.
[85] Lee JW, Krasnodembskaya A, McKenna DH, Song Y, Abbott J, Matthay MA. Therapeutic effects of human mesenchymal stem cells in ex vivo human lungs injured with live bacteria. Am J Respir Crit Care Med 2013;187(7):751–60.
[86] Monsel A, Zhu YG, Gennai S, Hao Q, Hu S, Rouby JJ, Rosenzwajg M, Matthay MA, Lee JW. Therapeutic effects of human mesenchymal stem cell-derived microvesicles in severe pneumonia in mice. Am J Respir Crit Care Med 2015;192(3):324–36.
[87] Rajarshi K, Chatterjee A, Ray S. Combating COVID-19 with mesenchymal stem cell therapy. Biotechnol Rep 2020;26, e00467.
[88] Song N, Scholtemeijer M, Shah K. Mesenchymal stem cell immunomodulation: mechanisms and therapeutic potential. Trends Pharmacol Sci 2020;41(9):653–64.
[89] Petersen TH, Calle EA, Niklason LE. Strategies for lung regeneration. Mater Today 2011;14(5):196–201.

7

Immunomodulatory properties of mesenchymal stem cells and hematopoietic stem cells—Potential therapeutic target for COVID-19

Josna Joseph[a] and Annie John[b]

[a]Department of Clinical Immunology and Rheumatology, Christian Medical College, Vellore, Tamil Nadu, India, [b]Advanced Centre for Tissue Engineering, Department of Biochemistry, University of Kerala, Thiruvananthapuram, Kerala, India

Introduction

Immunomodulatory therapies hold great promise in COVID-19 treatment as the pathomechanism of the disease itself involves a plethora of changes in the inflammatory pathway. Though various immunomodulatory agents like Interleukin-6, Interleukin-1 beta, kinase inhibitors, Interferon therapies, etc. have been recommended for COVID-19 treatment (https://www.covid19treatmentguidelines.nih.gov), they have their own limitations and contra-effects. Emergency Use Authorization of stem cells by Food and Drug Administration [1] points out to the fact that stem cell-based cellular therapies could offer alternate and safer treatment strategies.

Mesenchymal stromal cells are a stem cell source of choice, due to their HLA compatibility, wide immunomodulatory properties, and already established clinical protocols that aid in a rapid streamlining of administration processes. As of 31st January 2021, there were 70 registered clinical trials (Clinicaltrials.gov), out of which 8 were completed. Out of the 8 completed, two were on administration of MSC-derived exosomes through inhalation, for treatment of SARS-Cov-2 associated pneumonia (Table 1). Another earlier review (April 30, 2020) [2] of registered clinical trials of cellular therapy including three studies of total 1129 enrolled subjects, have shown that MSCs (41 studies, 76%)/MSC-derived extracellular vesicles or “exosomes” stood

Stem Cells and COVID-19. https://doi.org/10.1016/B978-0-323-89972-7.00005-2

Table 1 Completed clinical trials on MSC/MSC-derived exosomes for SARS-Cov-2 treatment.

Sl. no	Trial no./ identification	Duration	Dosage and route of administration	Nature of study	Outcome measured
1	NCT04713878	8 weeks	Intravenous 1 million cell/kg^{-3} doses	Interventional (open label RCT)	Recovery of patient from mechanical and oxygen support Cytokine storm parameters Oxygen saturation > 93% and pulmonary imaging of focus within 24–48 h > 50% progression Arterial pressure of oxygen/the fraction of inspired oxygen > 300 mmHg
2	NCT04288102	28 days	Intravenous 4×10^7 cells-3 doses UC-MSCs	Phase II (multicenter, randomized, double-blind, placebo-controlled trial)	Change in lesion proportion (%) of full lung volume from baseline to day 2 Evaluation of Pneumonia Improvement
3	NCT04573270	30 days	Intravenous UC-MSC	Phase I (double-blind, placebo-controlled, multiarm, multisite study)	Survival rates, contraction rates
4	NCT04355728	90 days	Intravenous 100×10^6 cells/in fusion—2 doses UC-MSC + heparin	Phase I and II, controlled clinical trial, double blinded, compared with placebo	Incidence of prespecified infusion associated adverse events (5,90 days)
5	NCT04535856	28 days	Intravenous 5×10^7 cells—low dose 1×10^8 cells—high dose	Phase I (double-blind, and placebo-controlled clinical trial)	Incidence of TEAE (treatment emergent adverse events)
6	NCT04492501	28 days	Intravenous Single dose of 2×10^6 cells/kg alone or in combinatorial therapy	Interventional (open label, nonrandomized)	Survival—death or recovery
7	NCT04491240	10–30 days	Inhalation-intranasal $0.5–2 \times 10^{10}$ of nanoparticles (exosomes)	Interventional (Phase I, Phase II)	Nonserious and serious adverse events during trial
8	NCT04276987	28 days	Inhalation—2.0×10^8 nano vesicles/3 mL	Interventional (Phase I)	Adverse reaction (AE) and severe adverse reaction (SAE)

first among other administered cell types (NK cells, mononuclear cells, antigen-presenting cells). Hence, this chapter aims to provide a comprehensive review of the clinical situation of COVID-19 that necessitates cellular therapeutic intervention, employing mesenchymal stromal cells (MSC) in context of their diversified and specific immunomodulatory properties.

COVID-19 and the cytokine storm

Manifold studies have shed light on the immune dysregulation involved in the pathomechanism of COVID-19, including cytokine storm, T cell exhaustion, myeloid cell exaggerated activation, dysregulated adaptive immunity, etc. Cytokine storms are the main cause of Acute Respiratory Distress Syndrome (ARDS) in COVID-19 patients observed in many critically ill patients [3,4]. SARS-Cov-2 through its spike protein (S protein) do engage with ACE2 receptors [5] present on type II alveolar epithelial cells/target cells. The alarmin chemokines secreted by these infected cells will cause an influx of inflammatory cells like neutrophils, monocytes, and T cells. This cellular influx in turn leads to release of a variety of proinflammatory cytokines constituting the cytokine storm [6]. It was shown that upon cellular entry, the virus is enabled to stimulate a terrible cytokine storm in the lungs and thereby increase the levels of interleukin (IL)-2, IL-6, IL-7, granulocyte colony-stimulating factor (GSCF), interferon-induced protein 10 (IP10), monocyte chemoattractant protein-1 (MCP1), macrophage inflammatory protein (MIP1A), and tumor necrosis factor-alpha (TNF) [3]. This probably lead to the pathologic changes of edema with prominent proteinaceous exudates (similar to those described in patients with severe acute respiratory syndrome, SARS-1), hyperplasia of pneumocytes, vascular congestion, and inflammatory clusters of fibrinoid material and multinucleated giant cells [7].

The immunological, serological, and histopathological profile of patients with COVID-19 is detailed elsewhere [8]. Cytokine storm predominantly comprise a release of cascade of proinflammatory cytokines, growth factors, and chemoattractants to bring about acute inflammatory marker uprise and cellular infiltration to affected organs. Of these, notable is the organ associated changes including pathological manifestations like desquamation of pneumocytes, hyaline membrane formation, alveolar hemorrhage, fibrinoid necrosis, interstitial accumulation of monocytes and T cells, and endothelial dysfunction due to cell apoptosis in lungs.

There were also reported cases with change in the T cell counts of COVID-19 patients, with a decrease in the number of T lymphocytes, including T helpers (CD4 +), cytotoxic lymphocytes (CB8 +), and a change in CD4 +/CD8 + ratio [9]. Levels of serum ferritin, D-dimer, lactate dehydrogenase, and IL-6 are increased during the worsening of the disease, providing an indication of the risk to mortality. Thus the key role of MSC treatment in COVID-19 is antiinflammatory as they could reduce the production of proinflammatory cytokines, through various mechanisms, as detailed [10,11] elsewhere.

MSC as a cellular therapy

MSCs are multipotent, nonspecialized cells, with self-renewal and differentiation capacity to specific lineages. Though MSCs can be derived from different sources like bone marrow, peripheral blood, adipose tissue (abdominal fat, buccal fat, and infrapatellar fat pad), human placental tissue, umbilical cord, Wharton's jelly, and dental tissues [12], most of the clinical trials or experimental research make use of MSCs that are derived from umbilical cord, adipose tissue, bone marrow, etc. Since MSCs are present in different tissues due to their close association with blood vessels, every piece of vascularized tissue could be used as a source of MSCs.

MSCs are considered as potent drug stores releasing biologically active substances collectively known as secretome aiding their paracrine mechanism of action [13]. MSC-secretome consists of soluble factors like cytokines, growth factors, epigenetic modulators like miRNA, lipids and extracellular vesicles (EVs) of micro- and nano-sizes [14]. They could modulate cellular responses after being delivered intracellularly by ligand–receptor interaction or by internalization. Secretome can activate endogenous stem cells and progenitor cells, suppress apoptosis, regulate the inflammatory response, stimulate the remodeling of the extracellular matrix and angiogenesis, reduce fibrosis, and mediate chemoattraction [15]. MSCs have been hypothesized to neutralize free virus particles through the production of antibiotic proteins like LL37 which bind to virus and lung cell binding sites [16].

These MSC-secretome-based therapeutic products are more preferred over mesenchymal stem cells due to safety, technology, and economic factors. Safety points include their nonself-replicating property, nil chance of endogenous tumor formation, low immunogenicity, low chance of emboli formation upon intravenous injection [17], stability in blood flow [18] and distribution to different tissues promoting immune modulation, resolution of inflammation, restoration of capillary barrier function, and enhanced bacterial clearance [19]. Technically too, secretome could be manipulated and stored more easily than cells, as a ready to use product suitable for emergency interventions and on-site delivery, specifically in resource-limited countries [20]. Lastly, in comparison with monoclonal antibody treatment, MSC-based secretome has both synergistic and simultaneous action on several cytokines [21] and is comparatively cost effective [22]. It will emerge sooner or later as a promising cell-free therapeutic tool for the treatment of acute and chronic lung diseases, with their proven effects in ARDS (acute respiratory distress syndrome), ex vivo and in vivo [23].

Recently, Bari and colleagues have suggested, for the first time, that MSC-secretome can be formulated as inhalable/injectable dosage forms [20]. Inhalation administration will lead to faster onset of action in lower doses and simultaneously being noninvasive will avoid side effects and pain associated with parenteral therapy. Hence MSC-secretome formulated as freeze-dried stable powder product may be easily administered by intravenous injection/inhalation. This will be an ideal approach for treatment of patients with COVID-19 pneumonia, especially for those in critically severe conditions [24].

Specific immunomodulatory changes

MSCs effect their modulatory action on both cellular and humoral arms of immunity. MSCs have majorly two major modes of cellular modulation, one via direct cell-cell contact and the other by release of extracellular vesicles/exosomes loaded with immunomodulatory molecules such as cytokines, growth factors, signaling molecules, mRNAs, and epigenetic miRNA modulators [24,25].

They have multimodal actions like suppressing T cell proliferation, regulating the balance of Th1/Th2 ratio, T regulatory cell proliferation, and controlling antigen presentation by dendritic cells [26,27]. MSCs can release prostaglandin E2 (PGE2) to induce production of antiinflammatory cytokines from macrophages along with suppressing ongoing T cell-dependent inflammation by releasing transforming growth factor beta (TGF-b), nitric oxide (NO), and indoleamine 2,3-dioxygenase (IDO). MSCs can also induce the proliferation of regulatory T cells to inhibit proliferation and activation of effector T cells [28], increase the phagocytic activity of neutrophils, promote differentiation of macrophages into the M2 phenotype (for producing antiinflammatory cytokines), inhibit neutrophil intravasation and proinflammatory activation by TGF-β release.

On the contrary, MSCs are reported to have wide immunomodulatory, antiapoptotic, antiscarring, chemoattractant, angiogenetic properties [29]. They directly inhibit the proliferation of T cells in vitro [30,31], suppress γδ T cells, and could escape cytotoxic T cell-mediated lysis [32]. MSCs can inhibit or promote B cell proliferation [33,34], suppress NK cell activation [35,36], involve in generation and maintenance of regulatory T cells, and modulate the cytokine secretion profile of dendritic cells [37] and macrophages [38]. Mediatory molecules that are involved in immunomodulatory effects of MSCs include prostaglandin E2, nitric oxide, indoleamine 2,3-dioxygenase

1(IDO-1), inducible nitric oxide synthase (iNOS), interleukin (IL)-10, TGF β1, Hepatocyte Growth Factor (HGF), TNF-α induced protein 6 (TSG-6), human leukocyte antigen G (HLA-G), and leukemia inhibitory factor (LIF) [30,37,39–44]. These immunomodulatory properties are summarized in Fig. 1. MSCs' behavior to immune system is contra specific as the stimulating agents could initiate opposing actions [29] dependent on their doses, e.g., low doses of MSCs could induce dendritic cells mediated T cell activation whereas high numbers of MSCs have an opposite effect [45].

In addition to their immunomodulatory properties that could create an antiinflammatory environment in the alveolar space, MSCs can directly differentiate to type II alveolar (ATII) epithelial cells (cell type constituting 60% of pulmonary alveolar epithelium) [46–48] contributing to overcome the COVID-19 infection associated cell loss. Furthermore, as MSCs lack the angiotensin-converting enzyme II (ACE2) receptor [49] through which SARS-Cov-2 gains cellular entry, they act as a safer form of cellular therapy source.

As results of COVID-19 associated MSCs/MSC-derived exosome therapy are yet to be reported and registered in clinical trial registry

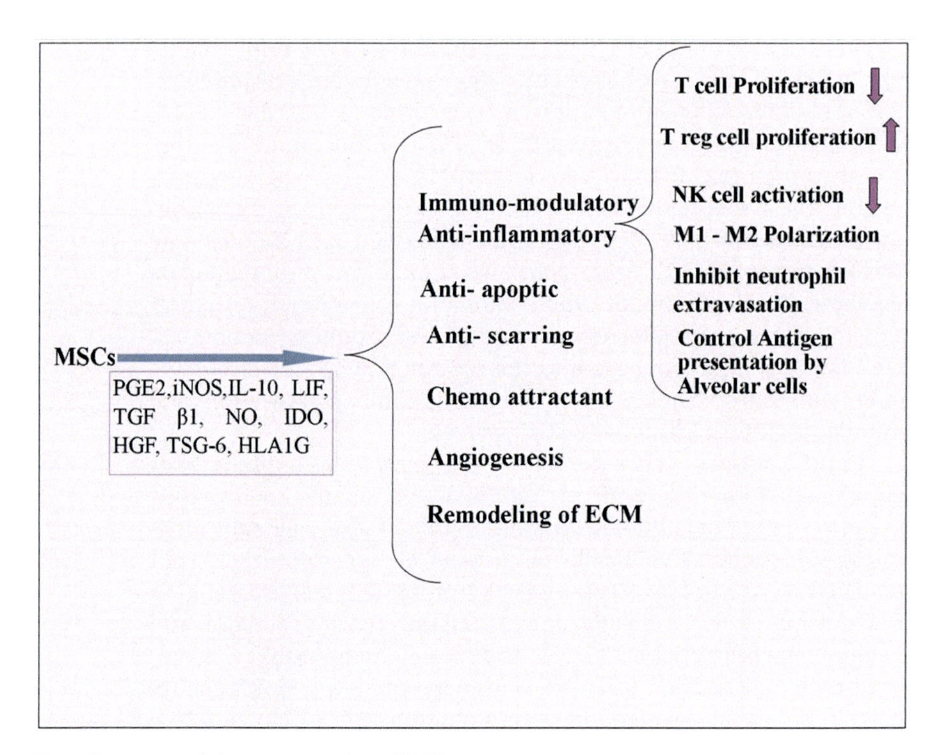

Fig. 1 Immunomodulatory properties of MSCs.

(clinicaltrials.gov.in), some clues could be gathered from earlier trials in related viral infections/diseases. Though several mediators, like IL-1β, IL-10, IL-12p70, and IL-17A, were undetectable in a cohort of nine stable Chronic obstructive pulmonary disease (COPD) patients treated with MSC administration (2×10^6 MSCs per kg, 1 week apart) [50], soluble CD163—a biomarker for macrophage activation that correlates with disease severity in COPD [51]—was found decreased. Again, there is a observed reduction in IL-6 levels, a major mediator in COPD which stimulates the secretion of matrix metalloproteinases and T cell responses [52] which can contribute to airway remodeling. Furthermore, an increase in the antiinflammatory circulating sTNFR1, a decoy noncell associated receptor that binds to and sequesters excess circulating TNF-α could occur, thus reducing systemic inflammation [53]. Similarly, upregulation of sTNFR1 in endotoxemic mice upon MSC administration has been reported [54]. Cell sources of MSCs and their respective immunomodulatory changes are briefly summarized in Table 2.

Lung damage and restoration by MSC administration

SARS-CoV-2 primarily affects the lungs [58] and incidentally the distribution of MSCs in the peripheral blood is mainly concentrated in the lungs after intravenous infusion [59], rendering MSCs a promising treatment option for patients with COVID-19 pneumonia.

Table 2 Different cell sources of MSCs and their immunomodulatory actions.

Sl. no.	Cells/sources	Infection	Immunomodulatory changes	Ref.
1	Allogenic menstrual blood-derived MSCs	Avian influenza A (H7N9) virus	CRP, procalcitonin, serum creatinine, and creatine kinase	[55]
2	ACE-2 negative MSCs	SARS-CoV-2	Increased peripheral lymphocytes, decreased TNF-α and CRP levels, enhanced IL-10 levels, disappearance of overactivated cytokine-secreting CXCR3 + CD4 + T cells, CXCR3 + CD8 + T cells, and NK CXCR3 + cells, CD14 + CD11c + CD11b mid regulatory DC cell populations increased	[56]
3	Umbilical cord MSCs	COVID-19 pneumonia	Serum bilirubin, C-reactive protein (CRP), and alanine transaminase/aspartate transaminase (ALT/AST) decreased	[57]

Lung damage caused by COVID-19 has been restored significantly as evident from CT scan in 67% of MSC administered patients and symptomatic improvement was reported from a retrospective studyin Chinese cohort.

It was reported that SARS-CoV-2 can increase the production of IL-10 to reduce the inflow and aggregation of neutrophils in the lung and reduce the production of TNF-α [60]. KGF (Keratinocyte growth factor) secreted by MSCs [61] can reduce injury and promote proliferation and repair of alveolar epithelial cells by increasing surface-active substances, matrix metalloprotein (MMP)-9, IL-1Ra, GM-CSF, etc. MSCs by the release of VEGF and HGF work together to stabilize endothelial barrier function by restoring pulmonary capillary permeability [62], pulmonary vascular endothelial cell apoptosis, enhancing the recovery of VE cadherin, and reducing proinflammatory factors enabling lung endothelial cells as a potential target for COVID-19 treatment.

In the injured lung, MSCs are involved in improving pathological resolution, functional restoration, and tissue repair in alveolar space, contributing to an overall antiinflammatory environment [63,64]. They could also regenerate alveolar bioenergy, by transferring mitochondria to the damaged alveolar epithelium and increase the concentration of alveolar ATP [65,66]. The immunomodulatory properties of MSCs in lung are represented in Fig. 2. Islam et al. [65] have revealed that bone marrow stromal cells formed connexin 43 (Cx43) containing gap junctional channels (GJCs) with the alveolar epithelium, released mitochondria-containing microvesicles that the epithelium engulfed which in turn increased the alveolar ATP.

In a clinical study in China involving seven COVID-19 pneumonia patients including elderly, MSC transplantation/injection improved the outcome of all patients by 14 days without apparent side effects. Patient pulmonary function improved alongside reduction of cytokine secreting immune cells, including CXCR3 + CD4 + T cells, CXCR3 + CD8 + T cells, CXCR3 + NK cells, and proinflammatory cytokine TNF-α levels [56].

Hematopoietic stem cells

Although there are reports of improved outcome or lack of worsening of symptoms in COVID-19 affected patients [67,68] who had prior hematopoietic stem cell administration, there are few/nil clinical studies/trials that showed efficacy of hematopoietic stem cell administration for COVID-19 treatment. Another promising study shows the antiviral effects of a peptide, thymalin, via its immunomodulatory property in peptide preparation, to influence the differentiation of human hematopoietic stem cells (HSCs) [69]. It also increases the differentiation of T cells from HSC and expression of mature T cell

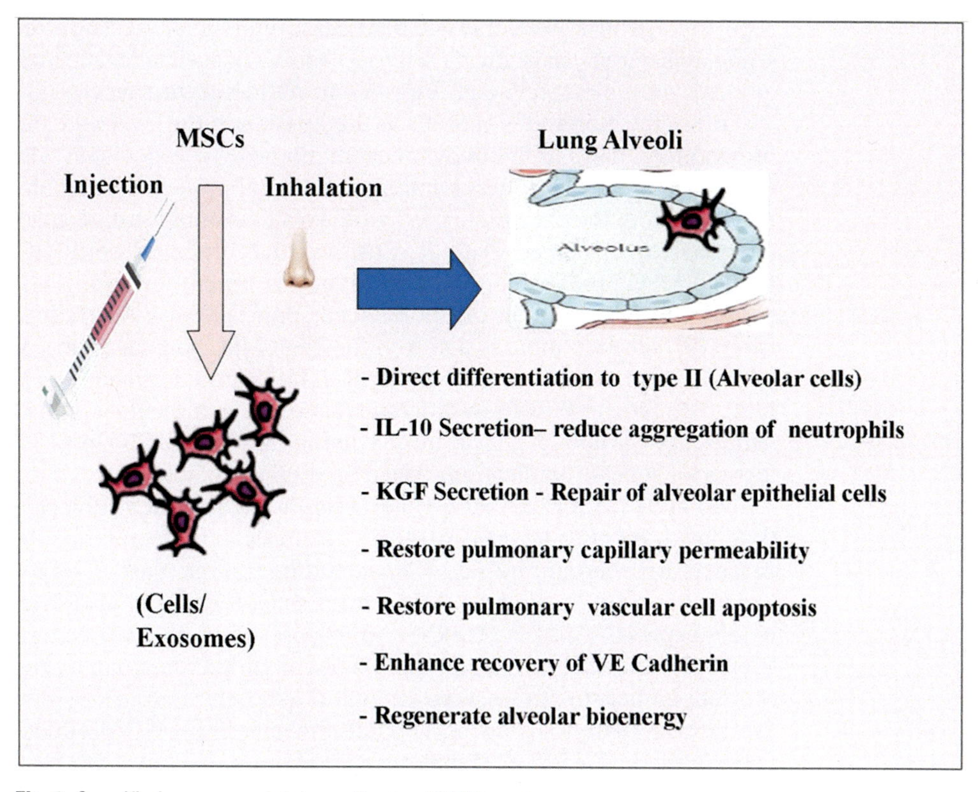

Fig. 2 Specific Immunomodulatory effects of MSCs on lungs.

marker—CD28, proving its augmenting effect on T cell differentiation, thus contributing to antiviral immunity in general and immunoprotective effect in COVID-19 en route hemostasis in the situation of a cytokine storm [70].

Though hematopoietic cellular therapy includes allogeneic, autogenic, and CART cells, CART cell consortium investigators have warned a cautious and careful administration in the context of COVID-19 due to its potential of triggering off a cytokine storm [71]. In a retrospective study between March 15, 2020 and May 7, 2020 conducted in 77 patients with SARS-Cov-2 who were recipients of cellular therapies, an overall survival at 30 days was 78% and they were able to recover from COVID-19 infection and simultaneously mount an antibody response with similar overall survival rate to the general hospitalized population [72].

Translational potential of MSCs

Based on their proven efficiency, MSCs are cells of choice in cellular therapy for COVID-19. Major manufacturers of advanced

MSC therapeutics like Athersys, Mesoblast, Pluristem, etc. have ongoing registered trials for COVID-19 (https://ipscell.com/2020/03/athersys-mesoblast-stem-cell-drugs-for-novel-coronavirus-covid-19/).

One of the limitations of MSCs as therapeutics is the low frequency of *bona fide* MSCs in the body, necessitating ex vivo expansion with its associated time lag, necessitating scale up approaches using bioreactors. From the preliminary 597 studies (status as of April 14, 2020) compiled by Cell-Trials-Data, it is evident that umbilical cord (UC) derived MSC products are the anticipated treatment for critically ill patients (https://celltrials.org/public-cells-data/all-covid-19-clinical-trials/79). Further, there is a search for fresh MSC sources from cell pools like pericytes which are thought to be disputed progenitors of MSC [73]. Since MSCs from different tissues are not identical, future studies are warranted to determine whether MSCs could be more effective for specific applications based on their site of origin.

Cao et al. [74] imply that it is necessary to select MSCs subject to their low level of ACE2 and TMPRESS2 expression in the therapeutic practice, to ascertain the safety by excluding the rare case of SARS-CoV-2 infection. Most of the MSCs from different organs or donors barely express ACE2 or TMPRSS2 which proved their exclusiveness from Sars-Cov-2 entry. Several reports documented cases of DIC and thromboembolism and associated multiorgan failure occurring after the intravascular infusion of TF/CD142-expressing MSC products, particularly in preactivated patients [75]. Also, in patients with metabolic acidosis or coronary heart disease, more cautious administration of MSCs is essential [76].

Thus there exists a need for reliable and extensive meta-analysis studies pooling in observed data from multiple controlled trials, to analyze and establish the safety and efficacy of MSC-based therapeutics for COVID-19 infection, as currently there is a paucity of reliable and concordant data regarding the same. Also, the variables such as manufacturing site, age of the donor, source of cells (whether umbilical cord, bone marrow, or adipose tissue), the extent of in vitro expansion, storage conditions, and release criteria prior to administration should be considered as basis of heterogeneity [77,78] while creating a cell bank for advanced therapeutic applications.

Not the least, there should be a good deal of concern about clinics offering unproven stem cell treatment for COVID-19 [11]. Hence a more cautious strategy should be strictly adopted toward the general implementation of MSC-based therapeutics.

Conclusion

The diversified modulatory properties of MSCs render them an irreplaceable position amid the cellular armamentarium in COVID-19 therapeutics. Possessing regulatory power over both the innate and adaptive arms of immunity, MSCs comprise a safer and controllable cellular choice. Nonetheless, together with an established and practiced administration protocol, MSCs become definitely the right choice for therapy. Though there is data scarcity on the applicability of hematopoietic stem cells for COVID-19 treatment, debut research data on the anvil highlights the modulatory and protective effects of HSC's on administered patients.

Acknowledgments

Authors express sincere thanks to the HOD, Department of Biochemistry and Director, Advanced Centre for Tissue Engineering, University of Kerala; and Head, Department of Clinical Immunology & Rheumatology, CMC Vellore for their valuable support; Ms. Niveditha K. for technical assistance in the graphical illustration; and lastly AJ gratefully acknowledges support from ICMR Emeritus Scientist Scheme.

References

[1] Choudhery MS, Harris DT. Stem cell therapy for COVID-19: possibilities and challenges. Cell Biol Int 2020;44(11):2182–91.

[2] Liao G, Zheng K, Lalu MM, Fergusson DA, Allan DS. A scoping review of registered clinical trials of cellular therapy for COVID-19 and a framework for accelerated synthesis of trial evidence-FAST evidence. Transfus Med Rev 2020;34(3):165–71.

[3] Huang C, Wang Y, Li X, Ren L, Zhao J, Hu Y, et al. Clinical features of patients infected with 2019 novel coronavirus in Wuhan, China. Lancet 2020;395(10223):497–506.

[4] Mehta P, McAuley DF, Brown M, Sanchez E, Tattersall RS, Manson JJ. COVID-19: consider cytokine storm syndromes and immunosuppression. Lancet 2020;395(10229):1033–4.

[5] Hoffmann M, Kleine-Weber H, Schroeder S, Krüger N, Herrler T, Erichsen S, et al. SARS-CoV-2 cell entry depends on ACE2 and TMPRSS2 and Is blocked by a clinically proven protease inhibitor. Cell 2020;181(2):271–280.e8.

[6] Xiao K, Hou F, Huang X, Li B, Qian ZR, Xie L. Mesenchymal stem cells: current clinical progress in ARDS and COVID-19. Stem Cell Res Ther 2020;11(1):305.

[7] Tian S, Hu W, Niu L, Liu H, Xu H, Xiao S-Y. Pulmonary pathology of early-phase 2019 novel coronavirus (COVID-19) pneumonia in two patients with lung cancer. J Thorac Oncol 2020;15(5):700–4.

[8] Rukavishnikova S, Akhmedov T, Pushkin A, Saginbaev U. Hematological indicators as predictors of the outcome of a new coronavirus infection Covid-19 patients of different age groups. Vrach 2020;31(7):33–6.

[9] Kermali M, Khalsa RK, Pillai K, Ismail Z, Harky A. The role of biomarkers in diagnosis of COVID-19—a systematic review. Life Sci 2020;254:117788.

[10] Liu S, Peng D, Qiu H, Yang K, Fu Z, Zou L. Mesenchymal stem cells as a potential therapy for COVID-19. Stem Cell Res Ther 2020;11(1):169.

[11] Can A, Coskun H. The rationale of using mesenchymal stem cells in patients with COVID-19-related acute respiratory distress syndrome: what to expect. Stem Cells Transl Med 2020;9(11):1287–302.

[12] Al-Nbaheen M, Vishnubalaji R, Ali D, Bouslimi A, Al-Jassir F, Megges M, et al. Human stromal (mesenchymal) stem cells from bone marrow, adipose tissue and skin exhibit differences in molecular phenotype and differentiation potential. Stem Cell Rev Rep 2013;9(1):32–43.

[13] Caplan AI. Mesenchymal stem cells: time to change the name!: mesenchymal stem cells. Stem Cells Transl Med 2017;6(6):1445–51.
[14] Crivelli B, Chlapanidas T, Perteghella S, Lucarelli E, Pascucci L, Brini AT, et al. Mesenchymal stem/stromal cell extracellular vesicles: from active principle to next generation drug delivery system. J Control Release 2017;262:104–17.
[15] Di Rocco G, Baldari S, Toietta G. Towards therapeutic delivery of extracellular vesicles: strategies for in vivo tracking and biodistribution analysis. Stem Cells Int 2016;2016:1–12.
[16] Krasnodembskaya A, Song Y, Fang X, Gupta N, Serikov V, Lee J-W, et al. Antibacterial effect of human mesenchymal stem cells is mediated in part from secretion of the antimicrobial peptide LL-37. Stem Cells 2010;28(12):2229–38.
[17] Zhu X, Badawi M, Pomeroy S, Sutaria DS, Xie Z, Baek A, et al. Comprehensive toxicity and immunogenicity studies reveal minimal effects in mice following sustained dosing of extracellular vesicles derived from HEK293T cells. J Extracell Vesicles 2017;6(1):1324730.
[18] Morishita M, Takahashi Y, Nishikawa M, Takakura Y. Pharmacokinetics of exosomes—an important factor for elucidating the biological roles of exosomes and for the development of exosome-based therapeutics. J Pharm Sci 2017;106(9):2265–9.
[19] Shah TG, Predescu D, Predescu S. Mesenchymal stem cells-derived extracellular vesicles in acute respiratory distress syndrome: a review of current literature and potential future treatment options. Clin Transl Med 2019;8(1). [Internet]. [cited 2021 February 15]. Available from: https://onlinelibrary.wiley.com/doi/abs/10.1186/s40169-019-0242-9.
[20] Bari E, Ferrarotti I, Torre ML, Corsico AG, Perteghella S. Mesenchymal stem/stromal cell secretome for lung regeneration: the long way through "pharmaceuticalization" for the best formulation. J Control Release 2019;309:11–24.
[21] Kyurkchiev D. Secretion of immunoregulatory cytokines by mesenchymal stem cells. World J Stem Cells 2014;6(5):552.
[22] Anon. Tocilizumab (Actemra): adult patients with moderately to severely active rheumatoid arthritis [Internet]. Ottawa (ON): Canadian Agency for Drugs and Technologies in Health; 2015. [cited 2021 February 15]. (CADTH Common Drug Reviews). Available from http://www.ncbi.nlm.nih.gov/books/NBK349521/.
[23] Abraham A, Krasnodembskaya A. Mesenchymal stem cell-derived extracellular vesicles for the treatment of acute respiratory distress syndrome. Stem Cells Transl Med 2020;9(1):28–38.
[24] Bari E, Ferrarotti I, Saracino L, Perteghella S, Torre ML, Corsico AG. Mesenchymal stromal cell secretome for severe COVID-19 infections: premises for the therapeutic use. Cell 2020;9(4):924.
[25] Huh JW, Kim WY, Park YY, Lim C-M, Koh Y, Kim M-J, et al. Anti-inflammatory role of mesenchymal stem cells in an acute lung injury mouse model. Acute Crit Care 2018;33(3):154–61.
[26] Martin-Rufino JD, Espinosa-Lara N, Osugui L, Sanchez-Guijo F. Targeting the immune system with mesenchymal stromal cell-derived extracellular vesicles: what is the Cargo's mechanism of action? Front Bioeng Biotechnol 2019;7:308.
[27] Duman DG, Zibandeh N, Ugurlu MU, Celikel C, Akkoc T, Banzragch M, et al. Mesenchymal stem cells suppress hepatic fibrosis accompanied by expanded intrahepatic natural killer cells in rat fibrosis model. Mol Biol Rep 2019;46(3):2997–3008.
[28] Genç D, Zibandeh N, Nain E, Arığ Ü, Göker K, Aydıner EK, et al. IFN-γ stimulation of dental follicle mesenchymal stem cells modulates immune response of CD4+ T lymphocytes in Der p1+ asthmatic patients in vitro. Allergol Immunopathol (Madr) 2019;47(5):467–76.

[29] da Silva ML, Fontes AM, Covas DT, Caplan AI. Mechanisms involved in the therapeutic properties of mesenchymal stem cells. Cytokine Growth Factor Rev 2009;20(5–6):419–27.
[30] Di Nicola M, Carlo-Stella C, Magni M, Milanesi M, Longoni PD, Matteucci P, et al. Human bone marrow stromal cells suppress T-lymphocyte proliferation induced by cellular or nonspecific mitogenic stimuli. Blood 2002;99(10):3838–43.
[31] Krampera M, Glennie S, Dyson J, Scott D, Laylor R, Simpson E, et al. Bone marrow mesenchymal stem cells inhibit the response of naive and memory antigen-specific T cells to their cognate peptide. Blood 2003;101(9):3722–9.
[32] Rasmusson I, Uhlin M, Le Blanc K, Levitsky V. Mesenchymal stem cells fail to trigger effector functions of cytotoxic T lymphocytes. J Leukoc Biol 2007;82(4):887–93.
[33] Corcione A, Benvenuto F, Ferretti E, Giunti D, Cappiello V, Cazzanti F, et al. Human mesenchymal stem cells modulate B-cell functions. Blood 2006;107(1):367–72.
[34] Traggiai E, Volpi S, Schena F, Gattorno M, Ferlito F, Moretta L, et al. Bone marrow-derived mesenchymal stem cells induce both polyclonal expansion and differentiation of B cells isolated from healthy donors and systemic lupus erythematosus patients. Stem Cells 2008;26(2):562–9.
[35] Sotiropoulou PA, Perez SA, Gritzapis AD, Baxevanis CN, Papamichail M. Interactions between human mesenchymal stem cells and natural killer cells. Stem Cells 2006;24(1):74–85.
[36] Spaggiari GM, Capobianco A, Abdelrazik H, Becchetti F, Mingari MC, Moretta L. Mesenchymal stem cells inhibit natural killer–cell proliferation, cytotoxicity, and cytokine production: role of indoleamine 2,3-dioxygenase and prostaglandin E2. Blood 2008;111(3):1327–33.
[37] Aggarwal S, Pittenger MF. Human mesenchymal stem cells modulate allogeneic immune cell responses. Blood 2005;105(4):1815–22.
[38] Németh K, Leelahavanichkul A, Yuen PST, Mayer B, Parmelee A, Doi K, et al. Bone marrow stromal cells attenuate sepsis via prostaglandin E2–dependent reprogramming of host macrophages to increase their interleukin-10 production. Nat Med 2009;15(1):42–9.
[39] Sato K, Ozaki K, Oh I, Meguro A, Hatanaka K, Nagai T, et al. Nitric oxide plays a critical role in suppression of T-cell proliferation by mesenchymal stem cells. Blood 2007;109(1):228–34.
[40] Nasef A, Chapel A, Mazurier C, Bouchet S, Lopez M, Mathieu N, et al. Identification of IL-10 and TGF-β transcripts involved in the inhibition of T-lymphocyte proliferation during cell contact with human mesenchymal stem cells. Gene Expr 2006;13(4):217–26.
[41] Nasef A, Mathieu N, Chapel A, Frick J, François S, Mazurier C, et al. Immunosuppressive effects of mesenchymal stem cells: involvement of HLA-G. Transplantation 2007;84(2):231–7.
[42] Nasef A, Mazurier C, Bouchet S, François S, Chapel A, Thierry D, et al. Leukemia inhibitory factor: role in human mesenchymal stem cells mediated immunosuppression. Cell Immunol 2008;253(1–2):16–22.
[43] Meisel R, Zibert A, Laryea M, Göbel U, Däubener W, Dilloo D. Human bone marrow stromal cells inhibit allogeneic T-cell responses by indoleamine 2,3-dioxygenase-mediated tryptophan degradation. Blood 2004;103(12):4619–21.
[44] Choi H, Lee RH, Bazhanov N, Oh JY, Prockop DJ. Anti-inflammatory protein TSG-6 secreted by activated MSCs attenuates zymosan-induced mouse peritonitis by decreasing TLR2/NF-κB signaling in resident macrophages. Blood 2011;118(2):330–8.
[45] Le Blanc K, Tammik L, Sundberg B, Haynesworth SE, Ringden O. Mesenchymal stem cells inhibit and stimulate mixed lymphocyte cultures and mitogenic responses independently of the major histocompatibility complex. Scand J Immunol 2003;57(1):11–20.

[46] Gotts JE, Matthay MA. Mesenchymal stem cells and acute lung injury. Crit Care Clin 2011;27(3):719–33.
[47] Harrell CR, Sadikot R, Pascual J, Fellabaum C, Jankovic MG, Jovicic N, et al. Mesenchymal stem cell-based therapy of inflammatory lung diseases: current understanding and future perspectives. Stem Cells Int 2019;2019:4236973.
[48] Harrell CR, Jovicic N, Djonov V, Arsenijevic N, Volarevic V. Mesenchymal stem cell-derived exosomes and other extracellular vesicles as new remedies in the therapy of inflammatory diseases. Cell 2019;11:8(12).
[49] Zhou P, Yang X-L, Wang X-G, Hu B, Zhang L, Zhang W, et al. A pneumonia outbreak associated with a new coronavirus of probable bat origin. Nature 2020;579(7798):270–3.
[50] Armitage J, Tan DBA, Troedson R, Young P, Lam K, Shaw K, et al. Mesenchymal stromal cell infusion modulates systemic immunological responses in stable COPD patients: a phase I pilot study. Eur Respir J 2018;51(3):1702369.
[51] Baines KJ, Backer V, Gibson PG, Powel H, Porsbjerg CM. Impaired lung function is associated with systemic inflammation and macrophage activation. Eur Respir J 2015;45(2):557–9.
[52] Jarvinen L, Badri L, Wettlaufer S, Ohtsuka T, Standiford TJ, Toews GB, et al. Lung resident mesenchymal stem cells isolated from human lung allografts inhibit T cell proliferation via a soluble mediator. J Immunol 2008;181(6):4389–96.
[53] Tan DBA, Fernandez S, Price P, French MA, Thompson PJ, Moodley YP. Impaired CTLA-4 responses in COPD are associated with systemic inflammation. Cell Mol Immunol 2014;11(6):606–8.
[54] Yagi H, Soto-Gutierrez A, Navarro-Alvarez N, Nahmias Y, Goldwasser Y, Kitagawa Y, et al. Reactive bone marrow stromal cells attenuate systemic inflammation via sTNFR1. Mol Ther 2010;18(10):1857–64.
[55] Chen J, Hu C, Chen L, Tang L, Zhu Y, Xu X, et al. Clinical study of mesenchymal stem cell treatment for acute respiratory distress syndrome induced by epidemic influenza a (H7N9) infection: a hint for COVID-19 treatment. Engineering 2020;6(10):1153–61.
[56] Leng Z, Zhu R, Hou W, Feng Y, Yang Y, Han Q, et al. Transplantation of ACE2- mesenchymal stem cells improves the outcome of patients with COVID-19 pneumonia. Aging Dis 2020;11(2):216.
[57] Liang B, Chen J, Li T, Wu H, Yang W, Li Y, et al. Clinical remission of a critically ill COVID-19 patient treated by human umbilical cord mesenchymal stem cells: a case report. Medicine (Baltimore) 2020;99(31), e21429.
[58] Guzik TJ, Mohiddin SA, Dimarco A, Patel V, Savvatis K, Marelli-Berg FM, et al. COVID-19 and the cardiovascular system: implications for risk assessment, diagnosis, and treatment options. Cardiovasc Res 2020;116(10):1666–87.
[59] Fischer UM, Harting MT, Jimenez F, Monzon-Posadas WO, Xue H, Savitz SI, et al. Pulmonary passage is a major obstacle for intravenous stem cell delivery: the pulmonary first-pass effect. Stem Cells Dev 2009;18(5):683–92.
[60] Mei SHJ, McCarter SD, Deng Y, Parker CH, Liles WC, Stewart DJ. Prevention of LPS-induced acute lung injury in mice by mesenchymal stem cells overexpressing angiopoietin 1. Singer M, editor, PLoS Med 2007;4(9):e269.
[61] Shyamsundar M, McAuley DF, Ingram RJ, Gibson DS, O'Kane D, McKeown ST, et al. Keratinocyte growth factor promotes epithelial survival and resolution in a human model of lung injury. Am J Respir Crit Care Med 2014;189(12):1520–9.
[62] Yang Y, Chen Q, Liu A, Xu X, Han J, Qiu H. Synergism of MSC-secreted HGF and VEGF in stabilising endothelial barrier function upon lipopolysaccharide stimulation via the Rac1 pathway. Stem Cell Res Ther 2015;6(1):250.
[63] English K, Ryan JM, Tobin L, Murphy MJ, Barry FP, Mahon BP. Cell contact, prostaglandin E_2 and transforming growth factor beta 1 play non-redundant roles in human mesenchymal stem cell induction of $CD4^+$ $CD25^{High}$ forkhead box $P3^+$ regulatory T cells. Clin Exp Immunol 2009;156(1):149–60.

[64] Wang M, Yuan Q, Xie L. Mesenchymal stem cell-based immunomodulation: properties and clinical application. Stem Cells Int 2018;2018:3057624.

[65] Islam MN, Das SR, Emin MT, Wei M, Sun L, Westphalen K, et al. Mitochondrial transfer from bone-marrow-derived stromal cells to pulmonary alveoli protects against acute lung injury. Nat Med 2012;18(5):759–65.

[66] Monsel A, Zhu Y-G, Gennai S, Hao Q, Liu J, Lee JW. Cell-based therapy for acute organ injury: preclinical evidence and ongoing clinical trials using mesenchymal stem cells. Anesthesiology 2014;121(5):1099–121.

[67] Sultan AM, Mahmoud HK, Fathy GM, Abdelfattah NM. The outcome of hematopoietic stem cell transplantation patients with COVID-19 infection. Bone Marrow Transplant 2020. Internet. [cited 2021 February 26]. Available from: http://www.nature.com/articles/s41409-020-01094-9.

[68] Maurer K, Saucier A, Kim HT, Acharya U, Mo CC, Porter J, et al. COVID-19 and hematopoietic stem cell transplantation and immune effector cell therapy: a US cancer center experience. Blood Adv 2021;5(3):861–71.

[69] Khavinson VK, Kuznik BI, Trofimova SV, Volchkov VA, Rukavishnikova SA, Titova ON, et al. Results and prospects of using activator of hematopoietic stem cell differentiation in complex therapy for patients with COVID-19. Stem Cell Rev Rep 2021;17(1):285–90.

[70] Khavinson V, Linkova N, Dyatlova A, Kuznik B, Umnov R. Peptides: prospects for use in the treatment of COVID-19. Molecules 2020;25(19):4389.

[71] Maziarz RT, Schuster SJ, Ericson SG, Rusch ES, Signorovitch J, Li J, et al. Cytokine release syndrome and neurotoxicity by baseline tumor burden in adults with relapsed or refractory diffuse large b-cell lymphoma treated with tisagenlecleucel. Hematol Oncol 2019;37:307.

[72] Shah GL, DeWolf S, Lee YJ, Tamari R, Dahi PB, Lavery JA, et al. Favorable outcomes of COVID-19 in recipients of hematopoietic cell transplantation. J Clin Invest 2020;130(12):6656–67.

[73] Lv F-J, Tuan RS, Cheung KMC, Leung VYL. Concise review: the surface markers and identity of human mesenchymal stem cells: markers and identity of MSCs. Stem Cells 2014;32(6):1408–19.

[74] Cao Y, Wu H, Zhai W, Wang Y, Li M, Li M, et al. A safety consideration of mesenchymal stem cell therapy on COVID-19. Stem Cell Res 2020;49:102066.

[75] Moll G, Hoogduijn MJ, Ankrum JA. Editorial: safety, efficacy and mechanisms of action of mesenchymal stem cell therapies. Front Immunol 2020;11:243.

[76] Chen X, Shan Y, Wen Y, Sun J, Du H. Mesenchymal stem cell therapy in severe COVID-19: a retrospective study of short-term treatment efficacy and side effects. J Infect 2020;81(4):647–79.

[77] Rizk M, Monaghan M, Shorr R, Kekre N, Bredeson CN, Allan DS. Heterogeneity in studies of mesenchymal stromal cells to treat or prevent graft-versus-host disease: a scoping review of the evidence. Biol Blood Marrow Transplant 2016;22(8):1416–23.

[78] Liu S, de Castro LF, Jin P, Civini S, Ren J, Reems J-A, et al. Manufacturing differences affect human bone marrow stromal cell characteristics and function: comparison of production methods and products from multiple centers. Sci Rep 2017;7:46731.

8

COVID-19 and acute myocardial injury: Stem cell driven tissue remodeling in COVID-19 infection

Jessy John[a], Mereena George Ushakumary[b], Soumya Chandrasekher[c], and Smitha Chenicheri[d,e]
[a]Division of Hematology and Oncology, Hillman Cancer Center, University of Pittsburgh, Pittsburgh, PA, United States, [b]Division of Pulmonary Biology, Cincinnati Children's Hospital Medical Center, Cincinnati, OH, United States, [c]Department of Zoology, KKTM Government College, Pullut, Trichur, India, [d]PMS College of Dental Sciences and Research, Thiruvananthapuram, India, [e]Biogenix Research Center for Molecular Biology and Applied Sciences, Thiruvananthapuram, India

Introduction

The emergence and ongoing COVID-19 pandemic has resulted in more than 19 million clinically confirmed cases claiming above 4 million deaths worldwide (as of July 20, 2021). The gush for vaccine research and clinical trials for taming COVID-19 continues expecting the most effective and efficient outcomes. Unfortunately, the virus retains the basic reproductive number (R_0) in the range of 2 to 6 since the onset challenging the global healthcare system and medical research [1]. Unfortunately, COVID-19 claimed lives of millions of people worldwide and it was the third leading cause of death in the US behind heart disease and cancer [2]. Being a respiratory virus, the mode of COVID-19 transmission is expected to occur through respiratory droplets/aerosols and the primary target of infection has been identified to be the respiratory system [3]. Despite the *acute respiratory distress syndrome* (ARDS), the clinical data unveiled the intimate association between COVID-19 infection and cardiovascular diseases (CVD) as evident from the increased infection and mortality rate in COVID-19 patients with the history of cardiovascular diseases. In addition, the preexisting CVD pathology significantly worsens the clinical situation, increasing the risk of myocardial ischemia, cardiac arrhythmia, and

Stem Cells and COVID-19. https://doi.org/10.1016/B978-0-323-89972-7.00001-5

coronary syndrome [4]. Furthermore, the CVD medication offers potential risk and CVD comorbidity became a serious clinical concern in this pandemic era. Moreover, COVID-19 infection displays abnormal coagulation chemistry leading to thromboembolic events [5]. Hence, the integration of the existing knowledge on CVDs with COVID-19 pathobiology is essential for development of effective preventative and management strategies. Interestingly, stem cells offer a promising management for the complications, especially the cytokine burst, associated with COVID-19 immunopathology. On this background, the present chapter throws novel insights perspectives into the molecular pathobiology associating COVID-19–CVD comorbidity and the potential applications of stem cells in clinical management.

COVID-19: Etiology, transmission, structure, and pathophysiology

Efforts to sequence the COVID-19 genome have been successful and resulted in the elucidation of viral structure. Generally, coronaviruses are a family of RNA viruses which were believed to originate in bats and the current understanding on COVID-19 owes to other members of the family, especially SARS-CoV [6]. The characteristic crown-like morphology of COVD-19 has been attributed to its structural proteins, especially the spike (S), envelope (E), membrane (M), and nucleocapsid (N) protein. The viral genome (~ 30 kb in length) encoding 24–27 genes is stabilized by the N protein. The 5′ terminal has been mapped with two polyproteins pp1a and pp1ab, which cleave to 16 nonstructural proteins essential for the survival of the virus. The 3′ terminal constitutes one-third of COVID-19 genome and encodes S, E, M, and N structural proteins [7]. S protein is pivotal for the attachment, binding, and entry of COVID-19 to the host cells via ACE2 receptors and is crucial in initiating the pathology [8]. Hence, the increased binding affinity between S protein and ACE2 receptors explains the severity of COVID-19 disease. The attachment of virus using the receptor-binding domain (S1) of S protein triggers the cleavage of S protein at the S1/2 and S2′ regions. This process is called S protein priming facilitated by the transmembrane serine protease TMPRSS2. S protein priming drives the fusion of viral membrane with the host, directing the entry of virus into the cytosol of the host cell [9]. The respiratory tract epithelial cells expressing both ACE2 and TMPRSS2 in abundance have been identified to be the initial route of COVID-19 entry. Alternatively, endosomal route of entry of ACE2–virus complex following the priming has been established which requires the endosomal cysteine proteases cathepsin B and cathepsin L [10]. Following the entry of COVID-19 genome to the cytosol, the translation of polyproteins pp1a and pp1ab is initiated which are cleaved by viral proteases into nonstructural proteins including the RNA-dependent RNA polymerase, the key to viral RNA replication. The structural proteins are subsequently translated exploiting the endoplasmic reticulum and Golgi complex of the host cell. Finally, the viral genome organization occurs along with the assembly of structural proteins leading to new viral particles which are released through exocytosis [4].

Clinically, COVID-19 infection presented profound lymphopenia with an increased apoptosis of T lymphocytes. In addition, the severity of infection compromises the integrity of epithelial-endothelial barrier aggravating the inflammatory responses by increasing the influx of monocytes and neutrophils [11]. The thickening of alveolar wall with increased infiltration of inflammatory macrophages was evident in autopsy specimen suggesting the aggravated inflammatory milieu. Also, edema of pulmonary system appearing as ground-glass opacities on computed tomographic imaging has been identified due to the uncontrolled deposition of hyaline membrane in the alveolar spaces [12]. In addition, bradykinin-dependent lung angioedema has been associated with severe COVID-19 infection. Importantly, the sudden activation of coagulation cascade and the increased consumption of clotting factors are common in severe COVID-19 pathology resulting in fatal thrombosis and sepsis [11,13]. Taken together, the disruption of endothelial barrier, pulmonary infection and dysfunction, impaired oxygen physiology, and hyperactivation of inflammatory episodes progress to extrapulmonary organs ultimately leading to multiorgan failure.

COVID-19-CVD comorbidity

Various studies reported the association between COVID-19 and comorbidities [14] and there is a strong interplay between COVID-19 and cardiovascular diseases. The major comorbidities reported in infected patients were diabetes, hypertension, cardiovascular and cerebrovascular diseases. A systematic meta-analysis by Li et al. [15] in 1527 patients in China reported that the prevalence of hypertension, cardiac-cerebrovascular diseases, and diabetes in patients with COVID-19 were 17.1%, 16.4%, and 9.7%, respectively, and these proportions increased twofold, threefold, and twofold, respectively, higher in ICU/severe cases than in their non-ICU/severe counterparts. Another case study with 1591 COVID-19 infected older patients from Italy reported hypertension as comorbid condition in 49% of patients [16] followed by cardiovascular disorders, hypercholesterolemia, and diabetes. Meta-analysis with 1576 infected patients reported hypertension (OR=2.36, 95% CI: 1.46–3.83) and cardiovascular disease (OR=3.42, 95% CI: 1.88–6.22) as the prevalent comorbidities in severe compared to nonsevere patients [17]. Similar studies from the United States identified hypertension, diabetes, and cardiovascular disease as risk factors for COVID-19 [18] and the persons with these risk factors are at a higher risk for severe disease from COVID-19 than are persons without these conditions [19]. Although CVD and its risk factors like hypertension and diabetes were significantly associated with severe outcomes in patients with COVID-19 across all ages, the relative risk for severity in young patients with hypertension, diabetes, and CVD was higher than in elderly patients [20].

Furthermore, COVID-19 severely affects cardiovascular system and early case studies reported cardiovascular complications like myocardial infarction, myocarditis, arrhythmia, heart failure, cardiogenic shock, and thromboembolism following infection, and which leads to elevated risk and mortality among these patients [21]. Mechanisms behind these COVID-19-associated complications include viral entry via ACE2 receptors leading to direct myocyte injury, elevated Ang II activity leading to oxidative stress and inflammatory damage to the myocardium, and a prothrombotic and proatherosclerotic state and cytokine storm induced by inflammation leading to plaque rupture and resulting in severe myocardial injury [22]. In summary, COVID-19 worsens underlying CVD and triggers newer cardiac complications, and the risk is high in patients with underlying CVD compared with those who do not [23].

Stem cells in cardiac remodeling

Cardiac remodeling refers to a series of events left ventricle undergoes postmyocardial infarction (MI). Activation of fibroblasts and inflammatory cells is necessary to initiate the healing response post-injury [24]. However, persistent activation of fibroblasts and the inflammatory cells leads to an adverse myocardial remodeling response that causes contractile dysfunction and eventually heart failure [25]. An orchestrated interplay between multiple cell types and adequate cues from the extracellular environment in the heart determines a physiological versus pathological remodeling response. In this regard, stem cells have been identified as a potential regenerative therapy for cardiac regeneration post-MI [26]. Cardiac resident stem cells are involved in the physiological remodeling of the myocardium [27]. They have the potential to support endogenous mechanisms ofmyocardial repair by either cell fusion with cardiomyocytes to make mature cardiomyocytes or through secreting paracrine factors required for the tissue repair [28]. Both mechanisms have proved effective in stem cell transplantation as the primary goal is to restore the functional mature myocytes. Besides the ability to restore cardiomyocytes, they also secrete cytokines and growth factors required for the initial inflammatory response as part of the reparative process [29].

Stem cells have been characterized depending on the source where they are derived from. The most studied stem cells include, but not limited to, adult stem cells, embryonic stem cells, and induced pluripotent stem cells [30]. Adult bone marrow-derived stem cells are the most used stem cells to date, and they have been used in both clinical and preclinical aspects [31]. Studies demonstrated that the c-kit positive cells derived from bone marrow were able to regenerate myocytes and coronary vessels in mice post-MI [32]. Tissue engineered cardiac grafts are in the frontline to overcome the limitations associated with direct transplantation of stem cells. Recent studies using stem cell-derived cell sheet technology opened a new realm for stem cell research [33]. This has the advantage of forming vascularized networks for fabricating thickened human cardiac tissue after transplantation [34]. Though the field is in its budding state, cell sheet engineering will bring novel insights into stem cell treatment strategies in future.

ACE2 signaling and cardiac pathology COVID-19 infection

COVID-19 virus enters the host cell through angiotensin-converting enzyme 2 (ACE2) receptor, which is expressed in various human organs like kidney, heart, lung, blood vessels, brain stem, stomach, liver, and gut [35]. ACE2 is negative regulator of the renin-angiotensin system (RAS). RAS network plays an important role in blood pressure, fluid, and electrolyte homeostasis and maintains the function of heart, blood vessels, and kidney (Fig. 1). In this system, the protease renin converts the precursor angiotensinogen to angiotensin I (Ang I) and subsequent conversion to angiotensin II (Ang II) mediated by angiotensin-converting enzyme (ACE).

Ang II, the most active peptide in this system, causes vasoconstriction, tissue fibrosis, cardiomyocyte hypertrophy, oxidative stress, and inflammation and blood coagulation. Pathological effects of Ang II are counteracted by ACE2 which converts Ang II to Ang-(1–7). Ang-(1–7) exerts antiinflammatory, antifibrotic, and vasodilating effects by binding to the G protein-coupled MAS receptors [36]. Thus ACE2 regulates the levels of Ang II which is involved in the development of hypertension and heart failure.

Binding of the glycosylated SPIKE protein (S-protein) of the COVID-19 viral coat to the extracellular domain of ACE2 mediates the viral entry into the host. Transmembrane protease seine 2 (TMPRSS2) mediated cleavage of the S-protein is essential for the membrane fusion of COVID-19 virus and its intake through the endocytic pathway. This viral entry leads to several mechanisms affecting ACE2 expression on

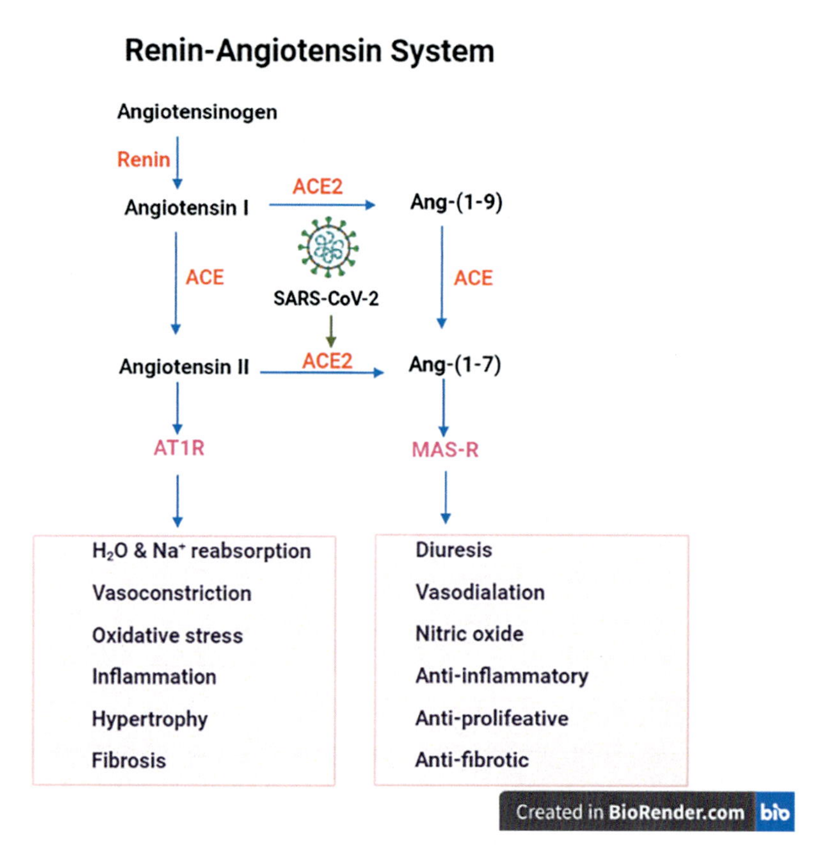

Fig. 1 Renin-Angiotensin signaling pathway: *Angiotensinogen is proteolytically cleaved to* angiotensin I by renin followed by conversion to angiotensin II by angiotensin-converting enzyme (ACE). Angiotensin II acts on angiotensin type I receptor (AT1R) to exert its cardiac actions which is counteracted by angiotensin-converting enzyme 2 (ACE2) which converts angiotensin I and angiotensin II to angiotensin-(1–7) [Ang 1–7)], which exerts cardioprotective actions by binding to MAS receptor. SARS-CoV-2 binding to the ACE2 receptor leads to exacerbation of angiotensin II-mediated signaling and subsequent cardiac damage.

cell surface, initially by receptor internalization upon viral entry leading to reduced cell surface expression. Following viral entry, unknown viral mediators inhibit ACE2 and induce ADAM metallopeptidase domain 17 (ADAM-17) gene expression. ADAM-17 sheddase mediates cleavage of ACE2 [37] and the release of various cytokines like TNFα, IL-4, and IFNγ which are involved in downregulation of ACE2 expression by autocrine pathways [38]. Expression of ACE2 by the cardiac myocytes leads to viral entry into the heart. Receptor internalization following viral entry and subsequent viral-mediated mechanisms induce downregulation of ACE2 which leads to elevated Ang II levels and RAS overactivation and loss of Ang-(1–7)/MAS receptor-mediated protective effects which exacerbate the cardiac damage. Numerous clinical studies reported cardiac manifestations like myocardial injury, myocarditis, heart failure, arrhythmias, etc. after COVID-19 infection [39]. Study by Guo et al. reported elevation in troponin T levels, an indicator of myocardial injury in infected patients [40]. Involvement of ACE2 in COVID-19 infection highlights the importance of monitoring the levels of myocardial enzymes, cardiac rhythm, and function, to decide the outcome of COVID-19 infection on heart.

Stem cell activation and signaling in ACE2 upregulation

Adult human heart is home to cardiac stem cells and multipotent progenitor cells [41]. They are inherently programmed to generate cardiomyocytes. Though studies provide a role for resident cardiac stem cells (CSCs) such as c-kit-positive (c-kit^{POS}) cells [42], Sca-1-positive (Sca-1^{POS}) cells [43], cardiosphere cells [44], and Isl1-positive (Isl1^{POS}) cells [45] in the endogenous self-renewal of cardiomyocytes [46], the need for transplantation of adult stem cells arise because of the time consuming and costly procedures to expand cardiac stem cells prior to cell transplantation. Understanding the molecular mechanism of action of stem cells paves the way for manipulating the cells for clinical application. The mechanism by which pluripotent stem cells and mesenchymal stem cells preserve myocardial homeostasis post-MI is through activation of Akt phosphorylation and thereby activating the downstream survival signaling cascade [47]. It prevents cardiomyocyte apoptosis.

Expression of ACE2 receptor has been reported on the surface of hematopoietic stem cells (HSCs) and endothelial progenitor cells (EPCs) [48]. Cardiac ACE2 expression is increased after MI and overexpression of ACE2 is reported to provide protection against ischemia-induced cardiac dysfunction [49]. In this regard, it is pertinent to note that COVID-19 virus enters human cells after binding to the ACE2 receptor [50]. Therefore ACE2 expression on stem cells raises a concern that the covid-19 virus may infect and damage the myocardial stem cell niche. Moreover, ACE2 expression is high in heart failure patients and therefore if infected with virus, these patients have a high chance of heart attack and critical illness [51]. ACE2 plays an important role as a member of the renin-angiotensin-aldosterone system (RAAS) in regulating the conversion of angiotensin II (Ang II) to other molecules that counteract the effects of Ang II [52] and at the same time acts as a receptor for COVID-19. These bidirectional aspects could impede the normal stem cell function and can lead to a total loss of stem cell niche after infection. However, studies on how COVID-19 virus infects and damages the stem cell niche post-MI are completely lacking. Understanding ACE2 dynamics in stem cells is key to the identification of viral entry mechanisms and possible damage to the niche.

Cytokine surge and stem cell activation in MI

Cardiac repair post-MI is tightly coupled to inflammatory response of the myocardium [24]. A surge in inflammatory cytokines contributes to cardiac remodeling during the acute phase of MI. The factors that trigger cytokine surge during MI include ischemia, reactive oxygen species (ROS), mechanical stress, etc. [53]. They trigger an inflammatory cytokine response to promote a normal repair process. In fact, these cytokines are prerequisites for wound healing and compensatory hypertrophy of myocytes [24]. The major cytokines released are TNFα, IL-6, and IL-1β [54]. Moderate release of these cytokines determines the cardioprotective role. Persistent activation of the inflammatory cytokines leads to pathological remodeling of myocardium [55]. Additionally, the profibrotic cytokine, transforming growth factor-β (TGF-β) triggers pathological repair by inducing myocardial fibrosis [56]. Stem cells have shown to produce different growth factors and cytokines after transplantation, and they act in tandem with immune cells to modulate the function of distinct immune cell population post-MI [57]. This paracrine action helps in modulating a pro-inflammatory phenotype to a pro-reparative immune cell phenotype to promote proper wound healing [58]. Several studies have demonstrated that mesenchymal stem cells produce chemotactic molecules and that recruit neutrophils to the site of injury [59]. However, it increases the antiinflammatory, pro-reparative cytokine expression in macrophages. This raises the concern that neutrophil infiltration could activate pro-inflammatory signaling and therefore may not be a good cell type for cell-based therapy [60]. Embryonic stem cells have shown to increase MHCI and MHCII and elicit a massive inflammatory response [59]. However, cardiac-derived stem cells have shown promise through paracrine and exosome-mediated mechanisms in reducing the pro-inflammatory response [61]. In addition, intramyocardial injection of cortical bone-derived stem cells has also shown cardioprotective effects by modulating the leukocyte inflammatory response [62].

Sheddases' response in cardiac stem cell differentiation (major focus to TMPRSS2, CTSL, and ADAM17)

Major host proteases involved in COVID-19 entry and infection include type II transmembrane serine protease (TMPRSS2), ADAM metallopeptidase domain-17 (ADAM-17), cysteine protease cathepsin L (CTSL), furin, etc. S-protein priming by TMPRSS2, proteolytic cleavage of S-protein by furin, and endosomal processing by CTSL facilitate viral entry into the host by binding to the ACE2 receptors. ADAM-17 mediates shedding of the ectodomain of ACE2 following viral entry. Various stem cells with cardiac regeneration potential include embryonic stem cells, induced pluripotent stem cells, cardiac stem cells, directly reprogrammed somatic cells, mesenchymal stem cells, bone marrow cells [63]. A recent preprint study using human embryonic stem cell-derived cardiomyocytes reported the presence of ACE2, TMPRSS2, cathepsin L, and furin at gene and protein level on these cardiomyocytes, which are the entry proteins for COVID-19 [64]. Another study reported the presence of cathepsin L not TMPRSS2 in induced pluripotent stem cell-derived cardiomyocytes suggesting cathepsin L dependent entry of COVID-19 into the cardiomyocytes [65].

Human mesenchymal stem (MSC) cells have been considered for treatment of COVID-19 infection owing to its immunoregulatory and differentiation potentials. Study by Avanzini et al. evaluated human MSCs derived from various adult and fetal origins for expression of ACE2 and TMPRSS2 and found that these cells do not express these proteins and showed failure of any cytopathic effect after exposure to SARS-CoV-2 infection in these MSCs which suggest that these human MSCs are resistant to SARS-CoV-2 infection. This makes MSCs as a safe treatment option for hyperactivation of the immune system and acute respiratory distress syndrome (ARDS) associated with COVOD-19 infection [66]. Another study by Hernández et al. also reported low expression of ACE2 and TMPRSS2 in human umbilical cord-derived MSCs from multiple donors which indicate human MSCs can evade SARS-CoV-2 infection and can be a treatment option in COVID-19 infection [67].

Perspectives in COVID-19 and cardiac responses

Elevated levels of ACE2 were reported in cardiac patients treated with ACE inhibitors (ACEIs) or Ang II receptor blockers (ARBs) [68,69]. ACE2 upregulation in these patients may make them more susceptible to severe COVID-19 infection [70]. This raised the concern of using ACEI/ARBs in infected patients with CVD, but clinical studies did not find evidence of risk for COVID-19 infection associated with the use of these drugs in patients. In a retrospective study with 4480 patients, prior use of ACEI/ARBs was not significantly associated with COVID-19 diagnosis or worse outcome in patients or with mortality [71]. Meta-analysis by Xu et al. also reported no association between prior use of ACEI/ARBs and risk of COVID-19 infection or severity [72]. Another study by **An et al. in a large cohort of patients with hypertension from US** also found no association between ACEI/ARBs use and COVID-19 infection [73]. These findings do not support the discontinuation of ACEI/ARBs in COVID-19 infected CVD patients.

Further studies also reported protective effect in the use of ACEI/ARBs in COVID-19 infected patients [74]. In a retrospective study among hospitalized COVID-19 infected patients with hypertension, in-patient use of ACEI/ARBs was associated with a lower risk of all-cause mortality compared with ACEI/ARB nonuser patients [75]. Also, ACE2 internalization by COVID-19 potentiates Ang II levels and leads to an increase in the pathological effects of Ang II and loss of cardioprotective effects of Ang-(1-7) and which can subsequently induce new or worsen the existing cardiac injury. Several studies reported elevated plasma Ang II levels after COVID-19 infection [76,77]. In this scenario, use of ACEi/ARBs in infected patients may protect from the deleterious effects of elevated Ang II activity which warrants further investigation. At present all guidelines recommend continuing the use of ACEI/ARBs in infected CVD patients and do not advocate initiating the use of these drugs to treat infected patients without CVD conditions.

Translational avenues and future

Though stem cells can differentiate into lineage committed cells and form specific tissues, its translational potential in a clinical setting needs to be improved to accelerate the process of regeneration postmyocardial injury. Numerous preclinical and clinical studies have identified the potential of stem cell transplantation post-MI. However, some of the questions remain are (a) the timing of stem cell transplantation, (b) appropriate stem cell type, (c) route of administration, and (d) how long the effect sustains. Intravenous injection of mesenchymal stem cells in a post-MI animal model showed very limited cells in the myocardium [78], most were sequestered in the lung [79]. This could raise the concern that these cells could form unpredicted lineages in different organs and can initiate an immune response apart of from the site of injury or at the site of injury. However, intracoronary and transendocardial routes of administration have shown improved outcome with the latter method being safer and more effective [80,81]. Stem cell preparation and storage is another key aspect regarding clinical/translational approach. Maintaining stem cell properties is important for the endogenous repair mechanisms post-MI. Injecting enough number of viable cells is important in a clinical setting. The media used must provide adequate environment for the differentiating into the functional lineage. There are ongoing efforts to standardize the means of stem cell preparation and storage. However, the process of long-term storage conditions and storage mediums is still lacking. Another factor to consider is increasing the number of preclinical and clinical trials to understand the reproducibility of the stem cell transplantation approach and thereby streamlining the process. Myocardial repair mechanisms post-MI and post-COVID-19 complications can be managed in future with more standardized ways and approaches of stem cell transplantation. Although there are more obstacles to consider and overcome, the capabilities of stem cell transplantation for various disease conditions are growing. An interdisciplinary collaboration is required to scale-up the process of stem cell therapy and overcoming the limitations explained before.

References

[1] Randolph HE, Barreiro LB. Herd immunity: understanding COVID-19. Immunity 2020;52(5):737–41.
[2] Ahmad FB, Anderson RN. The leading causes of death in the US for 2020. JAMA 2021;325(18):1829–30.
[3] Akin L, Gozel MG. Understanding dynamics of pandemics. Turk J Med Sci 2020;50(SI-1):515–9.
[4] Nishiga M, et al. COVID-19 and cardiovascular disease: from basic mechanisms to clinical perspectives. Nat Rev Cardiol 2020;17(9):543–58.
[5] Bikdeli B, et al. COVID-19 and thrombotic or thromboembolic disease: implications for prevention, antithrombotic therapy, and follow-up: JACC state-of-the-art review. J Am Coll Cardiol 2020;75(23):2950–73.
[6] Thankam FG, Agrawal DK. Molecular chronicles of cytokine burst in patients with coronavirus disease 2019 (COVID-19) with cardiovascular diseases. J Thorac Cardiovasc Surg 2021;161(2):e217–26.
[7] Cui J, Li F, Shi ZL. Origin and evolution of pathogenic coronaviruses. Nat Rev Microbiol 2019;17(3):181–92.
[8] Lurie N, et al. Developing Covid-19 vaccines at pandemic speed. N Engl J Med 2020;382(21):1969–73.
[9] Hoffmann M, et al. SARS-CoV-2 cell entry depends on ACE2 and TMPRSS2 and is blocked by a clinically proven protease inhibitor. Cell 2020;181(2):271–280 e8.
[10] Howley PMK, David M. Whelan, sean, fields virology: emerging viruses. Lippincott Williams & Wilkins (LWW); 2020.
[11] Wiersinga WJ, et al. Pathophysiology, transmission, diagnosis, and treatment of coronavirus disease 2019 (COVID-19): a review. JAMA 2020;324(8):782–93.
[12] Xu Z, et al. Pathological findings of COVID-19 associated with acute respiratory distress syndrome. Lancet Respir Med 2020;8(4):420–2.
[13] Tang N, et al. Abnormal coagulation parameters are associated with poor prognosis in patients with novel coronavirus pneumonia. J Thromb Haemost 2020;18(4):844–7.
[14] Zhou Y, et al. Comorbidities and the risk of severe or fatal outcomes associated with coronavirus disease 2019: a systematic review and meta-analysis. Int J Infect Dis 2020;99:47–56.
[15] Li B, et al. Prevalence and impact of cardiovascular metabolic diseases on COVID-19 in China. Clin Res Cardiol 2020;109(5):531–8.
[16] Grasselli G, et al. Baseline characteristics and outcomes of 1591 patients infected with SARS-CoV-2 admitted to ICUs of the Lombardy region, Italy. JAMA 2020;323(16):1574–81.
[17] Yang J, et al. Prevalence of comorbidities and its effects in patients infected with SARS-CoV-2: a systematic review and meta-analysis. Int J Infect Dis 2020;94:91–5.
[18] Goyal P, et al. Clinical characteristics of Covid-19 in new York City. N Engl J Med 2020;382(24):2372–4.

[19] Chow N, et al. Preliminary estimates of the prevalence of selected underlying health conditions among patients with coronavirus disease 2019-United States, February 12-march 28, 2020. Morb Mortal Wkly Rep 2020;69(13):382–6.

[20] Bae S, et al. Impact of cardiovascular disease and risk factors on fatal outcomes in patients with COVID-19 according to age: a systematic review and meta-analysis. Heart 2021;107(5):373.

[21] Samidurai A, Das A. Cardiovascular complications associated with COVID-19 and potential therapeutic~strategies. Int J Mol Sci 2020;21(18).

[22] Mahajan K, Chandra KS. Cardiovascular comorbidities and complications associated with coronavirus disease 2019. Med J Armed Forces India 2020;76(3):253–60.

[23] Guo T, et al. Cardiovascular implications of fatal outcomes of patients with coronavirus disease 2019 (COVID-19). JAMA Cardiol 2020;5(7):811–8.

[24] Prabhu SD, Frangogiannis NG. The biological basis for cardiac repair after myocardial infarction: from inflammation to fibrosis. Circ Res 2016;119(1):91–112.

[25] Humeres C, Frangogiannis NG. Fibroblasts in the infarcted, remodeling, and failing heart. JACC Basic Transl Sci 2019;4(3):449–67.

[26] Hare JM, Chaparro SV. Cardiac regeneration and stem cell therapy. Curr Opin Organ Transplant 2008;13(5):536–42.

[27] Leri A, Kajstura J, Anversa P. Role of cardiac stem cells in cardiac pathophysiology: a paradigm shift in human myocardial biology. Circ Res 2011;109(8):941–61.

[28] Lemcke H, et al. Recent Progress in stem cell modification for cardiac regeneration. Stem Cells Int 2018;2018:1909346.

[29] Broughton KM, et al. Mechanisms of cardiac repair and regeneration. Circ Res 2018;122(8):1151–63.

[30] Zakrzewski W, et al. Stem cells: past, present, and future. Stem Cell Res Ther 2019;10(1):68.

[31] Aly RM. Current state of stem cell-based therapies: an overview. Stem Cell Investig 2020;7:8.

[32] Orlic D, et al. Bone marrow stem cells regenerate infarcted myocardium. Pediatr Transplant 2003;7(Suppl 3):86–8.

[33] Masuda S, et al. Cell sheet engineering for heart tissue repair. Adv Drug Deliv Rev 2008;60(2):277–85.

[34] Guo R, et al. Stem cell-derived cell sheet transplantation for heart tissue repair in myocardial infarction. Stem Cell Res Ther 2020;11(1):19.

[35] Zamorano Cuervo N, Grandvaux N. ACE2: evidence of role as entry receptor for SARS-CoV-2 and implications in comorbidities. Elife 2020;9.

[36] South AM, Diz DI, Chappell MC. COVID-19, ACE2, and the cardiovascular consequences. Am J Physiol Heart Circ Physiol 2020;318(5):H1084–90.

[37] Lambert DW, et al. Tumor necrosis factor-alpha convertase (ADAM17) mediates regulated ectodomain shedding of the severe-acute respiratory syndrome-coronavirus (SARS-CoV) receptor, angiotensin-converting enzyme-2 (ACE2). J Biol Chem 2005;280(34):30113–9.

[38] Gross S, et al. SARS-CoV-2 receptor ACE2-dependent implications on the cardiovascular system: from basic science to clinical implications. J Mol Cell Cardiol 2020;144:47–53.

[39] Thakkar S, et al. A systematic review of the cardiovascular manifestations and outcomes in the setting of Coronavirus-19 disease. Clin Med Insights Cardiol 2020;14, 1179546820977196.

[40] Huang C, et al. Clinical features of patients infected with 2019 novel coronavirus in Wuhan, China. Lancet 2020;395(10223):497–506.

[41] Leri A. Human cardiac stem cells: the heart of a truth. Circulation 2009;120(25):2515–8.

[42] Marino F, et al. Role of c-kit in myocardial regeneration and aging. Front Endocrinol (Lausanne) 2019;10:371.

[43] Valente M, et al. Sca-1+ cardiac progenitor cells and heart-making: a critical synopsis. Stem Cells Dev 2014;23(19):2263–73.
[44] Dutton LC, et al. Cardiosphere-derived cells suppress allogeneic lymphocytes by production of PGE2 acting via the EP4 receptor. Sci Rep 2018;8(1):13351.
[45] Di Felice V, Zummo G. Stem cell populations in the heart and the role of Isl1 positive cells. Eur J Histochem 2013;57(2), e14.
[46] Wen Z, et al. Local activation of cardiac stem cells for post-myocardial infarction cardiac repair. J Cell Mol Med 2012;16(11):2549–63.
[47] Li H, et al. Myocardial survival signaling in response to stem cell transplantation. J Am Coll Surg 2009;208(4):607–13.
[48] Ratajczak MZ, et al. SARS-CoV-2 entry receptor ACE2 is expressed on very small CD45(−) precursors of hematopoietic and endothelial cells and in response to virus spike protein activates the Nlrp3 Inflammasome. Stem Cell Rev Rep 2021;17(1):266–77.
[49] Der Sarkissian S, et al. Cardiac overexpression of angiotensin converting enzyme 2 protects the heart from ischemia-induced pathophysiology. Hypertension 2008;51(3):712–8.
[50] Shang J, et al. Cell entry mechanisms of SARS-CoV-2. Proc Natl Acad Sci U S A 2020;117(21):11727–34.
[51] Chen L, et al. The ACE2 expression in human heart indicates new potential mechanism of heart injury among patients infected with SARS-CoV-2. Cardiovasc Res 2020;116(6):1097–100.
[52] Gheblawi M, et al. Angiotensin-converting enzyme 2: SARS-CoV-2 receptor and regulator of the renin-angiotensin system: celebrating the 20th anniversary of the discovery of ACE2. Circ Res 2020;126(10):1456–74.
[53] Gullestad L, et al. Inflammatory cytokines in heart failure: mediators and markers. Cardiology 2012;122(1):23–35.
[54] Kany S, Vollrath JT, Relja B. Cytokines in inflammatory disease. Int J Mol Sci 2019;20(23).
[55] Zhao W, Zhao J, Rong J. Pharmacological modulation of cardiac remodeling after myocardial infarction. Oxid Med Cell Longev 2020;2020:8815349.
[56] Frangogiannis N. Transforming growth factor-beta in tissue fibrosis. J Exp Med 2020;217(3), e20190103.
[57] Moore JB, Wysoczynski M. Immunomodulatory effects of cell therapy after myocardial infarction. J Cell Immunol 2021;3(2):85–90.
[58] Kang Y, et al. Administration of cardiac mesenchymal cells modulates innate immunity in the acute phase of myocardial infarction in mice. Sci Rep 2020;10(1):14754.
[59] Wagner MJ, Khan M, Mohsin S. Healing the broken heart; the immunomodulatory effects of stem cell therapy. Front Immunol 2020;11:639.
[60] Joel MDM, et al. MSC: immunoregulatory effects, roles on neutrophils and evolving clinical potentials. Am J Transl Res 2019;11(6):3890–904.
[61] Yuan Y, et al. Stem cell-derived exosome in cardiovascular diseases: macro roles of Micro particles. Front Pharmacol 2018;9:547.
[62] Kraus L, et al. Cortical bone derived stem cells modulate cardiac fibroblast response via miR-18a in the heart after injury. Front Cell Dev Biol 2020;8:494.
[63] du Pre BC, Doevendans PA, van Laake LW. Stem cells for cardiac repair: an introduction. J Geriatr Cardiol 2013;10(2):186–97.
[64] Williams TL, et al. Human embryonic stem cell-derived cardiomyocytes express SARS-CoV-2 host entry proteins: screen to identify inhibitors of infection. bioRxiv 2021. p. 2021.01.22.427737.
[65] Perez-Bermejo JA, et al. SARS-CoV-2 infection of human iPSC-derived cardiac cells reflects cytopathic features in hearts of patients with COVID-19. Sci Transl Med 2021;13(590).

[66] Avanzini MA, et al. Human mesenchymal stromal cells do not express ACE2 and TMPRSS2 and are not permissive to SARS-CoV-2 infection. Stem Cells Transl Med 2021;10(4):636–42.
[67] Hernandez JJ, et al. Dodging COVID-19 infection: low expression and localization of ACE2 and TMPRSS2 in multiple donor-derived lines of human umbilical cord-derived mesenchymal stem cells. J Transl Med 2021;19(1):149.
[68] Ferrario CM, et al. Effect of angiotensin-converting enzyme inhibition and angiotensin II receptor blockers on cardiac angiotensin-converting enzyme 2. Circulation 2005;111(20):2605–10.
[69] Sama IE, et al. Circulating plasma concentrations of angiotensin-converting enzyme 2 in men and women with heart failure and effects of renin-angiotensin-aldosterone inhibitors. Eur Heart J 2020;41(19):1810–7.
[70] Guo J, et al. Coronavirus disease 2019 (COVID-19) and cardiovascular disease: a viewpoint on the potential influence of angiotensin-converting enzyme inhibitors/angiotensin receptor blockers on onset and severity of severe acute respiratory syndrome coronavirus 2 infection. J Am Heart Assoc 2020;9(7), e016219.
[71] Fosbol EL, et al. Association of Angiotensin-Converting Enzyme Inhibitor or angiotensin receptor blocker use with COVID-19 diagnosis and mortality. JAMA 2020;324(2):168–77.
[72] Xu J, et al. The effect of prior angiotensin-converting enzyme inhibitor and angiotensin receptor blocker treatment on coronavirus disease 2019 (COVID-19) susceptibility and outcome: a systematic review and Meta-analysis. Clin Infect Dis 2020;72(11):e901–13.
[73] An J, et al. Angiotensin-converting enzyme inhibitors or angiotensin receptor blockers use and COVID-19 infection among 824 650 patients with hypertension from a US integrated healthcare system. J Am Heart Assoc 2021;10(3), e019669.
[74] Bavishi C, Maddox TM, Messerli FH. Coronavirus disease 2019 (COVID-19) infection and renin angiotensin system blockers. JAMA Cardiol 2020;5(7):745–7.
[75] Zhang P, et al. Association of Inpatient use of angiotensin-converting enzyme inhibitors and angiotensin II receptor blockers with mortality among patients with hypertension hospitalized with COVID-19. Circ Res 2020;126(12):1671–81.
[76] Liu Y, et al. Clinical and biochemical indexes from 2019-nCoV infected patients linked to viral loads and lung injury. Sci China Life Sci 2020;63(3):364–74.
[77] Wu Z, et al. Elevation of plasma angiotensin II level is a potential pathogenesis for the critically ill COVID-19 patients. Crit Care 2020;24(1):290.
[78] Toma C, et al. Human mesenchymal stem cells differentiate to a cardiomyocyte phenotype in the adult murine heart. Circulation 2002;105(1):93–8.
[79] Halkos ME, et al. Intravenous infusion of mesenchymal stem cells enhances regional perfusion and improves ventricular function in a porcine model of myocardial infarction. Basic Res Cardiol 2008;103(6):525–36.
[80] Wang X, et al. Stem cells for myocardial repair with use of a transarterial catheter. Circulation 2009;120(11 Suppl):S238–46.
[81] Amado LC, et al. Cardiac repair with intramyocardial injection of allogeneic mesenchymal stem cells after myocardial infarction. Proc Natl Acad Sci U S A 2005;102(32):11474–9.

9

Stem cell-driven tissue regeneration as treatment for COVID-19

Jane Joy Thomas[a], Jessy John[b], and Mereena George Ushakumary[c]
[a]Division of Nephrology and Hypertension, Feinberg Cardiovascular and Renal Research Institute, Northwestern University, Chicago, IL, United States, [b]Division of Hematology and Oncology, Hillman Cancer Center, University of Pittsburgh, Pittsburgh, PA, United States, [c]Division of Pulmonary Biology, Cincinnati Children's Hospital Medical Center, Cincinnati, OH, United States

Introduction

In December 2019, severe acute respiratory syndrome coronavirus 2 (SARS-CoV-2) was first identified as the cause of a respiratory illness designated coronavirus disease 2019, or COVID-19. Outbreaks of COVID-19 in China revealed patients exhibiting symptoms such as mild pneumonia and influenza-like illness. Other symptoms involved fever, cough, fatigue, shortness of breath, and muscle pain. Some people did not exhibit symptoms leading to suspecting exposure to SARS-CoV-2 [1–6]. Other complications of COVID-19 included acute respiratory distress syndrome, acute respiratory injury, acute renal failure, hypoxic respiratory failure of subacute onset evolving into ARDS, lymphopenia, and highly elevated C-reactive protein and proinflammatory cytokines [7]. Based on current knowledge, no treatment has been found to be effective in treating the SARS-CoV-2 infection. Several possible treatments with immunomodulatory and antiviral agents and inflammation inhibitors (anti-IL6, anti-IL1, inhibitors of Janus kinases) have been used to improve clinical outcomes in patients with critical COVID-19.

Management of COVID-19 by stem cells

Several COVID-19 treatments, including antiviral agents, lopinavir/ritonavir, remdesivir, new molecules, immunomodulatory drugs, tocilizumab, anakinra, baricitinib, corticosteroids, chloroquine and

Stem Cells and COVID-19. https://doi.org/10.1016/B978-0-323-89972-7.00002-7

hydroxychloroquine, anticoagulants, and therapeutic antibodies, have been used to inhibit the progression and complications of COVID-19 [8]. Recently, scientific studies identified stem cells as a promising therapy for COVID-19 infection based on their ability to self-renew, differentiate, and regenerate.

Tissue-resident stem cells

The major pathological outcomes of COVID-19, such as damage to epithelial cells and inflammatory and fibrotic complications [9–13], result from the depletion of the resident stem cell population [14]. This can be due to impairment of tissue regeneration and repair. One of the most effective strategies involves the alleviation of tissue stem cell loss caused by COVID-19 pathogenicity. Studies have shown that development of ex vivo human organoid experimental model systems of different organs including intestinal, pulmonary, and neuronal can help in investigating the effect of COVID-19 on tissue stem cells [15]. Also, understanding ways to develop host-specific resident stem cells with immunomodulatory properties can be useful in resolving site-specific COVID-19 damage.

Hematopoietic stem cells

Maintenance and differentiation of functional immune cells require a reserve from which the cells can evolve. The most important reserve consists of the pluripotent human hematopoietic stem cells, which differentiate into several lineages, giving rise to myeloid and lymphoid cells. Studies have shown that the effects of factors and stem cell markers, such as thymalin, CD44, and CD117, fight against the antiviral immunity in COVID-19 infection [16]. The fascinating plasticity of adult hematopoietic stem cells gives them the ability to act either as stem cells or to differentiate to become another tissue, which can be used to reprogram the fate of a cell [17]. This can be a rationale for treating COVID infection.

Embryonic stem cells

A challenge faced in the present COVID-19 scenario is the fact that coronaviruses have unique capabilities to maintain pathogenic genomes during replication. Developing a better understanding of viral susceptibility genes and genes necessary for responses such as chromatin modifiers and histone chaperones can help to combat the evolved changes of COVID-19 that make it more dangerous [18,19]. Here, stem cells can be susceptible to infection and can fail to fight and regenerate from viral infection [20]. Accumulating evidence has shown that use of stem cells with different lineages of differentiation potential can defend against viral infection. A major study by Abdulhasan et al. showed that, in mouse embryonic stem cells, a control hyperosmotic stress can decrease viral host resistance by increasing COVID-19 susceptibility [21].

Mesenchymal stem cells

A major association has been discovered between COVID-19 and inflammatory responses [22,23]. Several studies have shown that some stem cells such as mesenchymal stem cells (MSCs) can reduce inflammatory symptoms associated with COVID-19 [24]. Among all the different stem cells, MSCs are found to be more promising, as they have the capability to control most of the immune cells, release factors with immunosuppressive functions [25,26], and accelerate different activities like angiogenesis and antiapoptosis. They also play a role in immunomodulation, chemo-attraction, and tissue regeneration [27–31]. It has been shown that they can aid in regenerating damaged neurons or muscle fibers [32]. Studies by Leng et al. showed that use of MSCs resulted in the recovery of COVID-19 patients with pneumonia by targeting ACE2 mechanisms, which are essential for the pathogenesis [33]. MSCs have also been found to release extracellular vesicles (EVs) such as microvesicles and exosomes, which helped in curing ALI/ARDS [34]. The mechanism by which MSCs act as a therapy for COVID-19 is specifically based on their antibacterial, immunomodulatory, antiapoptotic, and regenerative effects.

Stem cell activation during pathology

Tissue stem cells reside in anatomically defined stem cell niches and play a role in tissue regeneration during disease pathogenesis. During pathological conditions, stem cell activation mechanisms are induced by cytokines, chemokines, growth factors, activation of complement cascade, bioactive phospholipids, and the interaction of stem cells with niche components [35,36]. In addition, the role of various metabolic pathways in stem cell differentiation was described by Hu et al. [37]. Various studies have reported the stem cell niche microenvironment as a critical regulator of stem cell activation. Yoshida et al. described the role of soluble factors, cell surface proteins, and extracellular matrixes in the rodent stem cell niche in pituitary [38]. Nguyen et al. reviewed the role of the local metabolic microenvironment in muscle stem cell activation [39]. COVID-19 causes multiorgan damage and can affect various tissue-resident stem cells, leading to impaired tissue regeneration followed by inflammatory and fibrotic events. Organoids containing tissue-specific stem cells are useful approaches to study the effect of COVID-19 on stem cells [15]. In COVID-19 patients, the virus causes depletion of type II alveolar cells (AT2 cells), which function as resident stem cells that can differentiate into the type I alveolar epithelial cells (AT1 cells). Studies in two COVID-19 patients reported the proliferation and differentiation of AT2 cells to AT1 cells with an intermediate stage [9]. Various studies using in vivo and ex vivo models of influenza virus-induced lung stem cell injury showed impaired stem cell regeneration due to virus-induced down-regulation of Wnt/β-catenin signaling [40,41] and the same scenario can occur with COVID-19 infection [42]. Hence targeting these signaling pathways may be beneficial for the mitigation of tissue stem cell injury. Researchers have used air-liquid interface pulmonary organoid [43], intestinal organoid [44], and kidney organoid [45] models to study the effect of COVID-19 on tissue-specific cells.

Growth factors as a vital target

A drawback of stem cells is that they can be susceptible to infection and can fail in tissue regeneration when exposed to the virus. Studies have shown that resident stem cells can be signaled by mesenchymal stem cells via their bioactive factors for damage repair. In this case, they do not require the use of differentiation [46]. Since MSCs can selectively differentiate, a major challenge lies in finding the right combination of growth factors and cytokines to orchestrate the differentiation towards a specific lineage. This also involves optimizing the controlled release and delivery of these factors and cytokines. Several other factors such as environmental factors, surrounding matrix, and cellular interactions affect the ability of MSC-secreted factors to promote tissue repair [47]. Studies have shown that signaling through growth factors and cytokines helps MSCs in immunosuppression by inhibition of immune cells [48].

The different stages of differentiation of MSCs involve a series of events that include mesenchymal cell recruitment, migration, proliferation, aggregation hypertrophy, and death. Several growth factors such as hedgehog proteins (Hhgs), Wnt proteins, Notch ligands, and FGFs are involved in this complex process. A number of studies have shown that growth factors modify surfaces of biomaterials to provide survival factors that aid in differentiation of stem cells [49,50]. These factors can be made to be released from biomaterials via proteolysis or diffusion to help in regeneration [51]. Works by Huang et al. have shown that PEI-condensed plasmid DNA encoding for bone morphogenetic protein-4 (BMP-4) can be used for bone regeneration in a rat with a cranial defect [52]. Several hematopoietic growth factors including ESAs, G-CSFs, and TPOm are used to mitigate complications like anemia, which result from cancer chemotherapy. This knowledge is important in expanding the use of growth factors in cancer patients exposed to COVID-19, to help in effective chemotherapy [53].

In contrast, certain factors such as LIF, VEGF, SCGF, and G-CSF were found to be upregulated in COVID-19 patients, whereas factors such as M-CSF, SCF, HGF, and basic FGF increased in patients with severe COVID-19 infection. Thus growth factors identified in COVID-19 patients can serve as biomarkers and provide instructive signals for the prognosis of COVID-19 patients. This in turn can be used in therapeutic targeting for treatment of mild, severe, and fatal COVID-19 patients.

COVID-19 clinical trials

Use of stem cells as a therapy for COVID-19 was first seen in reports from China, where a woman on ventilation was injected with MSCs. After several studies were conducted, the US Food and Drug Administration approved the use of MSCs as an investigational drug [54]. It was found that MSCs worked by reducing inflammation-inducing molecules, which led to reducing inflammation. Mesenchymal stem cells (MSCs) have since been widely investigated for the treatment of COVID-19. Their capacity to differentiate to a different cell lineage along with their immunosuppressive and immunomodulatory properties makes them a valuable candidate for treating lung injury and the cell-mediated immune response in COVID-19; however, there are limited data to assess the use of MSCs in COVID-19 infection. A study conducted with a small group of patients in China reported improved outcomes of patients after transplantation of $ACE2^{-}$ MSCs [33]. In a phase I clinical trial with human umbilical cord-derived MSCs in moderate to severely ill patients, the MSCs were well tolerated in the patients and no serious adverse events were reported [55]. A similar study using human umbilical cord Wharton's jelly-derived MSCs (hWJCs) reported the safety of the MSCs and the recovery of a severely ill COVID-19 patient [56]. Another double-blind randomized clinical trial evaluated the efficacy of human umbilical cord MSCs in treating COVID-19 ARDS and reported improved patient survival and decreased numbers of serious adverse events [57]. There are several ongoing clinical trials investigating the use of MSCs for treating COVID-19 and associated inflammatory syndromes. Most of the studies are limited by small sample size and patient inclusion criteria and these studies require long-term follow-up to evaluate the long-term benefits of MSC treatment in recovered patients. Other potential concerns regarding the use of MSCs include route, frequency, and dosage of administration of MSCs, differentiation to inappropriate cell types, potential contamination, chances of hypercoagulability [58], etc.

Owing to the higher risk of thrombus formation associated with MSC cell-free therapies, like MSC secretome or MSC extracellular vesicles, these therapies represent an interesting approach to treat COVID-19. New clinical trials with higher numbers of patients are required to evaluate the efficacy and safety of MSCs in treating COVID-19 infection and require sufficient data to assess the use of MSCs in children and during pregnancy as well.

Different trials have been carried out using MSCs isolated from various biological materials (Table 1). The two countries with the highest number of trials are the United States and China.

Potential drawbacks of the trials

(1) Some do not have a control group, whereas others lack controls involving different dosages of the drugs.
(2) Different critically ill patients have different complications, including inflammation and coagulopathy, which can lead to heterogeneity of treatment.
(3) Trials require time for symptom resolution and reduce the use of ventilation.
(4) Mortality still is an unresolved issue.
(5) Prevention of COVID-19 has not been addressed completely.
(6) Sharing of patient data is a requirement.
(7) Self-isolation of trial patients or investigators.

These drawbacks, if taken into consideration, could aid in prevention of a recurrence of the pandemic [59].

Table 1 Trial participants by country.

Si no:	Clinical trial no.	Number of patients	Country	Source
1	NCT0427698	30	Spain	Allogeneic AD-MSC exosomes
2	NCT0440003	9	Canada	BM-MSCs
3	NCT0438945	140	Israel	Placental mesenchymal-like adherent stromal cells
4	NCT0439830	70	United States	Allogeneic human UC-MSCs
5	NCT0430251	24	China	Dental pulp
6	NCT0442880	200	United States	Adipose tissue
7	NCT0438254	40	Belarus	Olfactory mucosa
8	NCT0431332	5	Jordan	Wharton's jelly

Several human trials also included usage of nutritional and dietary supplements like vitamin C, vitamin D, probiotics, and zinc (Zn) [60,61]. People are still curious regarding the nature of the vast number of trials that are being performed. Various factors such as patient, trial, geographical, and other characteristics are involved in COVID-19 trial mapping [62–64]. COVID-19 related mortality has been found to be sex and age dependent. These trials are meant to help researchers and clinicians to minimize morbidity and mortality based on evidence. The public and politicians also need to be aware of the much-needed clinical trials. The trials are registered on Clinical Trials.gov.

To consider [65]

(1) Obtaining consent from isolated patients.
(2) Requiring protocol modifications and specific visits.
(3) Protecting the well-being of patients, trial participants, and investigators.
(4) Reporting screening data.
(5) Delivery of Investigational Medicinal Products (IMPs) for self-administration.
(6) Missing information in the study report.

These trials are meant to ensure safety and efficacy and to bring out the greatest benefit in terms of disease regression. If the drawbacks and considerations stated are accounted for, it is possible to ensure high-quality research and patient care for COVID-19.

Summary and significance

Chronic disease conditions, serious infection, or injuries can compromise organ function and affect the quality of life of an individual. Recent developments in stem cell therapies show that stem cells can regenerate and repair diseased or damaged tissue [66–69]. The natural ability of stem cells to generate new cell types makes them important in regenerative medicine. Stem cells that are already in regenerative use include, but are not limited to, mesenchymal stem cells (MSCs), embryonic stem cells, and induced pluripotent stem cells (iPSCs) [70]. MSCs can self-renew by dividing and can differentiate into multiple types of tissues; iPSCs are reprogrammed adult cells that can be used instead of embryonic cells and can prevent immune rejection of the new stem cells. Studies show that researchers have been able to reprogram regular connective tissue cells to become functional heart cells. Animals with heart failure that were injected with new heart cells experienced improved heart function and survival rate [71].

An important aspect of stem cell therapy is treating respiratory disease caused by COVID-19. SARS-Cov-2 enters the body through binding onto the ACE2 receptor on AT1 cells [72]. This in turn damages the lung and promotes vasoconstriction, inflammation, and apoptosis of alveolar epithelial cells [73]. The virus also infects and damages organs including heart, kidney, pancreas, brain, and intestinal tract. Multiple studies have been carried out to see if stem cells can repair the damage caused by this serious viral infection. Studies show the efficacy of mesenchymal stem cells (MSCs) and MSC-derived exosomes in modulating the disease associated with COVID-19 infection [73]. They help reduce the immune cell infiltration and inflammatory cytokine production. Moreover, they secrete several growth factors that reduce the apoptosis of epithelial cells. The other advantage of MSCs is they lack the ACE receptor needed for viral entry and so they are resistant to infection [74]. The use of human amniotic epithelial cells (hAECs) and human mesenchymal stromal cells (hAMSCs) is promising, as they elicit antiinflammatory, immunoregulatory, and regenerative properties [75]. For example, the cytokine storm and severe inflammatory response could be managed with hAECs and hAMSCs, as these stem cells are isolated from human placenta and since they are easy to obtain, noninvasive, and have no ethical concerns [75]. They act by decreasing the pulmonary inflammatory microenvironment and increase the release of soluble antiinflammatory factors such as IL-10, IL-1β, and VEGF. Moreover, these cells reduce fibrotic responses through inhibiting TGF-β signaling. Another study used an embryonic stem cell-derived cardiomyocyte line to screen molecules to treat SARS-CoV-2 infection, as patients with cardiovascular comorbidities are more susceptible to severe infection due to COVID-19 [76]. Studies are ongoing using embryonic stem cell/induced pluripotent cell-derived somatic cells or organoids to screen molecules to study the drug response of COVID-19 patients in vitro [77]. Discovery of new medicines and screening of new molecules using stem cell approaches are critical for identifying new therapeutic approaches to protect patients from multiorgan damage caused by COVID-19. Transplantation of stem cells alone into injured tissues sometimes exhibits low therapeutic efficacy due to poor viability. However, combining the stem cell approach with tissue engineering aspects could help to improve the efficacy of transplantation in future.

References

[1] Chen N, Zhou M, Dong X, Qu J, Gong F, Han Y, Qiu Y, Wang J, Liu Y, Wei Y, Xia J, Yu T, Zhang X, Zhang L. Epidemiological and clinical characteristics of 99 cases of 2019 novel coronavirus pneumonia in Wuhan, China: a descriptive study. Lancet 2020;395(10223):507–13. https://doi.org/10.1016/S0140-6736(20)30211-7.

[2] Guan WJ, Ni ZY, Hu Y, Liang WH, Ou CQ, He JX, Liu L, Shan H, Lei CL, Hui DSC, Du B, Li LJ, Zeng G, Yuen KY, Chen RC, Tang CL, Wang T, Chen PY, Xiang J, Li SY, Wang JL, Liang ZJ, Peng YX, Wei L, Liu Y, Hu YH, Peng P, Wang JM, Liu JY, Chen Z, Li G, Zheng ZJ, Qiu SQ, Luo J, Ye CJ, Zhu SY, Zhong NS, China Medical Treatment Expert Group for, C. Clinical characteristics of coronavirus disease 2019 in China. N Engl J Med 2020;382(18):1708–20. https://doi.org/10.1056/NEJMoa2002032.
[3] Huang C, Wang Y, Li X, Ren L, Zhao J, Hu Y, Zhang L, Fan G, Xu J, Gu X, Cheng Z, Yu T, Xia J, Wei Y, Wu W, Xie X, Yin W, Li H, Liu M, Xiao Y, Gao H, Guo L, Xie J, Wang G, Jiang R, Gao Z, Jin Q, Wang J, Cao B. Clinical features of patients infected with 2019 novel coronavirus in Wuhan, China. Lancet 2020;395(10223):497–506. https://doi.org/10.1016/S0140-6736(20)30183-5.
[4] Wang D, Hu B, Hu C, Zhu F, Liu X, Zhang J, Wang B, Xiang H, Cheng Z, Xiong Y, Zhao Y, Li Y, Wang X, Peng Z. Clinical characteristics of 138 hospitalized patients with 2019 novel coronavirus-infected pneumonia in Wuhan, China. JAMA 2020;323(11):1061–9. https://doi.org/10.1001/jama.2020.1585.
[5] Wang Z, Yang B, Li Q, Wen L, Zhang R. Clinical features of 69 cases with coronavirus disease 2019 in Wuhan, China. Clin Infect Dis 2020;71(15):769–77. https://doi.org/10.1093/cid/ciaa272.
[6] Wu J, Liu J, Zhao X, Liu C, Wang W, Wang D, Xu W, Zhang C, Yu J, Jiang B, Cao H, Li L. Clinical characteristics of imported cases of coronavirus disease 2019 (COVID-19) in Jiangsu province: a multicenter descriptive study. Clin Infect Dis 2020;71(15):706–12. https://doi.org/10.1093/cid/ciaa199.
[7] Felsenstein S, Herbert JA, McNamara PS, Hedrich CM. COVID-19: Immunology and treatment options. Clin Immunol 2020;215. https://doi.org/10.1016/j.clim.2020.108448, 108448.
[8] Stasi C, Fallani S, Voller F, Silvestri C. Treatment for COVID-19: an overview. Eur J Pharmacol 2020;889. https://doi.org/10.1016/j.ejphar.2020.173644, 173644.
[9] Chen J, Wu H, Yu Y, Tang N. Pulmonary alveolar regeneration in adult COVID-19 patients. Cell Res 2020;30(8):708–10. https://doi.org/10.1038/s41422-020-0369-7.
[10] Chen X, Wu Y, Wang Y, Chen L, Zheng W, Zhou S, Xu H, Li Y, Yuan L, Xiang C. Human menstrual blood-derived stem cells mitigate bleomycin-induced pulmonary fibrosis through anti-apoptosis and anti-inflammatory effects. Stem Cell Res Ther 2020;11(1):477. https://doi.org/10.1186/s13287-020-01926-x.
[11] Huang K, Kang X, Wang X, Wu S, Xiao J, Li Z, Wu X, Zhang W. Conversion of bone marrow mesenchymal stem cells into type II alveolar epithelial cells reduces pulmonary fibrosis by decreasing oxidative stress in rats. Mol Med Rep 2015;11(3):1685–92. https://doi.org/10.3892/mmr.2014.2981.
[12] Nicolay NH, Lopez Perez R, Ruhle A, Trinh T, Sisombath S, Weber KJ, Ho AD, Debus J, Saffrich R, Huber PE. Mesenchymal stem cells maintain their defining stem cell characteristics after treatment with cisplatin. Sci Rep 2016;6:20035. https://doi.org/10.1038/srep20035.
[13] Nicolay NH, Ruhle A, Perez RL, Trinh T, Sisombath S, Weber KJ, Ho AD, Debus J, Saffrich R, Huber PE. Mesenchymal stem cells are sensitive to bleomycin treatment. Sci Rep 2016;6:26645. https://doi.org/10.1038/srep26645.
[14] Basiri A, Pazhouhnia Z, Beheshtizadeh N, Hoseinpour M, Saghazadeh A, Rezaei N. Regenerative medicine in COVID-19 treatment: real opportunities and range of promises. Stem Cell Rev Rep 2021;17(1):163–75. https://doi.org/10.1007/s12015-020-09994-5.
[15] Chugh RM, Bhanja P, Norris A, Saha S. Experimental models to study COVID-19 effect in stem cells. Cells 2021;10(1). https://doi.org/10.3390/cells10010091.
[16] Khavinson VK, Linkova NS, Kvetnoy IM, Polyakova VO, Drobintseva AO, Kvetnaia TV, Ivko OM. Thymalin: activation of differentiation of human hematopoietic stem cells. Bull Exp Biol Med 2020;170(1):118–22. https://doi.org/10.1007/s10517-020-05016-z.

[17] Abkowitz JL. Can human hematopoietic stem cells become skin, gut, or liver cells? N Engl J Med 2002;346(10):770–2. https://doi.org/10.1056/NEJM200203073461012.
[18] Cyranoski D. Profile of a killer: the complex biology powering the coronavirus pandemic. Nature 2020;581(7806):22–6. https://doi.org/10.1038/d41586-020-01315-7.
[19] Gordon DE, Jang GM, Bouhaddou M, Xu J, Obernier K, White KM, O'Meara MJ, Rezelj VV, Guo JZ, Swaney DL, Tummino TA, Huttenhain R, Kaake RM, Richards AL, Tutuncuoglu B, Foussard H, Batra J, Haas K, Modak M, Kim M, Haas P, Polacco BJ, Braberg H, Fabius JM, Eckhardt M, Soucheray M, Bennett MJ, Cakir M, McGregor MJ, Li Q, Meyer B, Roesch F, Vallet T, Mac Kain A, Miorin L, Moreno E, Naing ZZC, Zhou Y, Peng S, Shi Y, Zhang Z, Shen W, Kirby IT, Melnyk JE, Chorba JS, Lou K, Dai SA, Barrio-Hernandez I, Memon D, Hernandez-Armenta C, Lyu J, Mathy CJP, Perica T, Pilla KB, Ganesan SJ, Saltzberg DJ, Rakesh R, Liu X, Rosenthal SB, Calviello L, Venkataramanan S, Liboy-Lugo J, Lin Y, Huang XP, Liu Y, Wankowicz SA, Bohn M, Safari M, Ugur FS, Koh C, Savar NS, Tran QD, Shengjuler D, Fletcher SJ, O'Neal MC, Cai Y, Chang JCJ, Broadhurst DJ, Klippsten S, Sharp PP, Wenzell NA, Kuzuoglu-Ozturk D, Wang HY, Trenker R, Young JM, Cavero DA, Hiatt J, Roth TL, Rathore U, Subramanian A, Noack J, Hubert M, Stroud RM, Frankel AD, Rosenberg OS, Verba KA, Agard DA, Ott M, Emerman M, Jura N, von Zastrow M, Verdin E, Ashworth A, Schwartz O, d'Enfert C, Mukherjee S, Jacobson M, Malik HS, Fujimori DG, Ideker T, Craik CS, Floor SN, Fraser JS, Gross JD, Sali A, Roth BL, Ruggero D, Taunton J, Kortemme T, Beltrao P, Vignuzzi M, Garcia-Sastre A, Shokat KM, Shoichet BK, Krogan NJ. A SARS-CoV-2 protein interaction map reveals targets for drug repurposing. Nature 2020;583(7816):459–68. https://doi.org/10.1038/s41586-020-2286-9.
[20] Wu X, Kwong AC, Rice CM. Antiviral resistance of stem cells. Curr Opin Immunol 2019;56:50–9. https://doi.org/10.1016/j.coi.2018.10.004.
[21] Abdulhasan M, Ruden X, Rappolee B, Dutta S, Gurdziel K, Ruden DM, Awonuga AO, Korzeniewski SJ, Puscheck EE, Rappolee DA. Stress decreases host viral resistance and increases COVID susceptibility in embryonic stem cells. Stem Cell Rev Rep 2021. https://doi.org/10.1007/s12015-021-10188-w.
[22] Mehta P, McAuley DF, Brown M, Sanchez E, Tattersall RS, Manson JJ, Hlh Across Speciality Collaboration, UK. COVID-19: consider cytokine storm syndromes and immunosuppression. Lancet 2020;395(10229):1033–4. https://doi.org/10.1016/S0140-6736(20)30628-0.
[23] Stebbing J, Phelan A, Griffin I, Tucker C, Oechsle O, Smith D, Richardson P. COVID-19: combining antiviral and anti-inflammatory treatments. Lancet Infect Dis 2020;20(4):400–2. https://doi.org/10.1016/S1473-3099(20)30132-8.
[24] Khavinson VK, Kuznik BI, Trofimova SV, Volchkov VA, Rukavishnikova SA, Titova ON, Akhmedov TA, Trofimov AV, Potemkin VV, Magen E. Results and prospects of using activator of hematopoietic stem cell differentiation in complex therapy for patients with COVID-19. Stem Cell Rev Rep 2021;17(1):285–90. https://doi.org/10.1007/s12015-020-10087-6.
[25] Najar M, Raicevic G, Crompot E, Fayyad-Kazan H, Bron D, Toungouz M, Lagneaux L. The immunomodulatory potential of mesenchymal stromal cells: a story of a regulatory network. J Immunother 2016;39(2):45–59. https://doi.org/10.1097/CJI.0000000000000108.
[26] Shi Y, Wang Y, Li Q, Liu K, Hou J, Shao C, Wang Y. Immunoregulatory mechanisms of mesenchymal stem and stromal cells in inflammatory diseases. Nat Rev Nephrol 2018;14(8):493–507. https://doi.org/10.1038/s41581-018-0023-5.
[27] Carvalho AES, Sousa MRR, Alencar-Silva T, Carvalho JL, Saldanha-Araujo F. Mesenchymal stem cells immunomodulation: the road to IFN-gamma licensing and the path ahead. Cytokine Growth Factor Rev 2019;47:32–42. https://doi.org/10.1016/j.cytogfr.2019.05.006.

[28] Gao F, Chiu SM, Motan DA, Zhang Z, Chen L, Ji HL, Tse HF, Fu QL, Lian Q. Mesenchymal stem cells and immunomodulation: current status and future prospects. Cell Death Dis 2016;7. https://doi.org/10.1038/cddis.2015.327, e2062.
[29] Le Blanc K, Ringden O. Immunomodulation by mesenchymal stem cells and clinical experience. J Intern Med 2007;262(5):509–25. https://doi.org/10.1111/j.1365-2796.2007.01844.x.
[30] Saldanha-Araujo F, Haddad R, Farias KC, Souza Ade P, Palma PV, Araujo AG, Orellana MD, Voltarelli JC, Covas DT, Zago MA, Panepucci RA. Mesenchymal stem cells promote the sustained expression of CD69 on activated T lymphocytes: roles of canonical and non-canonical NF-kappaB signalling. J Cell Mol Med 2012;16(6):1232–44. https://doi.org/10.1111/j.1582-4934.2011.01391.x.
[31] Silva-Carvalho AE, Neves FAR, Saldanha-Araujo F. The immunosuppressive mechanisms of mesenchymal stem cells are differentially regulated by platelet poor plasma and fetal bovine serum supplemented media. Int Immunopharmacol 2020;79. https://doi.org/10.1016/j.intimp.2019.106172, 106172.
[32] Newman RE, Yoo D, LeRoux MA, Danilkovitch-Miagkova A. Treatment of inflammatory diseases with mesenchymal stem cells. Inflamm Allergy Drug Targets 2009;8(2):110–23. https://doi.org/10.2174/187152809788462635.
[33] Leng Z, Zhu R, Hou W, Feng Y, Yang Y, Han Q, Shan G, Meng F, Du D, Wang S, Fan J, Wang W, Deng L, Shi H, Li H, Hu Z, Zhang F, Gao J, Liu H, Li X, Zhao Y, Yin K, He X, Gao Z, Wang Y, Yang B, Jin R, Stambler I, Lim LW, Su H, Moskalev A, Cano A, Chakrabarti S, Min KJ, Ellison-Hughes G, Caruso C, Jin K, Zhao RC. Transplantation of ACE2(-) mesenchymal stem cells improves the outcome of patients with COVID-19 pneumonia. Aging Dis 2020;11(2):216–28. https://doi.org/10.14336/AD.2020.0228.
[34] Fujita Y, Kadota T, Araya J, Ochiya T, Kuwano K. Clinical application of mesenchymal stem cell-derived extracellular vesicle-based therapeutics for inflammatory lung diseases. J Clin Med 2018;7(10). https://doi.org/10.3390/jcm7100355.
[35] Bloch W. Stem cell activation in adult organisms. Int J Mol Sci 2016;17(7). https://doi.org/10.3390/ijms17071005.
[36] Budkowska M, Ostrycharz E, Wojtowicz A, Marcinowska Z, Wozniak J, Ratajczak MZ, Dolegowska B. A circadian rhythm in both complement cascade (ComC) activation and sphingosine-1-phosphate (S1P) levels in human peripheral blood supports a role for the ComC-S1P axis in circadian changes in the number of stem cells circulating in peripheral blood. Stem Cell Rev Rep 2018;14(5):677–85. https://doi.org/10.1007/s12015-018-9836-7.
[37] Hu C, Fan L, Cen P, Chen E, Jiang Z, Li L. Energy metabolism plays a critical role in stem cell maintenance and differentiation. Int J Mol Sci 2016;17(2):253. https://doi.org/10.3390/ijms17020253.
[38] Yoshida S, Kato T, Kato Y. Regulatory system for stem/progenitor cell niches in the adult rodent pituitary. Int J Mol Sci 2016;17(1). https://doi.org/10.3390/ijms17010075.
[39] Nguyen JH, Chung JD, Lynch GS, Ryall JG. The microenvironment is a critical regulator of muscle stem cell activation and proliferation [review]. Front Cell Dev Biol 2019;7(254). https://doi.org/10.3389/fcell.2019.00254.
[40] Hancock AS, Stairiker CJ, Boesteanu AC, Monzón-Casanova E, Lukasiak S, Mueller YM, Stubbs AP, García-Sastre A, Turner M, Katsikis PD. Transcriptome analysis of infected and bystander type 2 alveolar epithelial cells during influenza a virus infection reveals in vivo wnt pathway downregulation. J Virol 2018;92(21). https://doi.org/10.1128/jvi.01325-18.
[41] Hillesheim A, Nordhoff C, Boergeling Y, Ludwig S, Wixler V. β-catenin promotes the type I IFN synthesis and the IFN-dependent signaling response but is suppressed by influenza A virus-induced RIG-I/NF-κB signaling. Cell Commun Signal 2014;12:29. https://doi.org/10.1186/1478-811x-12-29.

[42] Trouillet-Assant S, Viel S, Gaymard A, Pons S, Richard JC, Perret M, Villard M, Brengel-Pesce K, Lina B, Mezidi M, Bitker L, Belot A. Type I IFN immunoprofiling in COVID-19 patients. J Allergy Clin Immunol 2020;146(1):206–208.e202. https://doi.org/10.1016/j.jaci.2020.04.029.

[43] Mulay A, Konda B, Garcia G, Yao C, Beil S, Sen C, Purkayastha A, Kolls JK, Pociask DA, Pessina P, Sainz de Aja J, Garcia-de-Alba C, Kim CF, Gomperts B, Arumugaswami V, Stripp BR. SARS-CoV-2 infection of primary human lung epithelium for COVID-19 modeling and drug discovery. bioRxiv 2020. https://doi.org/10.1101/2020.06.29.174623.

[44] Zang R, Gomez Castro MF, McCune BT, Zeng Q, Rothlauf PW, Sonnek NM, Liu Z, Brulois KF, Wang X, Greenberg HB, Diamond MS, Ciorba MA, Whelan SPJ, Ding S. TMPRSS2 and TMPRSS4 promote SARS-CoV-2 infection of human small intestinal enterocytes. Sci Immunol 2020;5(47). https://doi.org/10.1126/sciimmunol.abc3582.

[45] Allison SJ. SARS-CoV-2 infection of kidney organoids prevented with soluble human ACE2. Nat Rev Nephrol 2020;16(6):316. https://doi.org/10.1038/s41581-020-0291-8.

[46] Caplan AI. Mesenchymal stem cells: time to change the name! Stem Cells Transl Med 2017;6(6):1445–51. https://doi.org/10.1002/sctm.17-0051.

[47] Schuldiner M, Yanuka O, Itskovitz-Eldor J, Melton DA, Benvenisty N. Effects of eight growth factors on the differentiation of cells derived from human embryonic stem cells. Proc Natl Acad Sci U S A 2000;97(21):11307–12. https://doi.org/10.1073/pnas.97.21.11307.

[48] Glennie S, Soeiro I, Dyson PJ, Lam EW, Dazzi F. Bone marrow mesenchymal stem cells induce division arrest anergy of activated T cells. Blood 2005;105(7):2821–7. https://doi.org/10.1182/blood-2004-09-3696.

[49] Fan VH, Tamama K, Au A, Littrell R, Richardson LB, Wright JW, Wells A, Griffith LG. Tethered epidermal growth factor provides a survival advantage to mesenchymal stem cells. Stem Cells 2007;25(5):1241–51. https://doi.org/10.1634/stemcells.2006-0320.

[50] Suzuki Y, Tanihara M, Suzuki K, Saitou A, Sufan W, Nishimura Y. Alginate hydrogel linked with synthetic oligopeptide derived from BMP-2 allows ectopic osteoinduction in vivo. J Biomed Mater Res 2000;50(3):405–9. https://doi.org/10.1002/(sici)1097-4636(20000605)50:3<405::aid-jbm15>3.0.co;2-z.

[51] Kaigler D, Wang Z, Horger K, Mooney DJ, Krebsbach PH. VEGF scaffolds enhance angiogenesis and bone regeneration in irradiated osseous defects. J Bone Miner Res 2006;21(5):735–44. https://doi.org/10.1359/jbmr.060120.

[52] Huang YC, Simmons C, Kaigler D, Rice KG, Mooney DJ. Bone regeneration in a rat cranial defect with delivery of PEI-condensed plasmid DNA encoding for bone morphogenetic protein-4 (BMP-4). Gene Ther 2005;12(5):418–26. https://doi.org/10.1038/sj.gt.3302439.

[53] Griffiths EA, Alwan LM, Bachiashvili K, Brown A, Cool R, Curtin P, Geyer MB, Gojo I, Kallam A, Kidwai WZ, Kloth DD, Kraut EH, Lyman GH, Mukherjee S, Perez LE, Rosovsky RP, Roy V, Rugo HS, Vasu S, Wadleigh M, Westervelt P, Becker PS. Considerations for use of hematopoietic growth factors in patients with cancer related to the COVID-19 pandemic. J Natl Compr Cancer Netw 2020;1-4. https://doi.org/10.6004/jnccn.2020.7610.

[54] Shetty R, Murugeswari P, Chakrabarty K, Jayadev C, Matalia H, Ghosh A, Das D. Stem cell therapy in coronavirus disease 2019: current evidence and future potential. Cytotherapy 2021;23(6):471–82. https://doi.org/10.1016/j.jcyt.2020.11.001.

[55] Meng F, Xu R, Wang S, Xu Z, Zhang C, Li Y, Yang T, Shi L, Fu J, Jiang T, Huang L, Zhao P, Yuan X, Fan X, Zhang JY, Song J, Zhang D, Jiao Y, Liu L, Zhou C, Maeurer M, Zumla A, Shi M, Wang FS. Human umbilical cord-derived mesenchymal stem

cell therapy in patients with COVID-19: a phase 1 clinical trial. Signal Transduct Target Ther 2020;5(1):172. https://doi.org/10.1038/s41392-020-00286-5.
[56] Zhang Y, Ding J, Ren S, Wang W, Yang Y, Li S, Meng M, Wu T, Liu D, Tian S, Tian H, Chen S, Zhou C. Intravenous infusion of human umbilical cord Wharton's jelly-derived mesenchymal stem cells as a potential treatment for patients with COVID-19 pneumonia. Stem Cell Res Ther 2020;11(1):207. https://doi.org/10.1186/s13287-020-01725-4.
[57] Lanzoni G, Linetsky E, Correa D, Messinger Cayetano S, Alvarez RA, Kouroupis D, Alvarez Gil A, Poggioli R, Ruiz P, Marttos AC, Hirani K, Bell CA, Kusack H, Rafkin L, Baidal D, Pastewski A, Gawri K, Leñero C, Mantero AMA, Metalonis SW, Wang X, Roque L, Masters B, Kenyon NS, Ginzburg E, Xu X, Tan J, Caplan AI, Glassberg MK, Alejandro R, Ricordi C. Umbilical cord mesenchymal stem cells for COVID-19 acute respiratory distress syndrome: a double-blind, phase 1/2a, randomized controlled trial. Stem Cells Transl Med 2021;10(5):660–73. https://doi.org/10.1002/sctm.20-0472.
[58] Jeyaraman M, John A, Koshy S, Ranjan R, Anudeep TC, Jain R, Swati K, Jha NK, Sharma A, Kesari KK, Prakash A, Nand P, Jha SK, Reddy PH. Fostering mesenchymal stem cell therapy to halt cytokine storm in COVID-19. Biochim Biophys Acta Mol Basis Dis 2021;1867(2). https://doi.org/10.1016/j.bbadis.2020.166014, 166014.
[59] Bauchner H, Fontanarosa PB. Randomized clinical trials and COVID-19: managing expectations. JAMA 2020;323(22):2262–3. https://doi.org/10.1001/jama.2020.8115.
[60] Arrieta F, Martinez-Vaello V, Bengoa N, Jimenez-Mendiguchia L, Rosillo M, de Pablo A, Voguel C, Martinez-Barros H, Pintor R, Belanger-Quintana A, Mateo R, Candela A, Botella-Carretero JI. Serum zinc and copper in people with COVID-19 and zinc supplementation in parenteral nutrition. Nutrition 2021;91-92. https://doi.org/10.1016/j.nut.2021.111467, 111467.
[61] Skalny AV, Timashev PS, Aschner M, Aaseth J, Chernova LN, Belyaev VE, Grabeklis AR, Notova SV, Lobinski R, Tsatsakis A, Svistunov AA, Fomin VV, Tinkov AA, Glybochko PV. Serum zinc, copper, and other biometals are associated with COVID-19 severity markers. Metabolites 2021;11(4). https://doi.org/10.3390/metabo11040244.
[62] Cao B, Zhang D, Wang C. A Trial of Lopinavir-Ritonavir in COVID-19. Reply. N Engl J Med 2020;382(21). https://doi.org/10.1056/NEJMc2008043, e68.
[63] Chen PJ, Chao CM, Lai CC. Clinical efficacy and safety of favipiravir in the treatment of COVID-19 patients. J Infect 2021;82(5):186–230. https://doi.org/10.1016/j.jinf.2020.12.005.
[64] Yao X, Ye F, Zhang M, Cui C, Huang B, Niu P, Liu X, Zhao L, Dong E, Song C, Zhan S, Lu R, Li H, Tan W, Liu D. In vitro antiviral activity and projection of optimized dosing design of hydroxychloroquine for the treatment of severe acute respiratory syndrome coronavirus 2 (SARS-CoV-2). Clin Infect Dis 2020;71(15):732–9. https://doi.org/10.1093/cid/ciaa237.
[65] Singh AG, Chaturvedi P. Clinical trials during COVID-19. Head Neck 2020;42(7):1516–8. https://doi.org/10.1002/hed.26223.
[66] Afarid M, Sanie-Jahromi F. Mesenchymal stem cells and COVID-19: cure, prevention, and vaccination. Stem Cells Int 2021;2021:6666370. https://doi.org/10.1155/2021/6666370.
[67] Alanazi A. COVID-19 and the role of stem cells. Regen Ther 2021;18:334–8. https://doi.org/10.1016/j.reth.2021.08.008.
[68] Cuevas-Gonzalez MV, Garcia-Perez A, Gonzalez-Aragon Pineda AE, Espinosa-Cristobal LF, Donohue-Cornejo A, Tovar-Carrillo KL, Saucedo-Acuna RA, Cuevas-Gonzalez JC. Stem cells as a model of study of SARS-CoV-2 and COVID-19: a systematic review of the literature. Biomed Res Int 2021;2021:9915927. https://doi.org/10.1155/2021/9915927.

[69] Esquivel D, Mishra R, Soni P, Seetharaman R, Mahmood A, Srivastava A. Stem cells therapy as a possible therapeutic option in treating COVID-19 patients. Stem Cell Rev Rep 2021;17(1):144–52. https://doi.org/10.1007/s12015-020-10017-6.

[70] Zakrzewski W, Dobrzynski M, Szymonowicz M, Rybak Z. Stem cells: past, present, and future. Stem Cell Res Ther 2019;10(1):68. https://doi.org/10.1186/s13287-019-1165-5.

[71] Tompkins BA, Balkan W, Winkler J, Gyongyosi M, Goliasch G, Fernandez-Aviles F, Hare JM. Preclinical studies of stem cell therapy for heart disease. Circ Res 2018;122(7):1006–20. https://doi.org/10.1161/CIRCRESAHA.117.312486.

[72] Valyaeva AA, Zharikova AA, Kasianov AS, Vassetzky YS, Sheval EV. Expression of SARS-CoV-2 entry factors in lung epithelial stem cells and its potential implications for COVID-19. Sci Rep 2020;10(1):17772. https://doi.org/10.1038/s41598-020-74598-5.

[73] Jamshidi E, Babajani A, Soltani P, Niknejad H. Proposed mechanisms of targeting COVID-19 by delivering mesenchymal stem cells and their Exosomes to damaged organs. Stem Cell Rev Rep 2021;17(1):176–92. https://doi.org/10.1007/s12015-020-10109-3.

[74] Avanzini MA, Mura M, Percivalle E, Bastaroli F, Croce S, Valsecchi C, Lenta E, Nykjaer G, Cassaniti I, Bagnarino J, Baldanti F, Zecca M, Comoli P, Gnecchi M. Human mesenchymal stromal cells do not express ACE2 and TMPRSS2 and are not permissive to SARS-CoV-2 infection. Stem Cells Transl Med 2021;10(4):636–42. https://doi.org/10.1002/sctm.20-0385.

[75] Riedel RN, Perez-Perez A, Sanchez-Margalet V, Varone CL, Maymo JL. Stem cells and COVID-19: are the human amniotic cells a new hope for therapies against the SARS-CoV-2 virus? Stem Cell Res Ther 2021;12(1):155. https://doi.org/10.1186/s13287-021-02216-w.

[76] Williams TL, Colzani MT, Macrae RGC, Robinson EL, Bloor S, Greenwood EJD, Zhan JR, Strachan G, Kuc RE, Nyimanu D, Maguire JJ, Lehner PJ, Sinha S, Davenport AP. Human embryonic stem cell-derived cardiomyocyte platform screens inhibitors of SARS-CoV-2 infection. Commun Biol 2021;4(1):926. https://doi.org/10.1038/s42003-021-02453-y.

[77] Deguchi S, Serrano-Aroca A, Tambuwala MM, Uhal BD, Brufsky AM, Takayama K. SARS-CoV-2 research using human pluripotent stem cells and organoids. Stem Cells Transl Med 2021. https://doi.org/10.1002/sctm.21-0183.

10

Stem cell transplantation for COVID-19 management: Translational possibilities and future

Renjith P. Nair[a], P. Lekshmi[a], and Sunitha Chandran[b]

[a]Division of Thrombosis Research, Department of Applied Biology, Sree Chitra Tirunal Institute for Medical Sciences and Technology, Trivandrum, India, [b]Technology Business Incubator for Medical Devices and Biomaterials, Sree Chitra Tirunal Institute for Medical Sciences and Technology, Trivandrum, India

Introduction

COVID-19 disease attributable to severe acute respiratory syndrome corona virus II (SARS-COV II) is one of the longest pandemics the world has seen superseding cholera and Ebola. Although it was considered to be a zoonotic disease, it is now clarified as a highly contagious airborne disease with a zoonotic origin which is deleterious to human lives [1]. Manifestation of the disease is often encountered 2 to 14 days postinfection, which is contemplated as the incubation period of SARS-COV II in a human lung. A mild fever and cough accompanied by shortness of breath and pneumonia contributed to be the major symptoms of COVID-19, while the ailment could worsen in some patients owing to a condition called cytokine storm and ramificate in acute respiratory distress syndrome (ARDS) and acute cardiac ecchymosis, which eventually lead to demise [2,3]. Cytokine storm is associated with irregularity in immune system, ramping up the secretion of multiple pro-inflammatory cytokines counting tumor necrosis factor-α (TNF-α), interleukins (ILs), interferons (IFN), interferon γ-induced protein 10 (IP-10), granulocyte-colony stimulating factor (GCSF), granulocyte-macrophage colony-stimulating factor (GM-CSF), monocyte chemo-attractant protein-1 (MCP-1), and macrophage inflammatory protein 1-α (MIP1-α) resulting in a rush of these cytokines to vital tissues, and causing vandalization of those tissues and organs [3,4]. An observed combination of elevated levels

Stem Cells and COVID-19. https://doi.org/10.1016/B978-0-323-89972-7.00007-6

of iron-containing protein ferritin and a pro-inflammatory cytokine IL-6 in patients with chronic corona virus infection is believed to rationale the transience of life in association with impaired immune response [4].

As there was neither a specific drug nor vaccines available at the time of COVID-19 outbreak, vitamin supplements, certain antibiotics for handling secondary bacterial infections, tidal volume and provision of mechanical ventilator or extracorporeal membrane oxygenation (ECMO) or high-flow nasal oxygen (HFNOT) to support breathing, fluid management to ease respiration, care being given to patients with acute infection, while those with severe symptoms such as difficulty in breathing, pneumonia, ARDS, and other related discomforts had to be treated with an alternative [5,6]. Patients with chronic infection happen to have an increased CRS score for the cytokine IL-6. In this respect, these patients can be treated with any of the particular inhibitors for IL-6 receptors such as tocilizumab [4]. Human interferon beta-1 (IFN-β-1a) for treating multiple sclerosis, convalescent plasma and some of the broad spectrum antiviral drug remdesivir and HIV preventing drug ritonavir, and even antimalarial drug hydroxychloroquine are being attempted as antidotes against SARS-COV II [7].

Apart from the conventional treatment options, quite a few corrective as well as prophylactic courses of actions are being scrutinized to reduce the intensity of COVID-19-related chronic medical conditions; stem cell therapy gains attention as it is an engrossing method which is experimentally manifested to hamper pulmonary infection along with cytokine storm [8,9]. Excellent immunomodulatory effects, antiinflammatory property, decreased immunogenicity, and substantial property of regeneration are sole characteristics of stem cells which are being utilized for the treatment of ARDS and associated infections ensued from the pandemic [9,10]. In this chapter, we discuss the pros and cons of stem cells as a treatment alternative for chronic health issues related to COVID-19.

Sources of stem cells

Stem cells can be considered as self-recovering cells that are able to multiply into their inheritor cells and also possesses a capacity to proliferate and differentiate into specialized cells for specialized functions. They can exhibit their effect either directly through emission of certain molecules and exosomes or indirectly by exuding soluble secretome to the system which guarantees a systemic upshot rather than a localized effect [11]. It is particularly categorized as highly potential embryonic stem cells (ESCs) and adult stem cells; the former is distinctly demanding as their only source of origin are immature embryos (early embryos of a week old), whereas the latter is plentiful in umbilical cord and placenta of infants and several adult tissues such as adipose tissue, bone marrow, peripheral blood, and dental pulp [5,7,12]. Mesenchymal stem cells (MSCs) are one among the different types, which is being focused greatly on account of its immunomodulatory and regenerative properties along with its abundance in adipose tissue and bone marrow [7,13].

Use of different types of stem cells in translational research

Stem cells are known to be used in regenerative medicine in order to accomplish several difficult tasks of healing in a systemic and biological way. Fortunately, there are several varieties of stem cells which could be isolated from multiple sources of human body. Hematopoietic stem cells (HSCs) originate from the bone marrow and possesses a distinct function of regeneration into their successors as well as into myeloid and lymphoid blood cells [14]. Although HSCs are widely used in regenerative therapy for several medical conditions, there is no data available with a successful clinical trial against COVID-19. In fact, controversies exist that those who are already under treatment with HSCs are highly prone to COVID-19 as they are immunocompromised during the treatment [15].

MSCs are the most preferred among the stem cells which possess peculiar characteristics appropriate for performing standard translational researches. As of now, MSCs have been studied in the alleviation of various medical conditions including hyper and hypothyroidism, bone and cartilage related diseases, and neurological dysfunction; however, controversies subsist as MSCs bring about the promotion of cancer cells in several studies [11]. The crucial roles of MSCs are its high rate of proliferation, reproducibility after storage, and the ability to modulate body's immune system [16]. MSCs reduce the inflammatory stimulation by creating hindrance in the production of hydrogen peroxide by the immune cells [17,18]. Scientists started utilizing the advantages of MSCs from a long time considering its efficiency of being a multipotent stem cell [13]. The ultimate advantage of MSCs is its safety as a regenerative medicine which is proven through various studies in recovering the damaged alveolar epithelial cells in serious pulmonary complications [16]. The effectiveness of MSCs in healing lung tissues might later became the primary lead for considering these cells in treating intricacies due to COVID-19.

Amniotic membrane of human placenta is rich in amniotic epithelial cells (hAECs) as well as mesenchymal stem cells (hMSCs).

Both hAECs and hMSCs were used in clinical trials against pulmonary infections and they succeed in healing the infection and regained the injured tissues of lungs which made it successful in getting selected as an option for the treatment of COVID-19-related respiratory distresses [10]. Allogenic umbilical cord stem cells were studied and concluded to be a potential healing agent of COVID-19 infected lungs through clinical trials [3,19]. Stem cells isolated from endodontic pulp have also been experimentally proven to be effective as a potential regenerative therapy against several medical conditions including COVID-19; however, the therapeutic use of it has not been accepted yet [20]. An easy to isolate type of MSCs is adipose-derived MSCs which have a potential to be proliferated into adipocytes and differentiated into different types of cells like hepatocytes, pancreatic cells, neural cells, osteocytes, cardiomyocytes, and vascular endothelial cells. These therefore are being used lately for multiple regenerative studies [6,21].

Current COVID-19 related clinical trials using stem cells

Regenerative stem cell therapy could be successfully implemented in treating COVID-19-related comorbidities in reference with a previously published clinical data on successful treatment of serious pulmonary conditions like ARDS. It is always beneficial to instigate any treatment including stem cell therapy before the symptoms get worsen; that for COVID-19 it is believed to be effective within 14 days postinfection. Therapies applied with autologous cells are always prevailing over the allogenic one; however, experts prefer allogenic stem cells for the treatment of COVID-19-related discomforts due to the difficulty in isolating cells from the frail body of a patient [7].

A published clinical trial data includes the study of a group of seven patients injected with a single dose of stem cells along with three control patients who were given the conventional therapies. Both of the groups were under observation for 14 days, after which all the seven test patients recovered from its criticality, while that of the control group shoot up and one of them expired [3].

Stem cells can be injected as intravenous injections to achieve quick and a positive outcome. In the case of COVID-19, the intravenously injected MSCs interact with endothelial cells and get deposited in the vascular bed of lungs, which result in the release of secretome by MSCs thereby achieving an alleviation of damaged pulmonary tissues. There were several reported clinical trials with the intravenous administration of MSCs encountering zero side effects, which nullifies the dilemma in choosing stem cell therapy due to its unknown after-effects [22].

Innumerable studies are betiding with intravenous infusion of both autologous and allogenic MSCs to COVID-19 patients. Liang et al. studied a 65-year-old woman with COVID-19 symptoms. She was nursed to lessen the criticality with some of the conventional medicines (ritonavir, oseltamivir) and IFN-α inhalation prior to the supplementation of five doses of umbilical cord-derived MSCs: one million cells per dosage. An improvement was noticed from the second dosage of MSCs itself, which reinforces in relying MSCs to cure chronic effects of COVID-19 [19].

Another critical study was executed in China on injecting three doses of 50 million allogenic umbilical cord stem cells to seven patients aging 65 years and having COVID-19-related pneumonia. However, the chronic effect of viral pneumonia has been reduced after the second dose injection of cells; patients were observed for 14 days. The damaged lung tissues of the patients treated with umbilical cord stem cells recovered to its normal stage were observed through computed tomography [3].

Ye et al. describe an ongoing randomized clinical trial in assessing the safety and therapeutic effects of allogenic human dental pulp stem cells (DPSCs) in the treatment of COVID-19-related chronic pneumonia. The experimental group were intravenously injected by 30 million DPSCs in three shots as in 1st, 4th, and 7th day and were examined for a period of 28 days [23]. In a similar study, the dental pulp stem cell-treated patient lungs regain its normal physical feature [24].

Allogenic pooled olfactory mucosa-derived MSCs are also under clinical trial in treating pulmonary pneumonia ascribed to COVID-19 infection. Patients have to be injected with one million olfactory mucosa-derived MSCs followed by the injection of endodontic MSCs. Fifty milliliters normal saline has to be administered before and after the dosage of endodontic MSCs [25].

A similar method of study is going on with the intravenous infusion of around 200 million autologous adipose-derived MSCs in three doses for every 3 days. The test patients are monitored for 6 months from the onset of trial injection [26].

Not only the stem cells itself but also its components exert sufficient therapeutic efficacies in improving critical pulmonary disorders. Some such studies were carried out to assess the safety and coherence of the exosomes isolated from MSCs [27]. The efficiency of exosomes isolated from bone marrow-derived MSCs tested recently by Vikram et al. shows an improvement in the oxygenation and overall clinical status of the patients [5,28].

A total number of 88 clinical trials for the regenerative stem cells therapy in COVID-19 patients have been registered as of July 2021, of which 13 are reported to be completed, but results not yet submitted, 2 of them had already been withdrawn, one suspended, and another one no longer available. Seventy-one of them are active studies, where 35 of them currently recruit patients, while 11 of them are not recruiting at the moment. Twenty-two clinical trials didn't start recruiting patients yet. The details of the clinical trials are presented in Table 1. The majority of the study involves mesenchymal stem cells from different sources; the details are depicted in Fig. 1 and Fig. 2. (https://clinicaltrials.gov).

Challenges of using stem cells and future perspective

Stem cells have been successfully experimented against chronic symptoms connected to COVID-19; however, there are certain restrictions which trouble the vast usage of stem cell therapy. Despite the fact that stem cell therapies for various diseases and clinical conditions like COVID-19 are under clinical trial, there is no known FDA-approved stem cell treatment as a frontline of defense to any of the disorders; it could be used only as a compassionate drug which gives to a patient without hope or response to any other possible medications [7]. Although therapies using stem cells have a number of limitations like host rejection, hematology-based complexities, unnecessary immune reaction, limited sources, and tedious isolation procedure, it predominantly points out on the ethical issues related to the host selection and clinical trials [16]. Most of the clinical trials of stem cell therapy against COVID-19 are questionable because of the limited number of recruited patients for the study and the use of antiviral drugs and other conventional method for minimizing the symptoms. In addition, lack of proper controls for the trials also matters when it comes to the evaluation of effectiveness of the treatment [7,29,30].

It is always better to use autologous stem cells over allogenic one for aiding more protection and effectiveness, but it is not always possible to follow the norm for treatment of an emergency situation considering the fact that the number of stem cells isolated from the patient's body won't be sufficient for a single dosage [22]. Growing the isolated cells externally is a time-consuming effort and won't be useful for a patient in critical condition. This fact coerces the use of allogenic stem cells instead of autologous one for the trials. The safety and rejection are always a concern while using allogenic stem cells for any treatment because of its foreign origin and reactive

Table 1 A list of active clinical trials as of July 2021 (https://clinicaltrials.gov).

Cells used	Therapeutic area of research	Place of study	Clinical trial number
Multipotent adult progenitor cells	Treatment of ARDS induced by COVID-19 with MACoVIA	US	NCT04367077
Embryonic stem cells	Safety and efficacy study of CAStem for chronic COVID-19 related ARDS	Beijing	NCT04331613
Amnion epithelial stem cells	Heart patch for COVID-19 related myocardial infarction	Indonesia	NCT04728906
Mesenchymal stem cells	Treatment of COVID-19 related ARDS	Iran	NCT04366063
(source not specified)	Treatment of COVID-19 related ARDS	Mexico	NCT04416139
	Treatment of severe COVID-19 associated pneumonia	Spain	NCT04361942
	Safety and effectiveness of treatment of pneumonia associated with COVID-19	China	NCT04371601
	Secretome therapy on chronic COVID-19 patients	Indonesia	NCT04753476
	Therapy in COVID-19 patients with acute kidney injury	US	NCT04445220
	Safety and feasibility in the treatment of COVID-19	Brazil	NCT04467047
	Treating COVID-19 related severe pneumonia with NestaCell®	Brazil	NCT04315987
	A study with Descartes-30 on ARDS	US	NCT04524962
	MSC exosomes for the treatment of ARDS or pneumonia caused by COVID-19	US	NCT04798716
Hematopoietic stem cells	Effect of COVID-19 on hematopoietic stem cell (HSC) transplanted patients	UK	NCT04349540
	Impact of COVID-19 on HSC transplanted recipients	Sweden	NCT04760184
	Effect of COVID-19 vaccine in cellular therapy	US	NCT04723706
Adipose derived stem cells	IV treatment of ADSC for COVID-19	US	NCT04486001
	Treatment of severe COVID-19 associated pneumonia	Japan	NCT04888949
	Determine the safety and efficacy of ADSC-MSC from hope biosciences against COVID-19	US	NCT04348435
	Evaluation of efficacy and safety of AstroStem-V against COVID-19 related pneumonia	Republic of Korea	NCT04527224
	Study on treatment of COVID-19 associated acute respiratory distress	US	NCT04905836
	IV Administration of ADSC-MSC for COVID-19-induced acute respiratory distress	US	NCT04728698
	Treatment of COVID-19 using autologous ADSC-MSC	Texas	NCT04428801
	COVI-MSC IV against COVID-19 induced ARDS	US	NCT04903327
	MSC therapy on COVID-19 patients	Mexico	NCT04611256
	Efficacy and safety of ADSC-MSC for the COVID-19 treatment	US	NCT04362189
	Treatment of post COVID-19 "Long Haul" pulmonary compromise	US	NCT04909892
	IV treatment in critical patients with COVID-19 associated pneumonia	Spain	NCT04366323

Continued

Table 1 A list of active clinical trials as of July 2021—cont'd

Cells used	Therapeutic area of research	Place of study	Clinical trial number
Bone marrow derived mesenchymal stem cells	Collection of BM-MSC to treat severe COVID19 associated pneumonia	UK	NCT04397471
	MSCs study in adults with COVID-19 associated respiratory failure or another underlying cause	Australia	NCT04537351
	Study the safety of therapeutic Tx with MSC in critical COVID-19 patients	California	NCT04397796
	Extracellular vesicle injection for COVID-19 patients with associated ARDS	Texas	NCT04657458
	Infusion of MSC for COVID-19 infection	Pakistan	NCT04444271
	Inflammation-resolution programs of COVID-19 induced ARDS	Germany	NCT04377334
	Treatment for chronic COVID-19 patients	China	NCT04346368
	A phase 2 study of COVID-19 associated acute to chronic ARDS	US	NCT04780685
	Treatment of SARS-CoV-2-related acute respiratory failure	US	NCT04345601
	Therapy on COVID-19 related and Flu-Elicited ARDS	US	NCT04629105
	Treatment of COVID-19 related ARDS	Sweden	NCT04447833
	Treatment of COVID-19 related ARDS	US	NCT04371393
Allogenic placenta derived mesenchymal stem cells	Treatment of COVID-19 related pneumonia	Ukraine	NCT04461925
Dental pulp derived mesenchymal stem cells	COVID-19 induced severe pneumonia treatment	China	NCT04302519
	Safety and efficacy in treating severe COVID-19 patients	China	NCT04336254
Umbilical cord mesenchymal stem cells	Efficacy of stem cells IV in COVID-19 patients	Pakistan	NCT04437823
	Study on the treatment of severe COVID-19	China	NCT04273646
	COVID-19 induced pneumonia treatment	China	NCT04339660
	COVID-19 induced pneumonia treatment	Florida	NCT04398303
	COVID-19 induced pneumonia treatment	China	NCT04252118
	Adjuvant therapy for severe COVID-19 patients	Indonesia	NCT04457609
	Efficacy of hUC-MSC product (BX-U001) for the treatment ARDS associated with COVID-19	US	NCT04452097
	Comparison of UC treatment and placebo to treat COVID-19 related acute pulmonary inflammation	US	NCT04490486
	Extend of safety and efficacy in the management of severe COVID-19 related pneumonia	US	NCT04429763
	Treatment of COVID-19 patients with ARDS	Australia	NCT04494386
	Treatment of COVID-19 patients with ARDS	France	NCT04333368
	Repair of COVID-19 related ARDS using stromal cells	UK	NCT03042143

Cord blood derived mesenchymal stem cells	Treatment of COVID-19 cord blood and PRP	Egypt	NCT04393415
	Treatment of COVID-19 related ARDS	US	NCT04565665
	To treat the inflammation in COVID-19	US	NCT04299152
Wharton-Jelly mesenchymal stem cells	Treatment of COVID-19	Jordan	NCT04313322
	Treatment of COVID-19 related ARDS	Mexico	NCT04456361
	Treatment of COVID-19 related ARDS	China	NCT04625738
	Efficacy of IV treatment on COVID-19 associated ARDS	Colombia	NCT04390152
	Testing efficacy and safety of WJ-MSCs for the treatment of respiratory distress due to COVID-19	Spain	NCT04390139

MSC, mesenchymal stem cells; ***WJDSC***, Wharton-Jelly derived stem cells; ***HSC***, hematopoietic stem cells; ***DPDSC***, dental pulp-derived stem cells; ***ADSC***, adipose-derived stem cells; ***UCDSC***, umbilical cord-derived stem cells.

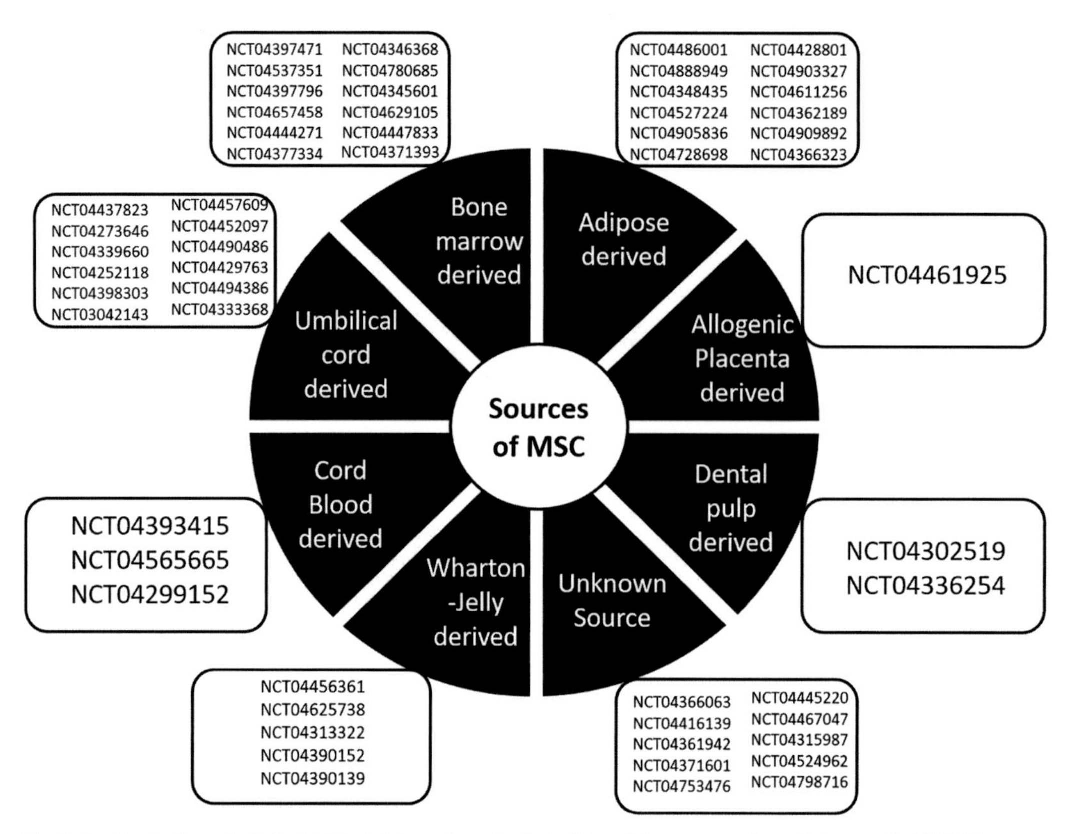

Fig. 1 An illustration of clinical trials (trial numbers included) involving mesenchymal stem cells (MSC) from different sources against COVID-19 (https://clinicaltrials.gov).

potential [7]. Stem cell culture should be processed in a completely sophisticated laboratory having a sterile environment to maintain GMP (Good Manufacturing Practices) standards of the product to be injected. Either autologous or allogenic, it is tedious to maintain cells without any contamination until injected to the patient; otherwise, it will augment the criticality of postinjection [18]. Another important factor to be noticed is the type of cells preferred for the therapy. Majority of the clinical trials use mesenchymal stem cells from multiple sources and the others remain countable (https://clinicaltrials.gov). The heterogeneity of same type of cells also adds up to the dilemma for appropriate cell selection criteria for therapies [31]. As of now, regenerative stem cell therapy is puzzle to be solved and there is a long way to go for the achievement of a successful era in the therapy.

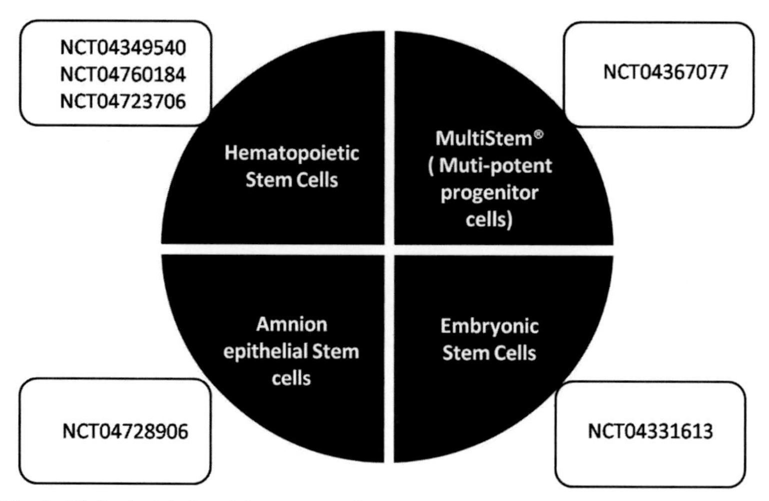

Fig. 2 Clinical trials involving stem cells other than mesenchymal cells (https://clinicaltrials.gov).

Conclusion

COVID-19 strike brought down the global health system and shattered all the aspects of human life. As it is utterly necessary to lessen the severity of the pandemic, several varieties of medications have been tried and the trend continues. Even though a number of vaccines and other antiviral and antibacterial drugs came up with a reduced number of intensities, scientists were making an effort in the development of a novel and more effective regenerative stem cell therapy to fight against COVID-19 pandemic. The effectiveness of the therapy was evident from the history of regenerative therapy in succeeding pulmonary disorders in diverse clinical conditions. However, a number of active clinical studies are going on with regenerative stem cells against the chronic symptoms of the pandemic, lack of strong experimental proofs, and proper controls to the therapy from being approved by the concerned authorities. Furthermore, victorious studies with proper controls and sufficient number of trials will pave the way to a glorious era of regenerative stem cell therapy.

References

[1] Haider N, Rothman-Ostrow P, Osman AY, Arruda LB, Macfarlane-Berry L, Elton L, Thomason MJ, et al. COVID-19—zoonosis or emerging infectious disease? Front Public Health 2020;8(November). https://doi.org/10.3389/fpubh.2020.596944, 596944.

[2] Can A, Coskun H. The rationale of using mesenchymal stem cells in patients with COVID-19-related acute respiratory distress syndrome: what to expect. Stem Cells Transl Med 2020;9(11):1287–302. https://doi.org/10.1002/sctm.20-0164.

[3] Musial C, Gorska-Ponikowska M. Medical progress: stem cells as a new therapeutic strategy for COVID-19; 2021. https://doi.org/10.1016/j.scr.2021.102239.
[4] Song N, Wakimoto H, Rossignoli F, Filippo R, Bhere D, Ciccocioppo R, Chen K-S, et al. Mesenchymal stem cell immunomodulation: in pursuit of controlling COVID-19 related cytokine storm. Stem Cells 2021;1–16. https://doi.org/10.1002/stem.3354.
[5] Jayaramayya K, Mahalaxmi I, Subramaniam MD, Raj N, Dayem AA, Lim KM, Kim SJ, et al. Immunomodulatory effect of mesenchymal stem cells and mesenchymal stem-cell-derived exosomes for COVID-19 treatment. BMB Rep 2020;53(8):400–12.
[6] Roudsari PP, Alavi-Moghadam S, Payab M, Sayahpour FA, Aghayan HR, Goodarzi P, Mohamadi-jahani F, Larijani B, Arjmand B. Auxiliary role of mesenchymal stem cells as regenerative medicine soldiers to attenuate inflammatory processes of severe acute respiratory infections caused by COVID-19. Cell Tissue Bank 2020;1–21. https://doi.org/10.1007/s10561-020-09842-3.
[7] Choudhery MS, Harris DT. Stem cell therapy for COVID-19: possibilities and challenges. Cell Biol Int 2020;44(11):2182–91. https://doi.org/10.1002/cbin.11440.
[8] Raza SS, Khan MA. Mesenchymal stem cells: a new front emerge in COVID19 treatment. Cytotherapy 2020. https://doi.org/10.1016/j.jcyt.2020.07.002.
[9] Rogers CJ, Harman RJ, Bunnell BA, Schreiber MA, Xiang C, Wang F-S, Santidrian AF, Minev BR. Rationale for the clinical use of adipose-derived mesenchymal stem cells for COVID-19 patients. J Transl Med 2020;18(1):203. https://doi.org/10.1186/s12967-020-02380-2.
[10] Riedel RN, Pérez-Pérez A, Sánchez-Margalet V, Varone CL, Maymó JL. Stem cells and COVID-19: are the human amniotic cells a new hope for therapies against the SARS-CoV-2 Virus? Stem Cell Res Ther 2021;12(1):155. https://doi.org/10.1186/s13287-021-02216-w.
[11] Sahu KK, Siddiqui AD, Cerny J. Mesenchymal Stem Cells in COVID-19: A Journey from Bench to Bedside. Lab Med 2021;52(1). https://doi.org/10.1093/LABMED/LMAA049.
[12] Chrzanowski W, Kim SY, McClements L. Can stem cells beat COVID-19: advancing stem cells and extracellular vesicles toward mainstream medicine for lung injuries associated with SARS-CoV-2 infections. Front Bioeng Biotechnol 2020;8(May):554. https://doi.org/10.3389/fbioe.2020.00554.
[13] Han Y, Li X, Zhang Y, Han Y, Chang F, Ding J. Mesenchymal stem cells for regenerative medicine. Cell 2019;8(8):886. https://doi.org/10.3390/cells8080886.
[14] Müller AM, Huppertz S, Henschler R. Hematopoietic stem cells in regenerative medicine: astray or on the path? Transf Med Hemother 2016;43(4):247–54. https://doi.org/10.1159/000447748.
[15] Sharma A, Bhatt NS, Martin AS, Abid MB, Bloomquist J, Chemaly RF, Dandoy C, et al. Clinical characteristics and outcomes of COVID-19 in haematopoietic stem-cell transplantation recipients: an observational cohort study. Lancet Haematol 2021;8(3):e185–93. https://doi.org/10.1016/S2352-3026(20)30429-4.
[16] Golchin A, Seyedjafari E, Ardeshirylajimi A. Mesenchymal stem cell therapy for COVID-19: present or future. Stem Cell Rev Rep 2020;April:1–7. https://doi.org/10.1007/s12015-020-09973-w.
[17] Basiri A, Pazhouhnia Z, Beheshtizadeh N, Hoseinpour M, Saghazadeh A, Rezaei N. Regenerative medicine in COVID-19 treatment: real opportunities and range of promises. Stem Cell Rev Rep 2020;1–13. https://doi.org/10.1007/s12015-020-09994-5.
[18] Galipeau J, Sensébé L. Mesenchymal stromal cells: clinical challenges and therapeutic opportunities. Cell Stem Cell 2018;22(6):824–33. https://doi.org/10.1016/j.stem.2018.05.004.

[19] Liang B, Chen J, Li T, Haiying W, Yang W, Li Y, Li J, et al. Clinical remission of a critically ill COVID-19 patient treated by human umbilical cord mesenchymal stem cells. Medicine 2020;99(31). https://doi.org/10.1097/MD.0000000000021429, e21429.
[20] Zayed M, Iohara K. Immunomodulation and regeneration properties of dental pulp stem cells: a potential therapy to treat coronavirus disease 2019. Cell Transplant 2020;29(August). https://doi.org/10.1177/0963689720952089. 0963689720952089.
[21] Konno M, Hamabe A, Hasegawa S, Ogawa H, Fukusumi T, Nishikawa S, Ohta K, et al. Adipose-derived mesenchymal stem cells and regenerative medicine. Dev Growth Differ 2013;55(3):309–18. https://doi.org/10.1111/dgd.12049.
[22] Harrell CR, Fellabaum C, Jovicic N, Djonov V, Arsenijevic N, Volarevic V. Molecular mechanisms responsible for therapeutic potential of mesenchymal stem cell-derived secretome. Cell 2019;8(5):E467. https://doi.org/10.3390/cells8050467.
[23] Ye Q, Wang H, Xia X, Zhou C, Liu Z, Xia Z-E, Zhang Z, et al. Safety and efficacy assessment of allogeneic human dental pulp stem cells to treat patients with severe COVID-19: structured summary of a study protocol for a randomized controlled trial (phase I/II). Trials 2020;21(1):520. https://doi.org/10.1186/s13063-020-04380-5.
[24] CAR-T (Shanghai) Biotechnology Co., Ltd. Clinical study of novel coronavirus induced severe pneumonia treated by dental pulp mesenchymal stem cells. Clinical trial registration NCT04302519. clinicaltrials.gov https://clinicaltrials.gov/ct2/show/NCT04302519; 2020.
[25] Institute of Biophysics and Cell Engineering of National Academy of Sciences of Belarus. Treatment of Covid-19 associated pneumonia with allogenic pooled olfactory mucosa-derived mesenchymal stem cells. Clinical trial registration NCT04382547. clinicaltrials.gov https://clinicaltrials.gov/ct2/show/NCT04382547; 2020.
[26] Celltex Therapeutics Corporation. Clinical study for the prophylactic efficacy of autologous adipose tissue-derived mesenchymal stem cells (AdMSCs) against coronavirus 2019 (COVID-19). Clinical trial registration NCT04428801 https://clinicaltrials.gov/ct2/show/NCT04428801; 2021.
[27] Gupta A, Kashte S, Gupta M, Rodriguez HC, Gautam SS, Kadam S. Mesenchymal stem cells and exosome therapy for COVID-19: current status and future perspective. Hum Cell 2020;33(4):907–18. https://doi.org/10.1007/s13577-020-00407-w.
[28] Sengupta V, Sengupta S, Lazo A, Woods P, Nolan A, Bremer N. Exosomes derived from bone marrow mesenchymal stem cells as treatment for severe COVID-19. Stem Cells Dev 2020;29(12):747–54. https://doi.org/10.1089/scd.2020.0080.
[29] Regmi S, Pathak S, Kim JO, Yong CS, Jeong J-H. Mesenchymal stem cell therapy for the treatment of inflammatory diseases: challenges, opportunities, and future perspectives. Eur J Cell Biol 2019;98(5–8). https://doi.org/10.1016/j.ejcb.2019.04.002, 151041.
[30] Yadav P, Vats R, Bano A, Bhardwaj R. Mesenchymal stem cell immunomodulation and regeneration therapeutics as an ameliorative approach for COVID-19 pandemics. Life Sci 2020;263(December). https://doi.org/10.1016/j.lfs.2020.118588, 118588.
[31] Li Z, Niu S, Guo B, Gao T, Wang L, Wang Y, Wang L, Tan Y, Jun W, Hao J. Stem Cell Therapy for COVID-19, ARDS and Pulmonary Fibrosis. Cell Prolif 2020;53(12). https://doi.org/10.1111/cpr.12939, e12939.

11

Therapeutic scale stem cell-derived exosomes for COVID-19: Models—Validation, management, and strategies

Francis Boniface Fernandez

Bioceramics Laboratory, Biomedical Technology Wing, Sree Chitra Tirunal Institute for Medical Sciences & Technology, Thiruvananthapuram, India

Introduction to Covid-19

Coronavirus disease (COVID-19) caused by the severe acute respiratory syndrome coronavirus 2 (SARS-CoV-2) [1] has cut a swathe of suffering and misery across the world. The virus has multifaceted transmittance mechanisms and is passed on by direct or indirect contact, aerosol from affected patients, and via body fluids that carry a high viral load. The outbreak of the virus that has been declared as a global pandemic [2] has affected day-to-day life and created a global economic challenge. This was based on the rapid rise of cases in China as well as the trend of disease transmission observed across the world. The outbreak as a pandemic was foreseen by quite a few with emphasis placed on the ability of health workers to track, trace, and isolate cases to preempt the pandemic [3]. This has slowly evolved to the idea of "COVID ZERO" [4] which was based on the long-term disability this disease may cause in affected people in large numbers. Agencies are slowly migrating away from this strategy currently with the help of vaccine availability and realistic economic costs associated with long-term movement restrictions. The approach to this situation varied from country to country with Sweden being a strong outlier in Europe. They followed a herd immunity approach with slight restrictions throughout, which has resulted in major outbreaks with a high number of deaths in the population [5].

From a reported cluster in Wuhan, China, COVID-19 has encircled the globe in its reach. The fifth reported pandemic [6] in human history will leave a mark forever on the research, development, and pharmaceutical industry. The 1918 flu pandemic infected about 500

Stem Cells and COVID-19. https://doi.org/10.1016/B978-0-323-89972-7.00010-6

million people and resulted in the deaths of around 50 million people worldwide with severe repercussions [7]. The tendency to draw parallels leads us to look back, as the current crisis approached unmanageable proportions in late 2020. Mortality rates of the flu pandemics are widely varying with older people with comorbidities more at risk from COVID-19. The Spanish flu caused affected people to be overcome by bacterial pneumonia wherein COVID-19 provokes an overactive immune response that causes multiorgan failure [8]. A proportion of people affected by COVID-19 show a downturn in condition due to complications from disease or development of acute respiratory distress syndrome that has been common with known coronavirus outbreaks [9]. The ability to understand both components of failure, develop markers and interventions tailored to the eventualities will help in management of COVID-19 cases in therapeutic settings. ARDS in COVID-19 is distinguished by involvement of alveolar epithelial cells, localized to the respiratory system and with a total inconsistency between severity of laboratory and imaging findings [9]. The onset time is also varying with reports of 8–12 days ([10,11]. Reports from retrospective analysis of data indicate that in hospital mortality ranged from 19.1% in men and 16.0% in women with advanced age, elevated CURB 65 scores, and not receiving azithromycin treatment. Active cancer and immune system complications affected survival rates along with cardiac disease and chronic lung disorders [12]. There has been considerable interest in the interaction of chronic conditions, medication, and disease outcome. A study in South Korea of 7713 COVID-19 patients admitted to in-hospital care indicated a strong correlation between use of strong opioids and in-hospital mortality [13]. A population of over 65-year-old Italians affected by COVID-19 was studied as per the PEO strategy (i.e. Population, Exposure, Outcome) as a systematic review. Relevant databases were used in the study and factors associated with mortality of senior citizens who were institutionalized or hospitalized due to diseases were evaluated. The group demonstrated that dementia, diabetes, chronic kidney disease, and hypertension were most associated with mortality in the abovementioned population [14].

Constant monitoring and supportive therapy has therefore been at the forefront of ensuring a positive prognosis in COVID-19. Symptomatic pharmacologic interventions are at the forefront of disease management, especially in populous developing countries. Long-term medical care and restorative therapies are currently out of the reach of the majority of the world's population.

Pathophysiology

The virus behind the current pandemic weighs in at a mere full length genome of 29 kb and is a single, positive-stranded RNA virus of the beta-subclass of the genus coronaviridae [15]. The virus via a spike glycoprotein on the surface (S protein) is able to invade human cells. The host cell ACE2 enzyme receptor (angiotensin-converting enzyme 2) is adsorbed and taken into the cells, which further on replicates, produces, and releases a host of viral particles [16]. The receptor is displayed on varied tissues in the human system and plays a key role in the rennin-angiotensin system. It can also deactivate angiotensin II and thus play a regulatory role control of pulmonary edema. The COVID-19 causative pathogen has about 10–20 times more affinity for the receptor than the SARS virus, thus enhancing its transmissibility [17]. SARS-CoV-2 is less fatal than SARS but has created more mayhem by its intrinsic transmissibility and affecting multiple systems via its axis of attack. This has also been enabled to a large extent by the asymptomatic and mild patients who act as active vectors of disease transmission.

Based on the 22nd June 2021 update of the World Health Organization, following are designated as the variants of concern and their approximate dates of detection [18].

Alpha (B.1.1.7): United Kingdom (UK) December 2020.

Beta (B.1.351): South Africa December 2020.

Gamma (P.1): Brazil in January 2021.

Delta (B.1.617.2): India in December 2020.

Patients who have been affected by the virus present with a wide range of symptoms. The incubation period has been presented as a wide range but commonly falls within 4 to 5 days [19]. Presenting symptoms include fever, cough, and shortness of breath. There are also associated symptoms of fatigue, general soreness, and headache. Gastrointestinal complaints and vomiting are also falling under the umbrella of COVID-19 symptoms [20,21]. A subset of the presenting patients progress via the clinical course to developing acute respiratory distress syndrome and multiorgan failure with the mortality rate rising in elderly and patients with significant comorbidities [11]. Cardiovascular disease, metabolic disease, diabetes, and weak immune systems render population vulnerable to complications from COVID-19 [22].

Severe illness associated with COVID-19 has been tracked to the cytokine mayhem unleashed. An excessive immune response that mediates cytokine release causes rapid disease progression and attendant mortality [23]. This has been directly correlated to the lung injury and multiorgan failure based on the cytokine profiles of afflicted patients with an unfavorable long-term prognosis [24,25]. The abnormal activation of inflammatory pathways that are associated with normal antiviral immune response results in severe disease [26].

A clinical manifestation in cytokine profiles has been investigated thoroughly. Levels of MIP1-B, PDGF, TNF-α IP-10, and VEGF were higher in COVID-19 affected patients regardless of their care status (C. [22]). Remarkable levels of CRP < GCSF and IL-6 have been recorded in COVID-19 patients [27]. The elated serum concentration of IL-6 and others was also earlier identified in MERS-CoV infections [28]. The cytokine release syndrome (CRS) also appears to precipitate respiratory failure, ARDS, and unfavorable clinical results [24]. Thrombosis promotion and thromboembolism resulting in fatalities are common in critically ill patients downstream of CRS [29]. Cytokine inhibitors and glucocorticoids have been handed out clinically for systemic immunosuppression with attendant risks of complications in viral pneumonia and sequelae such as osteonecrosis [30]. The inability to manage the CRS within the confines of disease progression and the inability of current therapies to act as an active modulator of a fickle immune system drive innovation in stem cell therapeutics to address the same.

Acute respiratory distress syndrome independent of any other constraints presents as hypoxia and bilateral infiltrates on chest imaging [31]. Management is prescribed as supportive with protective levels of artificial ventilation and fluid management. As per the Berlin definition, patients have acute lung injury when presenting with less severe hypoxemia and severe hypoxemia are considered to have acute respiratory distress syndrome [32]. ARDS in COVID-19 is a progressive pathology that generates rapid dyspnea, cough, nonproductive coughing, and in cases bloody sputum. Pulmonary interstitial fibrosis may occur in later stages with multiorgan failure a part of the disease progression. Patients on recovery from ARDS are able to lead a normal life but a significant portion progress to progressive pulmonary fibrosis that may be fatal [33]. Earlier work by George et al. places emphasis on lessons learned from MERS and SARS and advocates the rapid application of antifibrotics to head off the challenge faced by a large population currently dealing with the aftermath of COVID-19. Unregulated matrix metalloproteinases deal out damage to epithelial and endothelial systems causing the unregulated initiation and progression of fibrosis [34]. The activation of cellular components within the respiratory system in an irregular pattern promotes the release of profibrotic growth factors, procoagulatory factors, etc. which are collectively named as senescence-associated secretory phenotype (SASP) [35]. SASP has a context-based role and may drive tumor cell survival, promote fibrosis, and interferes at multiple levels ranging from chromatin remodeling to mRNA translation control and intracellular trafficking. This also leads to aberrated wound healing that will drive fibrosis in the respiratory system leading to long-term damage and unfavorable outcomes. The fibroblast and myofibroblast under adverse SASP influence will indicate elevated levels of stress and senescence markers, this will build resistance to apoptosis and nonregulated generation of ECM components [36]. Loss of regulatory control at the cellular level will result in varying matrix resistance, lack of microenvironment homeostasis, and disrupt cross talk between lining epithelial and circulatory endothelial cells which could drive nonrepairable damage and elevate fibrosis incidence [37]. The advent of pulmonary fibrosis and the lack of predictive clinical indicators as well as the ability to effectively disrupt inimical cytokine pathways to restore lung functionality have been an ongoing struggle in the recovery road from COVID-19.

Patients who recover from the acute stage while progressing to normal also may transition to a consistent side effects condition, being termed as long COVID. These persistent side effects are being denoted as Long COVID or postacute sequelae of COVD-19 [38]. The symptoms demonstrated include fatigue, weakness in major muscle groups, insomnia, sore throat, etc.(Y. [39]). Clinical studies have also demonstrated the presence of this loosely characterized symptom in young population [40], which gains prominence as schools and educational institutions rapidly cycle to offline sessions. This is especially prominent in countries where an effective vaccination program is not in place or implemented appropriately. In a particular cohort, more than 85% experienced relapse of symptoms with triggers that range from physical activity to stress. Nearly half the population required a reduced work schedule resulting in economic loss and related stressors. The participants also reported loss of existing work arrangements and inability to take up new assignments due to ongoing presentation of symptoms [41]. The Long COVID is expected to rise up as a public health challenge in the long term, especially in large population in underdeveloped countries and awareness is critical to preparation, management on a national and international level [42].

The ongoing pandemic is raising concerns over long-term prognosis of Long COVID patients as well as long-term complications associated with survival. Stem cell-based medicine offers a glimpse of hope in managing CRS during the acute phase and provides support in raising a nonfibrotic response in the respiratory tissue.

Interventions

The CDC recommends against the use of stem cell therapies in the treatment of COVID-19. This is based on the recommendation that stem cell therapies based on mesenchymal origin cells have to be evaluated within the auspices of clinical trials prior to their application in therapy mode. This also guards against the off-label application of preapproved therapies in this area [43,44]. The following therapies have been published as currently in use for COVID-19 and allied issues worldwide wherein the inflammatory cycle has been disrupted favorably. In cases where current therapeutic interventions were not sufficient, case studies of stem cell application have been carried out successfully [45]. The pandemic has caused a rapid increase in the registration of new trials in registries and open recruitment under multiple centers worldwide with more than 60 therapeutic and 32 safety trials recorded as of September 2020 [46].

Stem cell population/product condition addressed outcome

(1) Umbilical Cord Stem Cells [47] COVID-19 with ARDS Decrease in inflammatory cytokines (Clinical Trial).
(2) Whartons Jelly Derived MSC's Severe COVID-19 Decrease in Inflammatory Cytokines [48].
(3) Mesenchymal stem cells severe COVID-19 pneumonia decrease in inflammatory cytokines [49].
(4) Exosomes/bone marrow MSC severe COVID-19 decrease in inflammatory cytokines [50].
(5) Adipose tissue derived MSC severe COVID-19 decrease in inflammatory cytokines [51].
(6) Human umbilical cord stem cells severe COVID-19 decrease in inflammatory cytokines [52].
(7) Case study—Human umbilical cord stem cells severe COVID-19 decrease in inflammatory cytokines [45].
(8) Human umbilical cord derived stem cells severe COVID-19 decrease in inflammatory cytokines [53].

Interventions have lined up with the application of stem cells based on their self-renewal, differentiation potential rendering them an attractive tool for clinical application. Adult stem cells have been applied in multiple areas as depicted before as they are not burdened by ethical or legal restrictions that affect other categories in the clinic [54].

Exosome for management

Adult stem cells are being accessed over and over to provide blanket support in case of several refractory diseases. Viral infections with multiorgan involvement are key in being targeted [55]. This has been proposed extensively in case of acute lung injury due to a variety of viral insults ranging from influenza to cytomegalovirus [56]. They are hypothesized to play a moderate role in ensuring cytokine regulation, promotion of tissue repair, and enable production of beneficial cytokine pathways.

This application has been heralded by the wide use of MSCs in inflammatory disorders that originate in immune system malfunction such as graft versus host disease, Crohn's, inflammatory bowel disease, and ARDS. Modulatory and antiinflammatory properties of MSCs render an unique matchmaking in this application [57]. Several trials and positive indications have been generated based on this in the application of stem cells or stem cell products in the immune mediator role [58,59]. Their role in this process is via the direct or indirect interaction with the community of immune cells within the system. A broad outline is as follows.

MSCs are able to suppress proliferation and activation of T-cell populations via soluble- and contact-based mediators. It is via direct action on T-cells or via intermediary cell components [60]. Direct effects on T-cell effector populations prevent autoimmune activity and limit inflammation-related tissue damage. Their ability to directly promote and limit multiple T-cell subtypes is well documented ([61]. [62]). Inflammatory factor secretion and reducing dendritic cell activation can also be carried out via MSCs allowing a tamping down of the inflammatory response [63]. Macrophages are a key cell phenotype involved in antigen presentation and plays a role in ARDS [63]. Macrophage polarization, chronic inflammatory pathways, and tissue healing are influenced by secretory exosomes from MSCs [64]. Proinflammatory M1 macrophages can be polarized to antiinflammatory M2 macrophages by TSG-6 factor and IL-10 secretion by MSCs, thus modulating ARDS, antiviral immune response, and tissue healing in COVID-19 [65]. Neutrophils are dependent on reactive oxygen-based activity and phagocytosis to clear pathogens and perform its actions [66]. Excessive action of neutrophils has been contraindicated in COVID-19 immunopathology [67]. Neutrophil counts in lavage fluid generated from respiratory system of COVID-19 patients are directly correlated to disease severity and CRS [68]. The IV injection of exosomes derived from bone marrow-mesenchymal stem cells has demonstrated reduction of neutrophil count by 32% and improving oxygenation [50].

Mesenchymal stem cells can also directly inhibit B-cell proliferation and reduce excess immunoglobulin secretion [62]. NK cells that are overactivated in the $CXCR3^+$ phenotype are also removed within a few days of MSC application indicating a possible therapeutic target [49]. The many advantages of MSCs in the treatment, management, and long-term repair after COVID-19 necessitate the development of a solution that will be able to provide all the benefits in a cell-free solution. Cellular components, their generation, processing, and administration are fraught with several challenges [69].

Exosomes are a subset of cellular vesicles surrounded by lipid bilayer and secreted by most eukaryotic cells with a diameter range of 40–160 nm [70]. The functionality of exosomes depends on their cargo of lipids, nucleic acids, and proteins when presented appropriately to target cell populations [71,72].

Therapeutic scale exosome generation, its validation in clinical models, and navigation of regulatory challenges to reach a clinical setting are being outlined in the following paragraphs.

Exosome challenges

COVID-19 has caused a depletion in access to medical resources worldwide with hospitals carrying out triage or otherwise increasing thresholds for hospitalized disease management. This is based on the support-based therapy that necessitates intubation [73] and its attendant risks. Repurposed therapeutics, anti-HIV medication, and biologics (Remdesivir, tocilizumab) have not yielded reliable endpoint stability [74]. Prevention of infection based on vaccination depends on stable viral proteins that are undergoing mutation in the wild, rapidly with concurrent T-cell suppression on infection. Cell transplantation although quite promising is racked by safety, cell survivability, regulatory, and patient safety issues [75]. The exosomes derived from controlled stem cell populations can be a multitargeted, next generation therapeutic agent that can downregulate the cytokine storm and enable appropriate antiviral response from host [76]. There are several studies that have demonstrated reduced inflammatory in alveoli, higher edema clearance, improving membrane integrity, and better weathering of the cytokine storm [77–79].

There are several commercial preparations of exosomes that are available off the shelf and currently being tested in clinical trials in COVID-19 amelioration [50]. ExoFlo is a product (15 mL dose) that was evaluated in 24 patients with a survival rate of 83%. There was a demonstrated decrease in neutrophil count and reversal of lymphopenia. With a frank reduction of CRP, ferritin, the inflammatory landscape was rendered favorable.

Development of exosomal cargoes has followed various routes with significant risks and benefits associated with each one. The use of bioreactors has been widely advocated as they provide 100 times more yield than classic surfaces [80]. The intrinsic ability of exosome production is limited by the classical 2D cell culture techniques that introduce unnecessary perturbations in the end product. Natural 3D interactions are best fostered for ideal results [81]. The generation of size-controlled three-dimensional spheroids in a dynamic chamber would provide an enhanced production of therapeutic exosomes [82]. The group was able to detail the process by which mass production of exosome products via a custom designed cell culture system and prove results in various culture models.

The application of exosomes in treatment would require their production and inspection under strict regulatory conditions and processes that are conductive to this would be ideal. The most followed technique for exosome isolation is ultracentrifugation in excess of 120,000 × *g* [83]. MISEV 2018 [84] lays out the minimal information for the studies of the extracellular vesicles in a position statement that is widely referred to ensure harmonious application of standards. The requirement lays down three positive protein markers for EV identification and one negative marker depending on the application.

The harvesting of exosomes in large numbers would necessitate the setting up of a large-scale cell cultivation system with a downstream system similar to vaccine manufacturing. The diverse numbers of cellular systems from which exosomes are harvested are a major hurdle to harmonization. Exosomes are harvested from dendritic cells, MSCs, bone marrow-derived mesenchymal stem cells, adipose-derived mesenchymal stem cells, and umbilical cord-derived mesenchymal stem cells to name a few [85,86].

To generate therapeutic levels of exosomes, different groups have attempted varied approaches. Lee et al. working in an acute kidney injury rate model required the large-scale harvesting of a homogenous population of exosomes derived from adipose tissue-derived mesenchymal stem cells. Cells were characterized for their stemness and from 4 liters of ASC-CM, ExoSCRT technology was used to isolate exosomes in 4 h. The exosomes derived were of stable size, identity with negligible content of cellular impurities, and demonstrated up to 50% survival rate in treatment group overall [87]. The application of novel technical approaches results in a repeatable solution with long-term benefits.

The following studies have undertaken the manufacture of exosome components in compliance with good manufacturing practices. Monocyte-derived dendritic cells were cultured in T-175 flasks and exosomes purified via filtration using a 0.8um filter for debris removal. Five hundred kilodalton MWCO hollow fiber membrane was used as a secondary and final step was a sucrose/deuterium ultracentrifugation at 100,000 × *g*. Total protein content (ELISA) and characterization for tetraspanin proteins and adhesion proteins were carried out. The authors were able to achieve enhanced concentration of MHC class II compared to classical methods. Combining rapid and reproducible purification methods will be key for application [88]. Bone marrow stem cells have been cultured in T225 flasks offering enhanced cell numbers with primary clearance effected with a 0.22-μm filter. 30,000 g for 20 min completed the ultracentrifugation process and tertiary clearance was achieved by at 120,000 g for 3 h. The exosome fraction was characterized for total nucleic acids, 6 receptor proteins and provided confirmation that pooled human platelet lysate (10%)

based medium was ideal for human MSC-derived EV isolation [89]. Reduction in lysate concentration was detrimental to MSC characteristics and RNA content of released EV.

A typical dose in EV studies as described recently involves the amount collected from 2 million MSCs in 48h in a rat model [90], extrapolation to human would require more than half a billion cells. Generating applicable doses without changing the properties of the product has been a grueling challenge but has not been a focus of the field. The current pandemic demands a large-scale intervention that will dwarf current innovations. Cellular changes when moving away from classic labware to advanced techniques should be borne in mind. Large-scale manufacturing has moved toward the use of microcarriers and stirred reactors [91]. Hollow fiber reactors combined with well-studied process parameters offer a level of control [92] that is necessary for optimal production. The limitations faced by workers aiming at generating large-scale exosome production are not only biological. There is a distinct interference generated by the mechanical forces necessary to maintain large volumes in culture. Sparging of gas, agitation via mechanical instruments generate shear stress and other impacts on cell condition. One hundred and eighty revolutions per minute agitation causes T-cell expansion to be inhibited in a stirred system [93]. MSCs under mechanotransduction introduced by shear stress can be primed for osteogenic differentiation [94] which would in turn impact the quality of the exosome vesicles generated. When operated at capacity, impurities due to cell death will also rise in the exosome population which may signal unfavorable downstream events or alter its signaling [95].

As described, more innovations are necessitated to retrieve a concentrated dose of exosomes from the bioreactors or cell containers. Hollow fiber systems (packed bed) provide for development of systems where the end product can be retained with transfer of expended medium, enabling a simpler downstream process [96]. Tangential Flow Filtration systems [97], wherein a combination of 3D culture and TFF provides for an enriched harvest of exosome material from cell culture media. Proprietary systems like the CELLine from Integra provide for systems that provide retention of desired secretome and provide for replenishment of nutrients as required [98]. A 12-fold increase in exosome retention has been reported.

Phenomenal progress has been made in understanding the pathophysiology of COVID-19 and its effects on the human body. The remedial pathways identified could be well served by a cell-free therapeutic derived from an identified cell source as required. The production of this faces multiple challenges in its route from bench to bedside. Innovations in cell culture, media development, and biomaterials to support production and application will be critical to success in this struggle with the pandemic.

References

[1] Gorbalenya AE, Baker SC, Baric RS, de Groot RJ, Drosten C, Gulyaeva AA, Haagmans BL, et al. The species severe acute respiratory syndrome-related coronavirus: classifying 2019-NCoV and naming it SARS-CoV-2. Nat Microbiol 2020;5(4):536–44. https://doi.org/10.1038/s41564-020-0695-z.

[2] Cucinotta D, Vanelli M. WHO declares COVID-19 a pandemic. Acta Biomed 2020;91(1):157–60. https://doi.org/10.23750/abm.v91i1.9397.

[3] Thompson R. Pandemic potential of 2019-NCoV. Lancet Infect Dis 2020;20(3):280. https://doi.org/10.1016/S1473-3099(20)30068-2.

[4] McKee M. Achieving zero covid is not easy, but the alternative is far worse. BMJ 2020;371(October). https://doi.org/10.1136/bmj.m3859, m3859.

[5] Claeson M, Hanson S. COVID-19 and the Swedish enigma. Lancet 2021;397(10271):259–61. https://doi.org/10.1016/S0140-6736(20)32750-1.

[6] Madhav N, Oppenheim B, Gallivan M, Mulembakani P, Rubin E, Wolfe N. Pandemics: risks, impacts, and mitigation. In: Jamison DT, Gelband H, Horton S, Jha P, Laxminarayan R, Mock CN, Nugent R, editors. Disease control priorities: improving health and reducing poverty. 3rd ed. Washington (DC): The International Bank for Reconstruction and Development/The World Bank; 2017. http://www.ncbi.nlm.nih.gov/books/NBK525302/.

[7] Hardy A. The great influenza: the epic story of the deadliest plague in history. In: History: reviews of new books, vol. 34; 2006. p. 37. https://doi.org/10.1080/03612759.2006.10526765 [2].

[8] Wu C, Chen X, Cai Y, Xia J'a, Zhou X, Xu S, Huang H, et al. Risk factors associated with acute respiratory distress syndrome and death in patients with coronavirus disease 2019 pneumonia in Wuhan, China. JAMA Intern Med 2020;180(7):934–43. https://doi.org/10.1001/jamainternmed.2020.0994.

[9] Li X, Ma X. Acute respiratory failure in COVID-19: is it "typical" ARDS? Crit Care 2020;24(May):198. https://doi.org/10.1186/s13054-020-02911-9.

[10] Chen N, Zhou M, Dong X, Jieming Q, Gong F, Han Y, Qiu Y, et al. Epidemiological and clinical characteristics of 99 cases of 2019 novel coronavirus pneumonia in Wuhan, China: a descriptive study. Lancet 2020;395(10223):507–13. https://doi.org/10.1016/S0140-6736(20)30211-7.

[11] Zhou F, Ting Y, Ronghui D, Fan G, Liu Y, Liu Z, Xiang J, et al. Clinical course and risk factors for mortality of adult inpatients with COVID-19 in Wuhan, China: a retrospective cohort study. Lancet 2020;395(10229):1054–62. https://doi.org/10.1016/S0140-6736(20)30566-3.

[12] Fernández-Martínez NF, Ortiz-González-Serna R, Serrano-Ortiz Á, Rivera-Izquierdo M, Ruiz-Montero R, Pérez-Contreras M, Guerrero-Fernández I, de Alba Á, Romero-Duarte, Salcedo-Leal I. Sex differences and predictors of in-hospital mortality among patients with COVID-19: results from the ANCOHVID multicentre study. Int J Environ Res Public Health 2021;18(17):9018. https://doi.org/10.3390/ijerph18179018.

[13] Oh T-K, Song I-A, Lee J, Eom W, Jeon Y-T. Musculoskeletal disorders, pain medication, and in-hospital mortality among patients with COVID-19 in South Korea: a population-based cohort study. Int J Environ Res Public Health 2021;18(13):6804. https://doi.org/10.3390/ijerph18136804.

[14] Alves VP, Casemiro FG, Gedeon B, de Araujo M, de Souza A, Lima RS, de Oliveira F, de Souza T, Fernandes AV, Gomes C, Gregori D. Factors associated with mortality among elderly people in the COVID-19 pandemic (SARS-CoV-2): a systematic review and meta-analysis. Int J Environ Res Public Health 2021;18(15):8008. https://doi.org/10.3390/ijerph18158008.

[15] Cascella M, Rajnik M, Aleem A, Dulebohn SC, Di Napoli R. Features, evaluation, and treatment of coronavirus (COVID-19). In: StatPearls. Treasure Island (FL): StatPearls Publishing; 2021. http://www.ncbi.nlm.nih.gov/books/NBK554776/.

[16] Wrapp D, Wang N, Corbett KS, Goldsmith JA, Hsieh C-L, Abiona O, Graham BS, McLellan JS. Cryo-EM structure of the 2019-NCoV spike in the prefusion conformation. Science 2020;367(6483):1260–3. https://doi.org/10.1126/science.abb2507.

[17] Walls AC, Young-Jun Park M, Tortorici A, Wall A, McGuire AT, Veesler D. Structure, function, and antigenicity of the SARS-CoV-2 spike glycoprotein. Cell 2020;181(2):281–292.e6. https://doi.org/10.1016/j.cell.2020.02.058.

[18] Tracking SARS-CoV-2 Variants, https://www.who.int/emergencies/emergency-health-kits/trauma-emergency-surgery-kit-who-tesk-2019/tracking-SARS-CoV-2-variants; 2021. [Accessed 26 September 2021].

[19] Li Q, Guan X, Peng W, Wang X, Zhou L, Tong Y, Ren R, et al. Early transmission dynamics in Wuhan, China, of novel coronavirus-infected pneumonia. N Engl J Med 2020;382(13):1199–207. https://doi.org/10.1056/NEJMoa2001316.

[20] CDC. Coronavirus disease 2019 (COVID-19)—symptoms. Centers for Disease Control and Prevention; 2021. 22 February 2021 https://www.cdc.gov/coronavirus/2019-ncov/symptoms-testing/symptoms.html.

[21] Parasher A. COVID-19: current understanding of its pathophysiology, clinical presentation and treatment. Postgrad Med J 2021;97(1147):312–20. https://doi.org/10.1136/postgradmedj-2020-138577.

[22] Huang C, Wang Y, Li X, Ren L, Zhao J, Yi H, Zhang L, et al. Clinical features of patients infected with 2019 novel coronavirus in Wuhan, China. Lancet 2020;395(10223):497–506. https://doi.org/10.1016/S0140-6736(20)30183-5.

[23] Ye Q, Wang B, Mao J. The pathogenesis and treatment of the `cytokine storm' in COVID-19. J Infect 2020;80(6):607–13. https://doi.org/10.1016/j.jinf.2020.03.037.

[24] Chen G, Di W, Guo W, Cao Y, Huang D, Wang H, Wang T, et al. Clinical and immunological features of severe and moderate coronavirus disease 2019. J Clin Investig 2020;130(5):2620–9. https://doi.org/10.1172/JCI137244.

[25] Ruan Q, Yang K, Wang W, Jiang L, Song J. Clinical predictors of mortality due to COVID-19 based on an analysis of data of 150 patients from Wuhan, China. Intensive Care Med 2020;46(5):846–8. https://doi.org/10.1007/s00134-020-05991-x.

[26] Braciale TJ, Hahn YS. Immunity to viruses. Immunol Rev 2013;255(1):5–12. https://doi.org/10.1111/imr.12109.

[27] Moore JB, June CH. Cytokine release syndrome in severe COVID-19. Science 2020;368(6490):473–4. https://doi.org/10.1126/science.abb8925.

[28] Fehr AR, Channappanavar R, Perlman S. Middle East respiratory syndrome: emergence of a pathogenic human coronavirus. Annu Rev Med 2017;68(January):387–99. https://doi.org/10.1146/annurev-med-051215-031152.

[29] Beyrouti R, Adams ME, Benjamin L, Cohen H, Farmer SF, Goh YY, Humphries F, et al. Characteristics of ischaemic stroke associated with COVID-19. J Neurol Neurosurg Psychiatry 2020;91(8):889–91. https://doi.org/10.1136/jnnp-2020-323586.

[30] Robinson PC, Morand E. Divergent effects of acute versus chronic glucocorticoids in COVID-19. Lancet Rheumatol 2021;3(3):e168–70. https://doi.org/10.1016/S2665-9913(21)00005-9.

[31] Sweeney RM, McAuley DF. Acute respiratory distress syndrome. Lancet 2016;388(10058):2416–30. https://doi.org/10.1016/S0140-6736(16)00578-X.

[32] Ware LB, Matthay MA. The acute respiratory distress syndrome. N Engl J Med 2000;342(18):1334–49. https://doi.org/10.1056/NEJM200005043421806.

[33] George PM, Wells AU, Gisli Jenkins R. Pulmonary fibrosis and COVID-19: the potential role for antifibrotic therapy. Lancet Respir Med 2020;8(8):807–15. https://doi.org/10.1016/S2213-2600(20)30225-3.

[34] Zemans RL, Colgan SP, Downey GP. Transepithelial migration of neutrophils: mechanisms and implications for acute lung injury. Am J Respir Cell Mol Biol 2009;40(5):519–35. https://doi.org/10.1165/rcmb.2008-0348TR.

[35] Lopes-Paciencia S, Saint-Germain E, Rowell M-C, Ruiz AF, Kalegari P, Ferbeyre G. The senescence-associated secretory phenotype and its regulation. Cytokine 2019;117(May):15–22. https://doi.org/10.1016/j.cyto.2019.01.013.

[36] Barratt SL, Creamer A, Hayton C, Chaudhuri N. Idiopathic pulmonary fibrosis (IPF): an overview. J Clin Med 2018;7(8):201. https://doi.org/10.3390/jcm7080201.

[37] Sgalla G, Iovene B, Calvello M, Ori M, Varone F, Richeldi L. Idiopathic pulmonary fibrosis: pathogenesis and management. Respir Res 2018;19(1):32. https://doi.org/10.1186/s12931-018-0730-2.

[38] Proal AD, VanElzakker MB. Long COVID or post-acute sequelae of COVID-19 (PASC): an overview of biological factors that may contribute to persistent symptoms. Front Microbiol 2021;12:1494. https://doi.org/10.3389/fmicb.2021.698169.

[39] Huang Y, Pinto MD, Borelli JL, Mehrabadi MA, Abrihim H, Dutt N, Lambert N, et al. COVID symptoms, symptom clusters, and predictors for becoming a long-hauler: looking for clarity in the haze of the pandemic. MedRxiv 2021. https://doi.org/10.1101/2021.03.03.21252086. 2021.03.03.21252086.

[40] Buonsenso D, Munblit D, De Rose C, Sinatti D, Ricchiuto A, Carfi A, Valentini P. Preliminary evidence on long COVID in children. Acta Paediatr 2021;110(7):2208–11. https://doi.org/10.1111/apa.15870.

[41] Davis HE, Assaf GS, McCorkell L, Wei H, Low RJ, Re'em Y, Redfield S, Austin JP, Akrami A. Characterizing long COVID in an international cohort: 7 months of symptoms and their impact; 2021. https://doi.org/10.1101/2020.12.24.20248802.

[42] Garg M, Maralakunte M, Garg S, Dhooria S, Sehgal I, Bhalla AS, Vijayvergiya R, et al. The conundrum of 'long-COVID-19': a narrative review. Int J Gen Med 2021;14(June):2491–506. https://doi.org/10.2147/IJGM.S316708.

[43] Coronavirus disease 2019 (COVID-19) treatment guidelines. 2021, 360.

[44] Section_91.Pdf, https://files.covid19treatmentguidelines.nih.gov/guidelines/section/section_91.pdf; 2021. [Accessed 26 September 2021].

[45] Liang B, Chen J, Li T, Haiying W, Yang W, Li Y, Li J, et al. Clinical remission of a critically ill COVID-19 patient treated by human umbilical cord mesenchymal stem cells: a case report. Medicine 2020;99(31). https://doi.org/10.1097/MD.0000000000021429, e21429.

[46] Li Z, Niu S, Guo B, Gao T, Wang L, Wang Y, Wang L, Tan Y, Jun W, Hao J. Stem cell therapy for COVID-19, ARDS and pulmonary fibrosis. Cell Prolif 2020;53(12). https://doi.org/10.1111/cpr.12939, e12939.

[47] Lanzoni G, Linetsky E, Correa D, Cayetano SM, Alvarez RA, Kouroupis D, Gil AA, et al. Umbilical cord mesenchymal stem cells for COVID-19 acute respiratory distress syndrome: a double-blind, phase 1/2a, randomized controlled trial. Stem Cells Transl Med 2021;10(5):660–73. https://doi.org/10.1002/sctm.20-0472.

[48] Saleh M, Vaezi AA, Aliannejad R, Sohrabpour AA, Kiaei SZF, Shadnoush M, Siavashi V, et al. Cell therapy in patients with COVID-19 using Wharton's jelly mesenchymal stem cells: a phase 1 clinical trial. Stem Cell Res Ther 2021;12(1):410. https://doi.org/10.1186/s13287-021-02483-7.

[49] Leng Z, Zhu R, Hou W, Feng Y, Yang Y, Han Q, Shan G, et al. Transplantation of ACE2-mesenchymal stem cells improves the outcome of patients with COVID-19 pneumonia. Aging Dis 2020;11(2):216–28. https://doi.org/10.14336/AD.2020.0228.

[50] Sengupta V, Sengupta S, Lazo A, Woods P, Nolan A, Bremer N. Exosomes derived from bone marrow mesenchymal stem cells as treatment for severe COVID-19. Stem Cells Dev 2020;29(12):747–54. https://doi.org/10.1089/scd.2020.0080.

[51] Sánchez-Guijo F, García-Arranz M, López-Parra M, Monedero P, Mata-Martínez C, Santos A, Sagredo V, et al. Adipose-derived mesenchymal stromal cells for the treatment of patients with severe SARS-CoV-2 pneumonia requiring mechanical ventilation. A proof of concept study. EClinicalMedicine 2020;25(August). https://doi.org/10.1016/j.eclinm.2020.100454, 100454.

[52] Shu L, Niu C, Li R, Huang T, Wang Y, Huang M, Ji N, et al. Treatment of severe COVID-19 with human umbilical cord mesenchymal stem cells. Stem Cell Res Ther 2020;11(1):361. https://doi.org/10.1186/s13287-020-01875-5.

[53] Meng F, Ruonan X, Wang S, Zhe X, Zhang C, Li Y, Yang T, et al. Human umbilical cord-derived mesenchymal stem cell therapy in patients with COVID-19: a phase 1 clinical trial. Signal Transduct Target Ther 2020;5(1):172. https://doi.org/10.1038/s41392-020-00286-5.

[54] Esquivel D, Mishra R, Soni P, Seetharaman R, Mahmood A, Srivastava A. Stem cells therapy as a possible therapeutic option in treating COVID-19 patients. Stem Cell Rev Rep 2021;17(1):144–52. https://doi.org/10.1007/s12015-020-10017-6.

[55] Du J, Li H, Lian J, Zhu X, Qiao L, Lin J. Stem cell therapy: a potential approach for treatment of influenza virus and coronavirus-induced acute lung injury. Stem Cell Res Ther 2020;11(1):192. https://doi.org/10.1186/s13287-020-01699-3.

[56] Rubenfeld GD, Caldwell E, Peabody E, Weaver J, Martin DP, Neff M, Stern EJ, Hudson LD. Incidence and outcomes of acute lung injury. N Engl J Med 2005;353(16):1685–93. https://doi.org/10.1056/NEJMoa050333.

[57] Bian X-W, COVID-19 Pathology Team. Autopsy of COVID-19 patients in China. Natl Sci Rev 2020;7(9):1414–8. https://doi.org/10.1093/nsr/nwaa123.

[58] Galipeau J, Sensébé L. Mesenchymal stromal cells: clinical challenges and therapeutic opportunities. Cell Stem Cell 2018;22(6):824–33. https://doi.org/10.1016/j.stem.2018.05.004.

[59] Shi Y, Wang Y, Li Q, Liu K, Hou J, Shao C, Wang Y. Immunoregulatory mechanisms of mesenchymal stem and stromal cells in inflammatory diseases. Nat Rev Nephrol 2018;14(8):493–507. https://doi.org/10.1038/s41581-018-0023-5.

[60] Duffy MM, Ritter T, Ceredig R, Griffin MD. Mesenchymal stem cell effects on T-cell effector pathways. Stem Cell Res Ther 2011;2(4):34. https://doi.org/10.1186/scrt75.

[61] Fayyad-Kazan H, Faour WH, Badran B, Lagneaux L, Najar M. The immunomodulatory properties of human bone marrow-derived mesenchymal stromal cells are defined according to multiple immunobiological criteria. Inflamm Res 2016;65(6):501–10. https://doi.org/10.1007/s00011-016-0933-2.

[62] Uccelli A, Kerlero N, de Rosbo. The immunomodulatory function of mesenchymal stem cells: mode of action and pathways. Ann N Y Acad Sci 2015;1351(September):114–26. https://doi.org/10.1111/nyas.12815.

[63] Đokić JM, Tomić SZ, Čolić MJ. Cross-talk between mesenchymal stem/stromal cells and dendritic Cells. Curr Stem Cell Res Ther 2016;11(1):51–65.

[64] Ti D, Hao H, Tong C, Liu J, Dong L, Zheng J, Zhao Y, Liu H, Xiaobing F, Han W. LPS-preconditioned mesenchymal stromal cells modify macrophage polarization for resolution of chronic inflammation via exosome-shuttled let-7b. J Transl Med 2015;13(September):308. https://doi.org/10.1186/s12967-015-0642-6.

[65] Merad M, Martin JC. Pathological inflammation in patients with COVID-19: a key role for monocytes and macrophages. Nat Rev Immunol 2020;20(6):355–62. https://doi.org/10.1038/s41577-020-0331-4.

[66] Schönrich G, Raftery MJ. Neutrophil extracellular traps go viral. Front Immunol 2016;7:366. https://doi.org/10.3389/fimmu.2016.00366.

[67] Didangelos A. COVID-19 hyperinflammation: what about neutrophils? MSphere 2020;5(3). https://doi.org/10.1128/mSphere.00367-20, e00367-20.

[68] Perrone LA, Plowden JK, García-Sastre A, Katz JM, Tumpey TM. H5N1 and 1918 pandemic influenza virus infection results in early and excessive infiltration of macrophages and neutrophils in the lungs of mice. PLoS Pathog 2008;4(8). https://doi.org/10.1371/journal.ppat.1000115, e1000115.

[69] Dai J, Yangzhou S, Zhong S, Cong L, Liu B, Yang J, Tao Y, He Z, Chen C, Jiang Y. Exosomes: key players in cancer and potential therapeutic strategy. Signal Transduct Target Ther 2020;5(1):1–10. https://doi.org/10.1038/s41392-020-00261-0.

[70] Ruivo CF, Adem B, Silva M, Melo SA. The biology of cancer exosomes: insights and new perspectives. Cancer Res 2017;77(23):6480–8. https://doi.org/10.1158/0008-5472.CAN-17-0994.

[71] Puhka M, Takatalo M, Nordberg M-E, Valkonen S, Nandania J, Aatonen M, Yliperttula M, et al. Metabolomic profiling of extracellular vesicles and alternative normalization methods reveal enriched metabolites and strategies to study prostate cancer-related changes. Theranostics 2017;7(16):3824–41. https://doi.org/10.7150/thno.19890.

[72] Skotland T, Sandvig K, Llorente A. Lipids in exosomes: current knowledge and the way forward. Prog Lipid Res 2017;66(April):30–41. https://doi.org/10.1016/j.plipres.2017.03.001.

[73] Gattinoni L, Coppola S, Cressoni M, Busana M, Rossi S, Chiumello D. COVID-19 does not Lead to a "typical" acute respiratory distress syndrome. Am J Respir Crit Care Med 2020;201(10):1299–300. https://doi.org/10.1164/rccm.202003-0817LE.

[74] Alzghari SK, Acuña VS. Supportive treatment with tocilizumab for COVID-19: a systematic review. J Clin Virol 2020;127(June). https://doi.org/10.1016/j.jcv.2020.104380, 104380.

[75] Metcalfe SM. Mesenchymal stem cells and management of COVID-19 pneumonia. Med Drug Discov 2020;5(March). https://doi.org/10.1016/j.medidd.2020.100019, 100019.

[76] Hessvik NP, Llorente A. Current knowledge on exosome biogenesis and release. Cell Mol Life Sci 2018;75(2):193–208. https://doi.org/10.1007/s00018-017-2595-9.

[77] Lee JH, Park J, Lee J-W. Therapeutic use of mesenchymal stem cell-derived extracellular vesicles in acute lung injury. Transfusion 2019;59(S1):876–83. https://doi.org/10.1111/trf.14838.

[78] Tang X-D, Shi L, Monsel A, Li X-Y, Zhu H-L, Zhu Y-G, Jie-Ming Q. Mesenchymal stem cell microvesicles attenuate acute lung injury in mice partly mediated by Ang-1 MRNA. Stem Cells 2017;35(7):1849–59. https://doi.org/10.1002/stem.2619.

[79] Zhu Y-G, Feng X-M, Abbott J, Fang X-H, Hao Q, Monsel A, Jie-Ming Q, Matthay MA, Lee JW. Human mesenchymal stem cell microvesicles for treatment of *Escherichia coli* endotoxin-induced acute lung injury in mice. Stem Cells 2014;32(1):116–25. https://doi.org/10.1002/stem.1504.

[80] Palviainen M, Saari H, Kärkkäinen O, Pekkinen J, Auriola S, Yliperttula M, Puhka M, Hanhineva K, Siljander PR-M. Metabolic signature of extracellular vesicles depends on the cell culture conditions. J Extracell Vesicles 2019;8(1):1596669. https://doi.org/10.1080/20013078.2019.1596669.

[81] Placzek MR, Chung I-M, Macedo HM, Ismail S, Blanco TM, Lim M, Cha JM, et al. Stem cell bioprocessing: fundamentals and principles. J R Soc Interface 2009;6(32):209–32. https://doi.org/10.1098/rsif.2008.0442.

[82] Cha JM, Shin EK, Sung JH, Moon GJ, Kim EH, Cho YH, Dal Park H, Bae H, Kim J, Bang OY. Efficient scalable production of therapeutic microvesicles derived from human mesenchymal stem cells. Sci Rep 2018;8(1):1171. https://doi.org/10.1038/s41598-018-19211-6.

[83] Kalra H, Drummen GPC, Mathivanan S. Focus on extracellular vesicles: introducing the next small big thing. Int J Mol Sci 2016;17(2):170. https://doi.org/10.3390/ijms17020170.

[84] Théry C, Witwer KW, Aikawa E, Alcaraz MJ, Anderson JD, Andriantsitohaina R, Antoniou A, et al. Minimal information for studies of extracellular vesicles 2018 (MISEV2018): a position statement of the International Society for Extracellular Vesicles and Update of the MISEV2014 guidelines. Journal of Extracell Vesicles 2018;7(1):1535750. https://doi.org/10.1080/20013078.2018.1535750.

[85] Doyle LM, Wang MZ. Overview of extracellular vesicles, their origin, composition, purpose, and methods for exosome isolation and analysis. Cell 2019;8(7):E727. https://doi.org/10.3390/cells8070727.

[86] Yang X-X, Sun C, Wang L, Guo X-L. New insight into isolation, identification techniques and medical applications of exosomes. J Control Release 2019;308(August):119–29. https://doi.org/10.1016/j.jconrel.2019.07.021.
[87] Lee JH, Ha DH, Go H-k, Youn J, Kim H-k, Jin RC, Miller RB, Kim D-h, Cho BS, Yi YW. Reproducible large-scale isolation of exosomes from adipose tissue-derived mesenchymal stem/stromal cells and their application in acute kidney injury. Int J Mol Sci 2020;21(13):4774. https://doi.org/10.3390/ijms21134774.
[88] Lamparski HG, Metha-Damani A, Yao J-Y, Patel S, Hsu D-H, Ruegg C, Le Pecq J-B. Production and characterization of clinical grade exosomes derived from dendritic cells. J Immunol Methods 2002;270(2):211–26. https://doi.org/10.1016/s0022-1759(02)00330-7.
[89] Pachler K, Lener T, Streif D, Dunai ZA, Desgeorges A, Feichtner M, Öller M, Schallmoser K, Rohde E, Gimona M. A good manufacturing practice-grade standard protocol for exclusively human mesenchymal stromal cell-derived extracellular vesicles. Cytotherapy 2017;19(4):458–72. https://doi.org/10.1016/j.jcyt.2017.01.001.
[90] Phinney DG, Pittenger MF. Concise review: MSC-derived exosomes for cell-free therapy. Stem Cells 2017;35(4):851–8. https://doi.org/10.1002/stem.2575.
[91] Chen AK-L, Chen X, Choo ABH, Reuveny S, Oh SKW. Critical microcarrier properties affecting the expansion of undifferentiated human embryonic stem cells. Stem Cell Res 2011;7(2):97–111. https://doi.org/10.1016/j.scr.2011.04.007.
[92] Gerlach JC, Lin Y-C, Brayfield CA, Minteer DM, Han Li J, Rubin P, Marra KG. Adipogenesis of human adipose-derived stem cells within three-dimensional hollow fiber-based bioreactors. Tissue Eng Part C Methods 2012;18(1):54–61. https://doi.org/10.1089/ten.tec.2011.0216.
[93] Carswell KS, Papoutsakis ET. Culture of human T cells in stirred bioreactors for cellular immunotherapy applications: shear, proliferation, and the IL-2 receptor. Biotechnol Bioeng 2000;68(3):328–38. https://doi.org/10.1002/(SICI)1097-0290(20000505)68:3<328::AID-BIT11>3.0.CO;2-V.
[94] Brindley D, Moorthy K, Lee J-H, Mason C, Kim H-W, Wall I. Bioprocess forces and their impact on cell behavior: implications for bone regeneration therapy. J Tissue Eng 2011;2011(August). https://doi.org/10.4061/2011/620247, 620247.
[95] Bollini S, Gentili C, Tasso R, Cancedda R. The regenerative role of the fetal and adult stem cell secretome. J Clin Med 2013;2(4):302–27. https://doi.org/10.3390/jcm2040302.
[96] Wen Y-T, Chang Y-C, Lin L-C, Liao P-C. Collection of in vivo-like liver cell Secretome with alternative sample enrichment method using a hollow fiber bioreactor culture system combined with tangential flow filtration for secretomics analysis. Anal Chim Acta 2011;684(1):81–8. https://doi.org/10.1016/j.aca.2010.10.040.
[97] Haraszti RA, Miller R, Stoppato M, Sere YY, Coles A, Didiot M-C, Wollacott R, et al. Exosomes produced from 3D cultures of MSCs by tangential flow filtration show higher yield and improved activity. Mol Ther 2018;26(12):2838–47. https://doi.org/10.1016/j.ymthe.2018.09.015.
[98] Mitchell JP, Court J, Mason MD, Tabi Z, Clayton A. Increased exosome production from tumour cell cultures using the Integra CELLine culture system. J Immunol Methods 2008;335(1):98–105. https://doi.org/10.1016/j.jim.2008.03.001.

12

Current strategies and future perspectives in COVID-19 therapy

S.R. Aravind[a], Krupa Ann Mathew[a], Bernadette K. Madathil[a], S. Mini, and Annie John

Advanced Centre for Tissue Engineering, Department of Biochemistry, University of Kerala, Thiruvananthapuram, Kerala, India

Introduction

Coronavirus Disease 2019, abbreviated as (COVID-19), is one of the largest public health emergencies that the current generation has witnessed. The first case of COVID-19 was reported in the city of Wuhan in the Republic of China in December 2019 [1]. After confirming the evidence of human-to-human transmission around the globe and the causative agent to be the newly found Severe Acute Respiratory Syndrome Coronavirus 2 (SARS-CoV-2), the World Health Organization (WHO) declared what emerged as a cluster of atypical 'viral pneumonia' in Wuhan, China [2] as a public health emergency of international concern (PHEIC) on January 30, 2020. In spite of the concerted efforts globally to contain the disease through social distancing, lockdowns, and quarantining patients and their contacts, the outbreak continued to spread across the globe rampantly, and WHO declared COVID-19 as a pandemic on March 11, 2020 [3]. As of May 1, 2021, while writing this chapter, majority of the countries have been affected, and the total number of COVID-19 cases across the globe stands at approximately 150.6 million, with death at 3.17 million. The countries worst affected include the United States of America, followed by India and then Brazil. In India, the scenario is 19.1 million positive cases with 2,11,853 deaths [4]. In India, the first case of COVID-19 was reported in the state of Kerala on January 27, 2020. Although Kerala had initially mitigated the spread of COVID-19, there was a surge in the number of cases when state borders opened. Currently, the cases in Kerala stand at 1.57 million, with 1.26 million recoveries and 5308 deaths [5].

[a] The authors have contributed equally.

Stem Cells and COVID-19. https://doi.org/10.1016/B978-0-323-89972-7.00011-8

The COVID-19 pandemic continues to challenge governments, public health infrastructure, and the scientific community worldwide. It has caused a huge burden on health, economy, and society at large. In this chapter, the etiology, structure, pathomechanisms of infection, and manifestation of SARS-CoV-2 have been described, and the different modes of diagnosis and currently available treatment modalities elaborated. The chapter also gives an outlook on the future perspectives that are being adopted to mitigate and manage COVID-19, which can also be extended to other respiratory diseases.

COVID-19: An overview

SARS-CoV-2 virus etiology

The global population had witnessed and overcome a plethora of pandemics over the centuries, namely smallpox, plagues, cholera, Spanish flu [6], Asian flu [7], swine flu [8], HIV/AIDS [9], and Ebola [10]. Most of the pandemics encountered are zoonotic in origin, be it HIV/AIDS with chimpanzee origin [11,12], Ebola with bat origin [13], or influenza with bird or pig origin. Zoonotic pandemics use animals as reservoirs before getting transmitted to humans and are not fatal to their animal hosts. The animal hosts have specific physiology which prevents them from contracting the disease, for instance, the particular interferon present in bats [14]. Recent times are witnessing an unprecedented increase in the population of zoonotic viruses due to urbanization, deforestation, and resultant recurrent contact with animals [15].

Corona virus is a zoonotic strain of virus known to infect animals and humans, with the first report being published as early as 1949 on murine corona virus strain JHM (named in remembrance of the renowned bacteriologist Dr. J. Howard Mueller) [16,17]. The name corona virus was coined by D. A. Tyrrell based on the crown-like appearance of the virus under an electron microscope [18]. Corona viruses are enveloped positive-sense RNA viruses belonging to the kingdom Orthornavirae, order Nidovirales, and family Coronaviridae [19,20]. They have an unusually large genome for an RNA virus and possess unique self-replicating machineries such as replicase gene for expression of structural genes and ribosomal frameshifting for expression of nonstructural genes [15,21]. Because of their high rate of mutation and recombination, they need to cross species barriers and enter new animal reservoir hosts so as to survive and multiply. As they migrate across species, they get evolved into novel strains with enhanced virulence and infectivity.

Starting from the first report in 2005 [22], linking bovine corona viruses to animal infections, there have been several reports stating the involvement of corona viruses in causing respiratory infections and severe complications in various animals such as cows, pigs, dogs, chicken, feral pigeon, mallards, graylag geese [23], as well as humans [24]. Corona viruses had been very successful in crossing the species barrier and causing fatal infections in humans. The alpha and beta genera of coronaviruses constitute the majority of all the known coronaviruses and infect mammals and humans primarily, while the delta and gamma genera attack birds and marine mammals like cetaceans [25]. To date, seven zoonotic Coronaviruses, two under the alphacoronavirus genus and five under the betacoronavirus genus, were able to traverse the species barrier and cause infections in humans (Fig. 1) [26]. Of these, three betacoronaviruses, namely Severe Acute Respiratory Syndrome Coronavirus (SARS-CoV), Middle East Respiratory Syndrome Coronavirus (MERS-CoV), and Severe Acute Respiratory Syndrome Coronavirus 2 (SARS-CoV-2), were able to cause severe and fatal respiratory illness in humans worldwide [19,27].

Bats are the primary animal host for all the coronaviruses which cause zoonotic infections in humans [28,29], with two exceptions, OC43 and HKU1, originating from rodents [30]. The 2003 SARS outbreak that emerged in Guangdong, China, having a case fatality rate (CFR) of 10%, was caused by SARS-CoV [31], with bats as the natural host and civet as the intermediate host [32]. On the other hand, MERS-CoV, the causative agent of Middle East Respiratory Syndrome of 2012, with a CFR of 34.4% [33], was a species barrier breach from the intermediate animal reservoir, the dromedary camels of Saudi Arabia [32]. The current COVID-19 pandemic too, with a CFR ranging from 0.1% to 9.26% [34], is thought to have originated from a spillover infection from an animal reservoir [35]. A phylogenetic genomic similarity of 96.2%, which the SARS-CoV-2 shares with the SARS-like bat betacoronavirus, RaTG13, suggests its possible origin from bats [36–38]; the inability of the latter to directly infect humans necessitates a mutational event to occur to SARS-CoV-2, in an intermediate animal reservoir, prior to infecting humans.

Even though the scientific community is yet to arrive at a conclusive opinion, sequence similarity analysis points toward the possibility that Malayan pangolins may be the intermediate animal reservoir of SARS-CoV-2 [39,40]. Though many candidate reservoir hosts were suggested by various predictive and systematic sequencing and genetic analyses such as turtles and snakes, presumably from a live animal market from China [41–43], a study revealed the genomic sequence similarity of 100%, 98.6%, 97.8%, and 90.7% of E, M, N, and S genes, respectively, which tilted the balance in favor of pangolins betacoronavirus [44]. Moreover, there is striking sequence similarity (98.7%) between the receptor-binding motif (RBM) of SARS-CoV-2 and pangolins betacoronavirus (pangolin-CoV); specifically, there is no difference in the five amino acids critical for host binding [45].

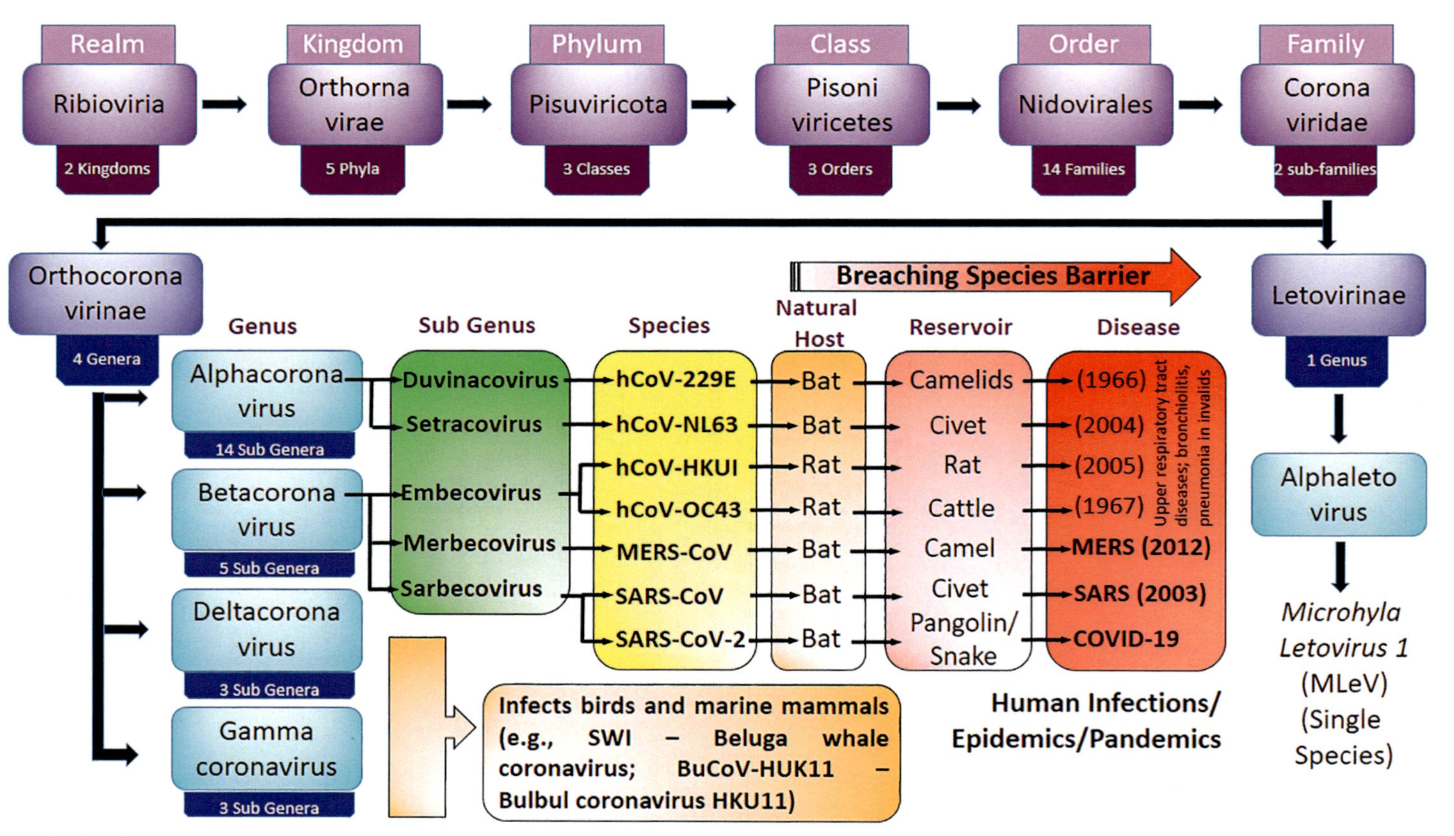

Fig. 1 Classification of coronaviruses. Modified from Hussein HA, Hassan RYA, Chino M, Febbraio F. Point-of-Care Diagnostics of COVID-19: From Current Work to Future Perspectives. Sensors. 20(15):4289. https://doi.org/10.3390/s2015428.

Though SARS-CoV-2 (COVID-19) exhibits similarity with SARS-CoV at the molecular level (79%) [46,47], they differ much in their pathophysiology. As per data published in the year 2020, COVID-19 possesses an increased transmission potential than SARS, quantified by an estimated R_0 value of 2.5 compared to an R_0 value of 1.9 for SARS [48]. This discrepancy can be explained by the lower extent of similarity (76%–78%) observed of the spike protein sequences of both the coronaviruses [49].

SARS-CoV-2 structure

The SARS-CoV-2 is an enveloped positive-sense, single-stranded RNA (+ ssRNA) virus. The virion is 50–200 nm in diameter, and the + ssRNA genome is approximately 29.9 kb in length with a 5′-cap structure and 3′-poly-A-tail and possess 14 putative open reading frames (ORFs) encoding 27 proteins [50]. SARS-CoV-2 comprises of structural proteins, namely the spike (S) protein, envelope (E) protein, membrane (M) protein, nucleocapsid (N) protein, and several nonstructural proteins (nsp) (Fig. 2). The N protein holds the RNA genome of the virus and plays a role in viral replication and transcription. This core is surrounded by an outer membrane envelope made of lipids with proteins inserted. The E and M proteins interact to form the viral envelope along with the S protein. The S protein is central to the binding to the host cell receptors to enable the entry of the virus into the host cell [51]. The S protein is a glycoprotein and comprises two functional subunits: the S1 subunit, which contains the receptor-binding domain (RBD), and the S2 subunit, which mediates the fusion machinery. The S protein is cleaved by the host cell proteases, which release the fusion peptide

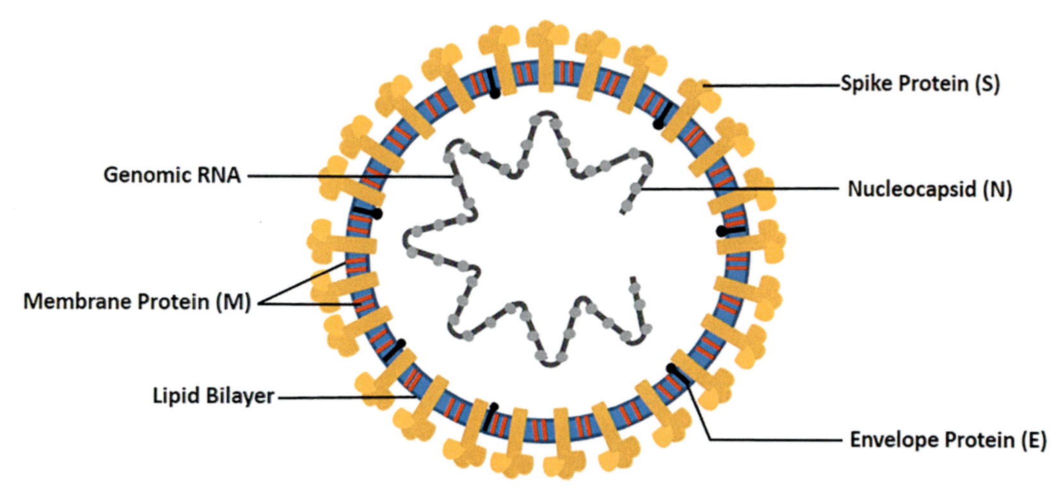

Fig. 2 Structure of the corona virus.

to facilitate membrane fusion and viral entry into the host cell [52,53]. The SARS-CoV-2 virus S protein binds to angiotensin-converting enzyme 2 (ACE2) receptors that are found on the surface of many human cells, including those in the lungs and intestine, via the RBD [54,55]. In the "down" state, these domains are usually hidden within the conformation of the S1 subunit and help immune evasion. The S protein in SARS-CoV-2 contains a polybasic cleavage sequence (PRRAR) at the S1–S2 boundary, which permits efficient cleavage by proprotein convertase furin. Preactivation of the S protein with proprotein convertase, furin, induces a conformational change in the trimeric S protein, exposing its RBD domains in an active "up" state and triggers the binding of the receptor-binding motif. Extensive conformational rearrangements of the S protein facilitate fusion of viral and cellular membranes and enables entry into the host cell via either cytoplasmic or endosomal membrane fusion. Furthermore, the use of furin reduces the dependence of the virus on host cell proteases, mainly cell surface serine protease TMPRSS2 (transmembrane serine protease 2) and lysosomal proteases cathepsin [56]. The furin-like cleavage site (FCS) in the spike protein (S) is absent in other βCoVs, such as SARS-CoV, and hence is speculated to be the reason for the high infectivity and transmissibility of the SARS-CoV-2 virus [57,58]. The virus on entry replicates via the replication and transcription complex (RTC) needed to synthesize the subgenomic RNAs as well as structural proteins (envelope and nucleocapsid). It hijacks the host ribosomes to replicate and then assembles in the endoplasmic reticulum. The full-length genome copies are incorporated into the newly produced viral particles. These new virions are exported from infected cells via the process of exocytosis and they go on to infect other cells. Meanwhile, the stress of viral production on the endoplasmic reticulum eventually leads to the death of the infected cell [52]. Hence the ability of the virus to bind to the host cell receptors is an important determinant of viral infectivity and pathogenesis.

Mutations occur in different regions of the virus, including the open reading frames (ORFs), the S protein, and the N protein. While most mutations are insignificant, few are reported to modulate viral transmissibility, replication efficiency, and virulence properties of the virus [59]. Currently, the reported mutants that might be more infectious are the Brazilian variant (P.1), the UK variant (B.1.1.7) [60], the South African variant (B.1.351), and the Double mutant Indian variant (B.1.617) [61,62].

Manifestation and transmission of COVID-19

COVID-19 affects different people in different ways. In some people, COVID-19 infection develops only mild to moderate illness and will recover without hospitalization. Most common symptoms include dry cough, tiredness, fever, and less common symptoms include sore throat, conjunctivitis, diarrhea, headache, aches and pain, loss of taste or smell, skin rashes, or discoloration of fingers or toes. Serious symptoms are difficulty breathing or shortness of breath, chest pain or pressure, loss of speech or movement. Those people who are healthy and with mild symptoms can manage their symptoms at home. When someone is infected with the virus, on average, it takes 5–6 days for symptoms to show and lasts up to 14 days.

Modes of transmission

Common modes of transmission for SARS-CoV-2 are contact, droplet, mother-to-child, animal-to-human transmission, airborne, fomite and fecal-oral. The SARS-CoV-2 infection causes respiratory illness, which ranges from mild to severe disease followed by death, and some people never develop symptoms even though they are infected with the virus.

Contact and droplet transmission

Transmission of SARS-CoV-2 can occur when an infected person coughs, sneezes, talks, or sings [63–65] or through infected secretions such as saliva and respiratory secretions or their respiratory droplets by direct, indirect, or close contact with infected people. When an infected person coughs or sneezes or comes in close contact (within 1 meter), they can transmit the virus and cause infection. When a person comes into contact with a susceptible host, with a contaminated object, or surface (fomite transmission), it is referred to as indirect contact transmission.

Airborne transmission

Airborne transmission: Spread of an infectious agent caused by the dissemination of droplet nuclei (aerosols) that remain infectious when suspended in air over long distances and time [66] is referred to as airborne transmission. Aerosols are generated during airborne transmission of SARS-CoV-2 in medical procedures [67]. In the absence of aerosol-generating procedures, SARS-CoV-2 can also spread through indoor settings with poor ventilation, as per WHO.

Fomite transmission

Respiratory secretions expelled by infected individuals which contaminate surfaces and objects are called fomites. In health care facilities where COVID-19 patients were being treated, viable SARS-CoV-2 virus can be found on those surfaces from hours to days, depending on the ambient environment (including temperature and humidity) and the type of surface, particularly at high concentrations [68,69].

Therefore transmission may also occur indirectly through objects contaminated with the virus from an infected person via a thermometer or stethoscope followed by touching eyes, nose, or mouth.

Other modes of transmission

SARS-CoV-2 RNA has also been detected in biological samples, including urine and feces of some patients [70]. Viral RNA has also been detected in either plasma or serum and is found to replicate in blood cells as per some studies. Low viral titers in plasma and serum suggest that blood-borne transmission remains uncertain, and the risk of transmission through this route may be low [70,71].

Prevention of transmission

The overall aim of the Strategic Preparedness and Response Plan for COVID-19 is to control COVID-19 by decreasing virus transmission and preventing death and associated illness. The virus primarily spreads through contact and respiratory droplets, and airborne transmission may occur, especially in indoor, crowded, and poorly ventilated settings. To prevent COVID-19 transmission, WHO recommends a comprehensive set of measures including,

- identifying suspected cases as early as possible, testing, and isolating all infected people to suitable facilities;
- identifying and quarantining all close contacts of infected people and testing those who develop symptoms so as to isolate them if they are infected and require care;
- use of fabric masks in public places where there is community transmission and prevention measures such as physical distancing is not possible;
- confirming that health workers caring for suspected and confirmed COVID-19 patients use contact/droplet and airborne precautions when aerosol-generating procedures are performed;
- strict and continuous use of a medical mask by health workers and caregivers working in all clinical settings, during all routine activities throughout the entire shift; and
- frequent practice of hand hygiene, avoiding close contact settings and crowded places, practicing physical distancing from others, using fabric masks in a closed environment, ensuring environmental cleaning and disinfection with good environmental ventilation.

The updated guidelines of US Centers for Disease Control and Prevention (CDC) recommend intensifying the protection by means of double masking to prevent the transmission and contracting COVID-19 from the newly emerging mutant strains. It is prescribed to wear a disposable mask underneath a multilayered cloth mask as well as knot and tie the ear loops of a 3-ply surgical mask while wearing it [72].

COVID-19 diagnosis

Effective management of the COVID-19 pandemic mandates timely detection and diagnosis of COVID-19 infection. It is critical for the treatment of the patient, to control transmission and for epidemiology analysis. The two main types of sample collected from the patient are (i) respiratory samples from the upper respiratory tract, e.g., nasopharynx or oropharynx, and less frequently from the lower respiratory tract, e.g., bronchoalveolar lavage fluid, and (ii) blood samples. The main testing strategies for COVID-19 are either molecular or serological and include (i) detecting viral RNA via amplification in nasopharyngeal or oropharyngeal swabs using the technique of real-time reverse transcriptase-polymerase chain reaction (RT-PCR), (ii) antigen testing, which detects viral proteins in the nasopharyngeal or oropharyngeal swabs, and (iii) antibody testing, which is a serological test that detects antibodies against the virus in the blood sample of the patient (Table 1) [73].

The diagnostic test method must have sufficient sensitivity and accuracy in order to avoid missing out on positive cases. The validity of a test is dependent on both its analytical and clinical sensitivity and specificity. Analytical sensitivity, i.e., the limit of detection, is the ability of the test to reliably detect the minimum concentration of a substance in a sample, and analytical specificity is the ability of the test to detect only the desired analyte in a specimen without cross-reacting with other substances. Clinical sensitivity measures how accurately a

Table 1 Comparison of different SARS-CoV-2 detection tests.

	Molecular	Antigen	Antibody
Test type	Nucleic acid (viral RNA) amplification	Detect viral proteins	Detects IgA, IgM, and IgG antibodies against SARS-CoV-2
Detection	Current infection with SARS-CoV-2	Current infection with SARS-CoV-2	Past exposure to SARS-CoV-2
Technology	RT-PCR, LAMP, CRISPR	Lateral flow	Lateral flow, ELISA, CIA, FIA
Sample type	Nasopharyngeal or oropharyngeal swab, bronchoalveolar lavage fluid	Nasopharyngeal or oropharyngeal swab	Plasma, Serum, Whole blood
Testing window	1–28 days after onset of symptoms. Optimal at 3–12 days	1–28 days after onset of symptoms. Optimal at 3–12 days	IgA, IgM: from day 5 to 21 after the onset of symptoms; IgG: from day 14 to 42 after the onset of symptoms

Modified from Chau CH, Strope JD, Figg WD. COVID-19 Clinical Diagnostics and Testing Technology. Pharmacotherapy: The Journal of Human Pharmacology and Drug Therapy. 40(8):857-868. https://doi.org/10.1002/phar.2439.

test identifies positive patients who are infected, i.e., a lower sensitivity test means higher false-negative results. Clinical specificity determines how accurately a test identifies negative patients who do not have COVID-19, i.e., a lower specificity test means higher false positives [73].

Molecular tests that amplify viral RNA are reported to have higher sensitivity than antigen tests [74]. Viral RNA extracted from the patient sample is used to make complementary DNA using the enzyme, reverse transcriptase. Using forward and reverse primers that flank the region of interest on the DNA strand, the desired gene sequence that is unique to the SARS-CoV-2 is amplified. RT-PCR is reliable and remains the gold standard for detection of the infection. The regions of interest include ORF1a/b, ORF1b-nsp14, non-structural RNA-dependent RNA polymerase (RdRp), S (Spike), E (Envelope), or N (Nucleocapsid) gene of SARS-CoV-2 [73]. The specificity of the test is high, while the sensitivity rate is around 66%–80% [75]. Alternate molecular test systems advocate the use of RT-LAMP (Reverse transcription loop-mediated isothermal amplification). For the amplification of target RNA, the reverse transcription (RT)-LAMP method can synthesize cDNA from target RNA and use LAMP technology to specifically amplify the resultant cDNA. LAMP exhibits high specificity and selectivity as it uses four primers (pairs of inner and outer primers), recognizing 6 distinct regions on the target base sequence. The technique exerts high amplification efficiency and can be completed in a short time as it performed under isothermal conditions as thermal denaturation of the DNA strand does not take place, hence the use of expensive thermal cyclers can be avoided [76]. CRISPR (clustered regularly interspaced short palindromic repeats) based molecular techniques have also been used for the detection of COVID-19. Although CRISPR/Cas is being used as a programmable tool for gene editing since 2013, the collateral cleavage activities of a unique group of Cas nucleases discovered recently are harnessed for in vitro nucleic acid detection. The most commonly adopted technique is that of CRISPER-based SHERLOCK (Specific High Sensitivity Enzymatic Reporter UnLOCKing) technique. The test also uses RNA purified from patient samples, as is used for other RT-PCR assays, and downstream processes involve isothermal amplification and detection of viral RNA using enzymes such as Cas13 [77,78]. The results are obtained in less than an hour and do not require elaborate instrumentation.

Antigen test kits enable rapid, point of care testing, which facilitates screening the population to enable reliable public health interventions to mitigate transmission and spread of the virus. These tests capture viral proteins in a rapid lateral flow format, and results are obtained within a short turnaround time of fifteen minutes and are qualitative. The nasopharyngeal or oropharyngeal swab is mixed with a solution that breaks the virus open and frees specific viral proteins. This is then added to a paper strip that contains an antibody tailored to bind to viral proteins. A positive test result can be detected either as a fluorescent glow or as a dark band on the paper strip. However, the test is less sensitive and specific than reverse transcriptase-polymerase chain reaction tests [74].

Antibody tests are serological tests that are used to detect IgA, IgM, IgG, or total antibody to COVID-19 in the patient. They are cheap, quick, and aid in rapid screening at points of care. Studies have shown that IgA and IgM antibodies can be detected as early as 5 days after a new infection, with an increase in the second and third week. Hence the test is not so effective in diagnosis during the initial stages of infection. IgG antibodies become detectable later in the infection course, around 7–14 days after symptom onset, and levels remained relatively high until 6 weeks [79]. The test kits use purified proteins of SARS-CoV-2, namely the N protein and S protein, to which antibodies in the patient's blood/serum/plasma bind. The kit operates on one of the following detection platforms: colloidal gold-based immunochromatographic assay (also known as lateral flow immunoassay LFIA), chemiluminescent immunoassay (CLIA), Fluorescence Immunoassays (FIA), and enzyme-linked immunosorbent assay (ELISA). Results are obtained within twenty minutes and are qualitative. The test also helps detect those who have had an infection and are now negative in the RT-PCR test. Kontoue et al. had reported in their meta-analysis of data from 38 studies involving 7848 individuals that antibody tests using the S antigen are more sensitive than those using the N antigen. They also highlighted that IgG tests perform better compared to IgM. A combined IgG/IgM test displayed better sensitivity than measuring either antibody type alone, with ELISA tests being more sensitive than lateral flow immunoassays. All methods had high specificity, while in terms of sensitivity ELISA- and CLIA-based methods perform better in terms of sensitivity (90%–94%) followed by LFIA and FIA with sensitivities ranging from 80% to 89% [80].

Radiological studies have shown that chest Computed Tomography (CT) scans are sensitive and specific and suggests their potential use in screening and diagnosing COVID-19. The characteristic CT findings of COVID-19 pneumonia include bilateral, peripheral, often-rounded ground-glass opacities predominantly located in the lower lobes and may be accompanied by consolidation [81]. The size and type of CT abnormalities are related to disease severity. Patients with mild type pneumonia have no obvious changes in the CT scans, while ground glass opacities occur in patients with common or severe pneumonia. Pulmonary consolidation is mainly found in severe and critical type patients, which can coexist with ground glass and fibrotic changes. However, the use of CT scans for the early diagnosis and screening of COVID-19 is still debatable. There is still a lacuna in the specificity of CT in differentiating COVID-19 pneumonia from other diseases with similar CT findings, thereby limiting the use of CT as a confirmatory diagnostic test.

Current treatment strategies

At the onset of the pandemic, there were no proven effective treatments or interventions for COVID-19. The rapid spread of COVID-19 across the globe, escalating mortality rates and nonavailability of any approved drugs or vaccines specifically for COVID-19, potentiated the need for adopting the strategies employed to tackle SARS-CoV and MERS-CoV to handle SAR-CoV-2 also. The initial efforts were aimed at containing the transmission of the disease by quarantining patients and their contacts at home, and monitoring for deterioration of the condition, if any. Symptomatic treatment was the main approach to handle the disease clinically upon weakening of the patient's health [82]. Oxygen supplementation is the standard mode of treatment, either with a nasal cannula or with mechanical ventilation for those with acute respiratory distress syndrome (ARDS). To avoid the possibility of generation of aerosols during the procedure and subsequent airborne transmission, nebulization is maximally avoided and done only with personal protection equipment (PPE) for the health care practitioners [83]. COVID-19 patients have an elevated risk of developing venous thromboembolism and related complications [84]. Hence it was recommended that anticoagulant pharmaceuticals be administered prophylactically in seriously ill hospitalized individuals [85]. Antiviral drugs such as Remdesivir and antiinflammatory drugs such as corticosteroids and acetaminophen were administered to alleviate symptoms of COVID-19 patients in hospital settings. Though the benefit of using both these drugs is arguable due to the possible risk of increasing viral replication [86,87], their use was not dismissed [88] and was recommended to be practiced in an individualized manner as part of clinical trials, taking into consideration the possible risks, the severity of the disease and comorbidities [82]. Other than these drugs, several bioactive compounds, nonantiviral drugs such as hydroxychloroquine, and monoclonal antibodies are also tried out to save critically invalid COVID-19 patients [89]. As per the International Clinical Trials Registry Platform (ICTRP), approximately 9400 clinical trials are being conducted to investigate different intervention methodologies such as repurposed drugs, novel therapeutics, and various vaccine candidates against COVID-19, as of April 29, 2021 [90].

Drug repurposing for pathogen targeting

The urgency of the pandemic necessitated the repurposing of drugs that had FDA approval and were already in use for other diseases [91]. Those drugs that have documented evidence of activity against viral diseases such as HIV, MERS, and SARS were the initial candidates. The repurposed drugs used against COVID-19 can be broadly classified into three based on the mode of action, namely cellular entry blockers, viral replication inhibitors, and immunotherapies [92]. Various aspects of the viral life cycle and host response pathways can be exploited to identify lead compounds that can be repurposed for anti-COVID-19 treatment strategies [19,82]. While most of the repurposed drugs come under the first two classes, immunoglobulins that can neutralize SARS-CoV-2 viral particles in the blood and inflammatory modulators come under the drugs/molecules repurposed for targeting COVID-19 through immunotherapy.

The drug classes used for repurposing against SARS-CoV-2 are varied, such as antivirals, antiparasites, antibiotics that inhibit viral replication, and small molecules inhibiting the cellular entry of the virus [92]. As a first step, WHO had initiated 'Solidarity Trial,' a concerted worldwide effort to carry out randomized clinical trials to evaluate the efficacy of 4 drugs, namely Remdesivir, Hydroxychloroquine, Lopinavir/Ritonavir combination, and Lopinavir/Interferon-beta combination against COVID-19, on March 18, 2020 [93]. Remdesivir is an adenosine analog developed against Ebola [94], targeting viral replication by competing with cellular nucleosides for the active site of RNA-dependent RNA polymerase (RdRp) [95]. Though the efficacy of Remdesivir against COVID-19 was unclear in the initial reports [96], it was permitted by the US FDA for use in hospitalized SARS-CoV-2 patients under emergency conditions [97]. Chloroquine and hydroxychloroquine are antimalarial drugs that can impair viral entry into cells by glycosylating the ACE2 receptor [98]. Lopinavir and Ritonavir are anti-HIV drugs that inhibit the 3C-like proteinase, a proteinase with a decisive role in the processing of viral polypeptides and assembly of coronaviruses and enteroviruses [99,100]. Clinical trials exploring the usage of a combination of these HIV drugs with interferon-beta, an antiinflammatory molecule, were also included in the initiative [101]. However, it was a very ambitious project, the interim results of the Solidarity Trial published by WHO report that all the studied treatment regimens didn't have any effect on the 28-day mortality rate of hospitalized COVID-19 patients [102].

Apart from the solidarity initiative, other RdRp inhibiting antiviral drugs such as Favipiravir [103] and ribavirin [104] are being investigated for their ability to inhibit SARS-CoV-2 replication [105]. Even though the results are yet inconclusive, it seems Favipiravir is effective for the treatment of mild-to-moderate disease and significantly reduced the time of clinical cure in COVID-19 patients [106]. Ritonavir, in combination with an HCV (Hepatitis C Virus) protease inhibitor, danoprevir, was assessed clinically for its antiviral effect and found promising without any adverse effects [107]. An antiviral drug that is known to inhibit viral entry into the cells is Arbidol. It is known to inhibit viral membrane fusion in the case of influenza virus and vesicle trafficking in the case of hepatitis C virus. A study by Wang et al.

has shown that it can prevent the mechanism of viral attachment and vesicular trafficking in the case of SARS-CoV-2 also [108]. Preliminary results from clinical studies show that Arbidol significantly improved clinical as well as laboratory parameters such as oxygen saturation, period of hospitalization, and ESR in COVID-19 patients [109]. K22 is a small antiviral molecule that inhibits the replication of a broad range of coronaviruses [110] via targeting membrane-bound RNA synthesis [111]. It is also hypothesized as a lead compound with potent antiviral activity against the COVID-19 virus [112].

Two related anticoagulant drugs used to treat pancreatitis, namely Camostat [113] and Nafamostat [114], are known to block the entry of SARS-CoV and hCoV-NL63 viruses into cells. They both are TMPRSS2 antagonists and inhibit their activity [115]. Recent works have shown that Camostat can inhibit the infection of the human lung by the SARS-CoV-2 virus [116] and possess antiviral activity [117,118]. Nafamostat mesylate also showed similar activity against the SARS-CoV-2 virus and is suggested as a treatment option for COVID-19 [119]. Another cellular entry blocker with potential application to treat COVID-19 is human recombinant soluble ACE2 (hrsACE2), developed initially to treat SARS-CoV by APEIRON Biologics AG in the trade name APN01 (alunacedase alfa) [120]. In vitro studies also confirmed that clinical-grade soluble ACE2 protected engineered cells from SARS-CoV-2 infection [121]. It has already advanced till Phase II of clinical trials for ARDS [122] and hence is being used for its ability to inhibit SARS-CoV clinically [123,124]. The phase II trial was conducted as a multicenter, randomized, placebo-controlled double-blind study involving 178 patients from 4 countries. Interim results released by the company state that APN01 treatment reduced the viral load and increased the ventilator-free days in COVID-19 patients [125].

Antiinflammatory molecules such as corticosteroids and interferons are also being repurposed to treat COVID-19 patient. NIH had recommended the usage of dexamethasone, a glucocorticoid, and other steroids on COVID-19 patients under artificial ventilation [126]. A randomized, open-label clinical trial named 'Randomised Evaluation of COVID-19 Therapy (RECOVERY)' was conducted to evaluate the effect of dexamethasone on hospitalized COVID-19 patients. Preliminary results suggest that the rate of mortality was lower among patients who received dexamethasone than those who received standard care only [127]. Clinical trials are being carried out by China to test the efficacy of INF-$\alpha 2\beta$ (NCT04293887) [128] and INF-α-ribavirin combination therapy (ChiCTR2000029387) [129] for COVID-19 treatment. Therapeutic combinations of repurposed statins and angiotensin receptor blockers (ARBs) are proposed to ameliorate host immune reaction to SARS-CoV-2 infection due to their immunomodulatory and inflammatory cytokine lowering characteristics [130].

Piyush et al. have explained in detail the prospects of using nucleic acid-based strategies to tackle COVID-19 [131]. Small interfering RNA (siRNA) against spike protein of SARS-CoV is reported to have virus replication-inhibiting activity in Vero-E6 cells [132]. Hence siRNAs and RNA interference can be exploited to develop treatment strategies against COVID-19 [133]. These siRNAs should be targeted against two different categories of molecules involved in the viral life cycle of SARS-CoV-2, namely (i) viral proteins involved in replication and maintenance such as protease-coding nsp5 [133], and (ii) host factors needed for viral entry and cellular trafficking such as ACE2 [134]. Fukushima et al. had developed a chimeric RNA-DNA hammerhead ribozyme against SARS, targeting the loop region and inhibiting RNA synthesis of the SARS-CoV virus [135]. This ribozyme can be exploited to tackle SARS-CoV-2 also [136]. Another chimeric protein, namely, dsRNA-activated caspase oligomerizer (DRACO), was identified as an antiviral therapeutic candidate against H1N1 by virtue of the presence of a dsRNA-binding domain which can attack viral dsRNAs [137]. It has been suggested as a potential treatment modality for COVID-19 in children [138].

Various antibiotics are also being investigated as candidate drugs for repositioning as a treatment modality for COVID-19. Results from a nonrandomized study indicate that hydroxychloroquine and azithromycin when used in combination in hospitalized patients can reduce mortality rate [139]. With earlier proven instances of effectiveness against dengue [140] and chikungunya virus [141], doxycycline, an antibiotic used to treat bacterial infections [142], was a primary candidate as the treatment option for high-risk COVID-19 patients. Various initial clinical experience and well-known safety profile prompted their widespread use [143]. There were earlier reports of azithromycin being used along with antiviral agents to manage MERS-CoV [144]. Another platform randomized, open-label, multiarm clinical trial (PRINCIPLE) conducted by the University of Oxford involving older COVID-19 patients showed that both azithromycin and doxycycline are ineffective against COVID-19. The study only recommends the use of antibiotics in cases where bacterial pneumonia is suspected [145]. Another clinical trial is in the pipeline using azithromycin in combination with antiparasite drug ivermectin and cholecalciferol to manage early stage COVID-19 disease [146]. A novel synthetic macrolide antibiotic showed antiviral activities against a broad spectrum of coronaviruses, including SARS-CoV-2 [147] and has been approved by FDA to carry out phase III trials in severe cases of COVID-19 [148]. Hence, clinicians should stick to judicious use of these drugs in the wake of the COVID-19 pandemic [149].

Various bioactive small molecules [89] are also considered for repurposing against COVID-19, such as luteolin and tetra-o-galloyl-beta-d-glucose (TGG), which find use in traditional Chinese Medicine

[150]. Published data shows that these compounds can inhibit SARS-CoV from entering Vero cells [151] and hence is considered as a treatment option for SARS-CoV-2 [152] and is suggested to alleviate the long-COVID-associated brain fog [153]. Similarly, several formulations currently in use for other ailments in Ayurvedic medicine were also investigated for their potential to treat and manage COVID-19 symptoms with considerable success [154,155]. AYUSH-64, a poly-herbal drug, formulated by Central Council of Research in Ayurvedic Sciences, India, is in use for the last four decades as an antimalarial and antiinflammatory drug [156]. A pilot study has already shown its efficacy and safety against influenza-like illness (ILI) [157]. In silico screening of the formulation revealed that it could be considered as a suitable candidate drug for repurposing against COVID-19 [156]. A recent report from a randomized control trial states that it can be used as an adjuvant therapy as part of standard care in the case of COVID-19 patients with mild-to-moderate symptoms [158,159].

Various drugs currently being used for other diseases are being repurposed to treat, alleviate symptoms, and manage early stage as well as severe COVID-19. Since these drugs are originally designated to be used for another ailment, their specific action against SARS-CoV-2 has not been studied in experimental animal models. Also, they may elicit potential toxicity when administered synergistically with other therapeutic agents. These factors should always be under consideration while evaluating the repurposing potential of any drugs against COVID-19 [19].

Immunomodulation in COVID-19

Immune response against SARS-CoV-2

SARS-CoV-2 invasion and pathogenesis are associated with the host immune response. The structure of the SARS-CoV-2 S protein and its binding affinity for ACE2 is very similar to that of SARS, although with minor differences when analyzed using cryogenic electron microscopy and surface plasmon resonance [54,160]. SARS-CoV-2 might transmit more readily from person to person since the affinity of SARS-CoV-2 S protein binding to ACE2 is 10 to 20 times higher than that of the SARS S protein [160]. The innate immunity is the first line of defense against during SARS-CoV-2 invasion. Cytolytic immune responses, mainly through the type I interferons (IFN) and natural killer cells, are recognized by pathogen-associated molecular patterns. During viral clearance via the adaptive immune response, activated cytotoxic T cells destroy virus-infected cells, while the antibody-producing B cells target virus-specific antigens. Patients with COVID-19, especially those with severe pneumonia, are reported to have higher plasma concentrations of inflammatory cytokines mainly, IL-6 and tumor necrosis factor (TNF), and substantially lower lymphocyte counts [64,161,162]. In critical patients, CD4 + T cells, CD8 + T cells, and natural killer cells were reduced when compared with those with mild disease symptoms [161]. Xu et al. have reported on the substantial reduction of CD4 + T cell, and CD8 + T cell counts in the peripheral blood in a patient who subsequently died of COVID-19. In patients with severe immune injury in the lungs, the proinflammatory subsets of T cells, including IL-17-producing CCR4 + CCR6 + CD4 + (T-helper 17 or Th17) cells and perforin and granulysin-expressing cytotoxic T cells, were increased [163].

Antiviral immune response might also cause massive production of inflammatory cytokines and damage to host tissues [164]. The overproduction of cytokines induced by aberrant immune activation is known as a cytokine storm. Cytokine storms are a major cause of disease progression and eventual death in the stages of coronavirus disease, including SARS, MERS, and COVID-19 [165]. Huang and colleagues found increased plasma concentrations of both Th1 (e.g., IL-1β and IFNγ) and Th2 (e.g., IL-10) cytokines [65]. Patients admitted to the intensive care unit (ICU) had higher plasma concentrations of TNF, macrophage inflammatory protein 1α, macrophage chemoattractant protein-1, IFNγ-induced protein-10 (IP-10), granulocyte-colony stimulating factor, IL-2, IL-7, IL-10 compared to those not admitted to the ICU. IL-6 concentrations were above the normal range in patients with severe symptoms of COVID-19 compared with healthy individuals and those with milder symptoms [161,166]. Secondary hemophagocytic lymphohistiocytosis (HLH) could be associated with severe COVID-19 cases [167]. The condition of uncontrolled cytokine storm and expansion of tissue macrophages or histiocytes that exhibit hemophagocytic activity is often referred to as hemophagocytic lymphohistiocytosis (HLH). HLH can result from other diseases such as infection, malignancy, and rheumatic disease (sHLH) or from genetic defects in cytolytic pathways (familial or primary HLH) [168]. In 1952 Farquhar and Claireaux first described cytokine storm in patients with HLH [169]. The characteristics of HLH, including multiorgan damage, hyperferritinemia cytopenias, unremitting fever, and hypercytokinemia, are commonly seen in seriously ill patients with COVID-19 [170]. It is suggested that alveolar macrophages expressing ACE2 are the primary target cells for SARS-CoV-2 infection. For reducing the mortality of patients with COVID-19, early identification and appropriate treatment of hyperinflammatory status is important [167].

Potential immunotherapy in COVID-19

Evidence has shown that in asymptomatic COVID-19 patients, the virus has a wider range of incubation time than initially thought (0–24 days), and these carriers can transmit the disease to others [171] with higher infectivity. Repurposing of approved drugs is commonly employed to fight against newly emerged diseases such as COVID-19, as these drugs have known pharmacokinetic and safety profiles. Several immune-modulating drugs that regulate different aspects of inflammation are being tested for their efficacy in the treatment of severe COVID-19 due to the importance of immune imbalance in the pathogenesis of SARS-CoV-2 infection. Immunosuppression might be beneficial to reduce mortality in patients with severe symptoms, and hyperinflammation is an important determinant of disease outcome in COVID-19 [163]. Patients at high risk of hyperinflammation can be screened using laboratory tests of sedimentation rate, erythrocyte counts, platelet counts, leukocyte counts, and ferritin. HScore can be applied for the evaluation of patients with sHLH to identify patients with COVID-19 at high risk of hyperinflammation. The HScore combines both laboratory and clinical parameters, including signs of immunosuppression, hemophagocytosis on bone marrow aspirate, body temperature, organomegaly, cytopenias, ferritin, fibrinogen, triglycerides, and serum aspartate aminotransferase [167]. For selecting appropriate immunosuppressants (e.g., tocilizumab could be considered in patients with high concentrations of serum IL-6), evaluation of cytokine profiles and immune cell subsets plays a vital role. The pros and cons of using an immunosuppressant on these patients should be carefully considered since antiviral immunity is required to recover from COVID-19. Viral load or replication and the severity of the hyperinflammation status need to be taken into consideration. Choosing selective instead of broad immunosuppressive drugs is one way to avoid the suppression of antiviral immunity. Unfortunately, there is not yet any definitive evidence with regard to the appropriate timing of administration of these agents since the timing of treatment is also crucial to reduce the side effects of immunosuppression. Further studies are required to determine the appropriate timing and routes of drug administration.

Biological immunomodulating drugs

IL-6 is elevated in patients with COVID-19, which is a key inflammatory cytokine that has a critical part in an inflammatory cytokine storm [161]. Tocilizumab, which is widely used in treating autoimmune diseases, such as rheumatoid arthritis, is a recombinant humanized monoclonal antibody against the IL-6 receptor [172]. IL-6-producing CD14 + CD16 + inflammatory monocytes were significantly increased in patients with COVID-19, and numbers of these cells were further increased in patients with COVID-19 admitted to the ICU [173]. IL-6 and GM-CSF might be potential therapeutic targets in patients with COVID-19 since hyperactivated Th1 cells producing granulocyte-macrophage colony stimulating factor (GM-CSF) and IFNγ in the lung promote IL-6-producing monocytes through the release of GM-CSF [174]. In patients with comorbidities, Tocilizumab is a first-line drug for the treatment of cytokine release syndrome (a rapid and massive release of cytokines into the blood from immune cells, usually caused by immunotherapy). Tocilizumab suppresses JAK-signal transducer and activates transcription (STAT) signaling pathway and production of downstream inflammatory molecules by binding to both the membrane and soluble forms of IL-6 receptor [175,176]. However, animal studies have shown that IL-6 is required for the clearance of viruses and control of pulmonary inflammation [177]. Hence clinicians can pay close attention to the possibility that blocking IL-6 could interfere with viral clearance or exacerbate lung inflammation. In severe COVID-19 cases administered with tocilizumab, improvements were seen in COVID-19 symptoms, peripheral oxygen saturation, and lymphopenia within a few days [178].

IL-1 pathway blockade is used for the treatment of some hyperinflammation conditions. Anakinra is an IL-1 receptor antagonist approved for cryopyrin-associated periodic syndrome, Still's disease, and rheumatoid arthritis. Anakinra administration significantly lowered 28-day mortality in patients who were septic with hyperinflammation, without increased adverse events as per phase 3 randomized controlled trial (RCT) for severe sepsis [179]. Anakinra treatment resulted in a 57% decrease of ferritin concentrations, and early initiation of anakinra was associated with reduced mortality in a retrospective analysis of 44 patients with sHLH [180]. Blockade of IL-1 seems a reasonable approach for the treatment of hyperinflammation in these patients since IL-1 was reported to be increased in some patients with COVID-19 [65,181]. Several trials of anakinra are currently underway, including anakinra and emapalumab (IFNγ inhibitor) in reducing hyperinflammation and respiratory distress in patients with COVID-19 and a two-third clinical trial evaluating the efficacy and safety.

Targeted synthetic immunosuppressants

Baricitinib is a small molecule compound that selectively inhibits the kinase activity of JAK1 and JAK2. It is approved for the treatment of rheumatoid arthritis [182] and psoriatic arthritis [183] and can be used in combination with one or more TNF inhibitors. SARS-CoV-2 binds to the ACE2 receptor on host cells and enters lung cells through receptor-mediated endocytosis. ACE2 is widely expressed in lung, heart, and renal cells. AP2-associated protein kinase 1 (AAK1) regulates endocytosis via phosphorylation of the clathrin adaptor protein AP2. Baricitinib can bind to cyclin G-related kinases, which also regulate receptor-mediated endocytosis and possess relatively mild side effects with the feasibility to achieve effective concentrations in the blood. In severe cases of COVID-19 with hyperactive immune status, the immunosuppressive function of baricitinib is beneficial where immune-mediated lung injury and ARDS might occur [184].

Ruxolitinib, which is used for the treatment of sHLH, is a promising inhibitor of JAK1 and JAK2. Ruxolitinib was shown to substantially improve liver function, fibrinogen, lactate dehydrogenase, and serum ferritin in a 38-year-old female patient with refractory Epstein-Barr virus-related sHLH [185] and improve hemodynamic function, respiratory and refractory HLH of the liver in an 11-year-old patient [186]. Ruxolitinib was well tolerated and manageable for treating sHLH with symptoms, and cytopenias improved in all patients within the first week of ruxolitinib treatment as per an open-label clinical trial. Concentrations of STAT1 phosphorylation, soluble IL-2 receptor, and ferritin were also reduced after the administration of ruxolitinib [187]. Animal studies showed that inhibition of JAK1 and JAK2 using ruxolitinib improved tissue inflammation, hypercytokinemia, thrombocytopenia, anemia, organomegaly, weight loss in animal models of both primary HLH and sHLH by reducing STAT1-dependent CD8 + T cell expansion [188]. Baricitinib and ruxolitinib could be potential treatments for the hyperinflammation seen in COVID-19 due to the similar hyperinflammatory nature of sHLH and severe COVID-19, JAK1 and JAK2 inhibitors [189]. Several registered RCTs are evaluating the efficacy of ruxolitinib and baricitinib in the treatment of COVID-19.

Role of anticoagulant, antiplatelet therapy, and statins in COVID-19

Anticoagulant therapy for patients admitted with COVID-19

Patients with SARS-CoV-2 infection are at increased risk of thromboembolic events, especially VTE (Venus thromboembolic disease), which is associated with the critical situation and immobilization entailed by this disease. Effective VTE prevention strategies become crucial since these ill patients are at increased thromboembolic risk. However, optimal therapeutic anticoagulation and prophylactic strategies during hospitalization are not clearly established.

Antiplatelet therapy for patients admitted with COVID-19

During the current SARS-CoV-2 infection pandemic, an apparent reduction in patients with the acute coronary syndrome is observed. To prevent the collapse of the health care system and avoid infections, scientific societies had recommended delaying nonurgent procedures [190]. In patients with ST-segment elevation acute myocardial infarction (STEMI), primary angioplasty is the preferred reperfusion strategy. However, in facilities where percutaneous coronary intervention is not available, fibrinolysis can be considered in non-infected patients with an estimated time of more than 120 minutes from diagnosis to coronary intervention. Similarly, in COVID-19 infected patients with poor clinical status or having a low risk of bleeding with symptom onset less than 3 h, fibrinolysis may be contemplated [190]. Under these considerations, there are a number of patients, with or without SARS-CoV-2 infection, who are admitted after an acute coronary syndrome and undergo coronary intervention.

In uninfected patients, current clinical practice guidelines are recommended with indications for antiplatelet therapy [191]. However, in patients with SARS-CoV-2 infection, there are two factors that lead to modifications for the antiplatelet therapy strategy. The primary factor is the prothrombotic component and high inflammation that this infection appeared to have, and the second factor is potential drug-drug interactions between COVID-19 drugs and antiplatelet agents.

Correlation between statins and COVID-19

Angiotensin-converting enzyme 2 (ACE-2) is the entry receptor of SARS-CoV-2 into host cells. Eighty-three percent of the cells account for Type II pneumocytes expressing ACE-2 in the lung. Extra pulmonary tissues such as the kidney, brain, heart, and liver express the ACE-2 receptor. Conversion of angiotensin II (AT-II) to angiotensin 1–7 (AT 1–7) by ACE-2 is an important regulatory enzyme in the renin-angiotensin system. AT 1–7 opposes the effects induced by AT-II, with vasodilating actions, antiinflammatory, antifibrous, and antioxidative stress. Downregulation of ACE-2 occurs when SARS-CoV-2 infection is in the most severe stages. This effect can increase the likelihood of lung injury, which can be fatal in some cases. ACE-2 plays a double role in COVID-19 infection, initially as a protector against the damaging effects of hyperinflammatory response and later as an entry receptor for SARS-CoV-2. Whether the use of drugs such as angiotensin receptor blockers (ARB) and ACE inhibitors (ACEi) may represent a COVID-19 risk factor, or on the contrary, protection has led the scientific world to investigate this issue. Little has been said about the effects of statins to modify ACE-2 concentrations. The presence of preexisting comorbidities, such as dyslipidemia, diabetes, and cardiovascular disease, are COVID-19 risk factors at an advanced age [55,192]. Statins have been the first-choice therapy in the treatment of hypercholesterolemia for years. In addition to their effect of reducing LDL concentrations, known pleiotropic actions are attributed, including antifibrotic actions [193]. Severe lung injury from COVID-19 occurs as a result of pulmonary fibrosis [194]. Some evidence associates pulmonary antifibrotic effects with statins [195]. ACE-2 expression has been increased in treatment with statins [196]. ACE-2 is the input receptor of SARS-CoV-2, but it also has a protective role against virus injury, especially in organs such as the lungs. To date, it is not clear how the clinical results in patients with COVID-19 are affected by the use of statins, alone or in combination with ACEi and ARB. In a clinical trial conducted in Cambridge, Massachusetts, a cholesterol-lowering drug called atorvastatin is currently pursued to check the effects of statins on COVID-19 [197].

Convalescent plasma therapy

Treatment modalities to manage COVID-19 also include the use of convalescent plasma, which is the plasma that is collected from individuals following resolution of infection and development of antibodies [198]. Historically, convalescent plasma had been used as a passive immunization strategy to treat viral diseases such as poliomyelitis, measles, mumps, Spanish flu, West Nile virus and have also been adopted recently to manage Ebola virus disease, SARS, and MERS [199–202] and therefore is expected to have potential also in the treatment of COVID-19. Convalescent plasma is a source of antiviral neutralizing antibodies which bind to the receptor-binding domain (RBD) of the virus, thus preventing them from entering the host cell and mitigates viral replication. Although RBD is the main target of interest, some neutralizing or blocking antibodies recognize other epitopes, including domains in the S1 subunit, S-ectodomain, HR1 and HR2 domains in the S2 subunit, nucleoprotein (NP), or envelope (E) protein [203]. Nonneutralizing antibodies in the plasma could play a role in various immune pathways. These immunoglobulins bind to the virus but do not affect its ability to enter the host cells. They induce cellular destruction via mainly antibody-dependent cellular cytotoxicity, antibody-dependent cellular phagocytosis, and complement activation. The plasma is also a store house of immune-modulatory proteins, including antiinflammatory cytokines, defensins, pentraxins which might have a role in alleviating systemic inflammatory response syndrome, which is the main cause for acute respiratory distress syndrome and COVID-19-related pneumonia [204].

The donors of convalescent plasma are usually COVID-19 survivors of age 18–65 years and have been tested COVID-19 negative after 14 days from recovery. The plasma is collected via apheresis wherein the selective blood components, mainly plasma, are separated from the blood by continuous centrifugation, and the remaining components are returned to the donor. Approximately 400–800 mL of plasma can be obtained from a single donor and is frozen within 24 h in single units of 200 to 250 mL. The survivor is initially tested for transfusion-transmitted infections, mainly human immunodeficiency virus (HIV), hepatitis B, hepatitis C, and syphilis and only those that are negative are chosen to be donors for convalescent plasma. Currently, the recommendation is to administrate 3 mL/kg with a titer of greater than 1:64 [198,205].

Li et al. conducted an open-label, multicenter, randomized clinical trial of 103 severe and critical COVID-19 patients, with convalescent plasma treatment performed in 7 medical centers in Wuhan, China. The treatment did not result in significant clinical improvement or survival within 28 days. This may be because treatment was initiated at least 14 days after the onset of symptoms, indicating the potential importance of early clinical intervention for more effective results [206]. In another multicenter, open-label study, patients receiving convalescent plasma treatment within 3 days or a delayed administration after 4 days of COVID-19 diagnosis showed a significant improvement in survival associated with earlier administration [207]. Agarwal et al. conducted a study across India to determine the effectiveness and safety of convalescent plasma therapy in patients with moderate COVID-19 who had been admitted to hospitals. The study was an open labeled, parallel-arm, phase II, multicenter, randomized controlled trial of 464 adults conducted in a total of 39 hospitals from both the public and private sectors (Clinical Trial Registry of India CTRI/2020/04/024775) [208]. The outcome of the trial, however, showed that convalescent plasma treatment had limited effectiveness. It did not cause a reduction in progression to severe COVID-19 or mortality. The treatment was, in some cases, able to effect an earlier resolution of shortness of breath and fatigue and higher negative conversion of SARS-CoV-2 RNA on the seventh day of enrollment. The authors also conclude that effective titers of antiviral neutralizing antibodies, optimal timing for donation, the commencement of convalescent plasma treatment, and the severity class of patients who are likely to benefit from convalescent plasma needs are important parameters that determine the treatment outcome and warrant further in-depth studies [209].

Cell-based therapy

SARS-CoV-2 infection can lead to pneumonia, multiorgan failure, severe acute respiratory syndrome, septic shock, and even death in severe cases. Patients with severe COVID-19 disease exhibit severe pneumonia, acute respiratory distress syndrome (ARDS), multiorgan failure, and death in extreme cases [19]. Regenerative medicine aims to repair, regrow, or replace damaged tissues and organs. In the perspective of the current COVID-19 pandemic, regenerative medicine has been adopted to offer various cellular therapeutics, mainly that of mesenchymal stem cell (MSC) therapy, natural killer (NK) cell therapy, exosomes, and tissue products [210].

Mesenchymal stem cells are regenerative cells that can be harvested from different adult tissue types, mainly bone marrow, umbilical cord, adipose, and endometrium. MSCs are reported to have immunomodulatory and antiinflammatory properties in addition to their self-renewal properties. They can modulate the cells of both the innate and the adaptive immunity and exert their effects by secreting trophic factors, cytokines, and chemokines [211]. MSCs have inhibitory effects on different immune cells, mainly T lymphocytes, B lymphocytes, NK natural killer cells, and dendritic cells. MSCs can modulate the cytokine storm by reducing the concentration of the proinflammatory cytokines IL1α, TNFα, IL6, IL12, and IFNγ [210,212]. Li et al. had demonstrated that intravenous injection of mouse bone marrow-derived MSC into H9N2 virus-infected mice could significantly attenuate pulmonary inflammation. This action was effected by lowering chemokine (GM-CSF, MCP-1, KC, MIP-1α, and MIG) and proinflammatory cytokine (IL-1α, IL-6, TNF-α, and IFN-γ) levels and reducing inflammatory cell recruitment into the lungs [213]. In another experiment using H5N1 infected aged mice, Chan et al. demonstrated that MSC administration 5 days postinfection reduced histological lung injury, lung edema, and pro-inflammatory cytokines [214]. Zheng et al. had conducted a randomized placebo-controlled pilot study of adipose-derived mesenchymal stem cells in patients suffering from ARDS. The results showed no adverse effect of MSC therapy, although the clinical outcomes in reducing cytokine level were not significant [215]. Considering the immunomodulatory and antiinflammatory properties of MSC, their use as a therapeutic in treating COVID-19 is extensively studied. In an early case study, a critically ill COVID-19 patient with severe pneumonia, respiratory, multiorgan failure, and on a ventilator was treated with 3 doses each of 50 million allogeneic umbilical cord stem cells, three days apart along with conventional therapy. The patient showed gradual improvement of vital signs with the treatment and had survived [216]. In a recent study, seven patients with severe COVID-19 were assessed for 14 days after intravenous MSC injections. The patients showed an increase in lymphocyte count, a decrease in inflammatory markers and cytokines, and an increase in antiinflammatory cytokine (IL-10). The authors reported that treatment with intravenous MSCs could cure or significantly improve the functional outcome of seven patients without observed adverse effects. Further, the peripheral lymphocytes were increased, the C-reactive protein decreased, and the overactivated cytokine-secreting immune cells CXCR3 + CD4 + T cells, CXCR3 + CD8 + T cells, and CXCR3 + NK cells disappeared in 3–6 days [217]. Currently, a number of clinical trials are ongoing, and there are many published reviews on the potential of mesenchymal stem cell therapies for treating COVID-19 [218,219].

Mesenchymal stem cell-derived products like extracellular vesicles, including microvesicles, nanovesicles, and exosomes have also been considered cell-free therapeutics as they exhibit immunomodulatory and reparative effects. Microvesicles arise from the blebbing of the cellular plasma membrane and are in the size range of 100–1000nm. Exosomes are smaller within 30–100nm and arise by the fusion of intravesicular bodies of the plasma membrane, while nanovesicles are artificially produced by serial extrusions through filters [218,220]. These extracellular vesicles carry bioactive molecules such as nucleic acids, microRNA, lipids, and proteins which are shuttled to recipient cells and modulate their behavior. Hence they act as modes of intracellular communication. Exosomal particles secreted by mesenchymal stem cells can be used for therapy, as such, and, alternatively, specific microRNAs, mRNA, and drugs can be incorporated and used as therapeutics [221]. Extracellular vesicles have demonstrated potential as acellular therapies in experimental models of ARDS and sepsis. Park et al. had generated nanovesicles from human bone marrow-derived MSCs by subjecting the cells to serial extrusion through micro-sized filters. The authors demonstrated in vitro that the nanovesicles are taken up by macrophages and could significantly reduce the production of pro-inflammatory cytokines from the cells activated by the gram-negative outer membrane vesicle (OMV). Further in vivo studies in a mice model of peritoneal sepsis caused by gram-negative outer membrane vesicle demonstrated that MSC nanovesicles are capable of attenuating the systemic response to sepsis with the maintenance of body temperature and a reduction in circulating cytokines [220].

Wang et al. demonstrated in an in vivo mice model of lipopolysaccharide-induced lung injury that systemic and intratracheal administration of mesenchymal stem cell-derived extracellular vesicles could alleviate lung injury via the downregulation of nuclear factor kappa B subunit 1 (NFKB1). The MSC-derived extracellular vesicles contained miR-27a-3p, a microRNA whose target is NFKB1 [222]. A few preclinical and clinical studies on MSC-derived exosomes for treating COVID-19, including management of associated cytokine storm, are being undertaken. Sengupta et al. have reported an open-labeled cohort study investigating exosomes derived from allogenic bone marrow MSCs (ExoFlo^{TM}) as a treatment for severe COVID-19 pneumonia. In the study, 24 patients with moderate to severe COVID-19 were administered a single 15mL of Exoflo through intravenous infusion. Their results indicated no adverse reactions, and the patients' clinical status and oxygenation improved after one treatment [223]. However, more multicenter, controlled, randomized trials will be needed to adequately assess the future of MSCs as well as exosomes in the treatment of COVID-19. Literature is replete with a detailed outline of the current clinical trials that are being undertaken in using MSC and their exosomes as therapeutics [221,224].

Natural killer (NK) cells have also been used as an alternative cellular therapy for treating COVID-19. NK cells play a pivotal role in bridging the innate and adaptive immune system and in modulating the body's response to viral infections. Physiologically NK cells are found in circulation and form 10%–15% of the total peripheral blood leukocytes in humans [225]. They are recruited to the site of infection via a chemokine gradient, where they facilitate viral clearance. The NK cells secrete cytokines and induce apoptosis of target cells through receptors such as TRAIL and Fas or induce cytotoxicity through Ca^{2+}-dependent exocytosis of cytolytic granules, mainly perforin and granzymes [226]. These cells also promote interferon production, which is essential for antiviral activity [227]. It has been reported that persons with NK cell dysfunction or compromise are more susceptible to viral infections [228]. Earlier studies evaluating the NK phenotype during a viral infection such as SARS and MERS have shown a reduction in circulating NK cell numbers and could be a reason for the progression of the disease [229,230]. Zheng et al. had evaluated the levels of NK cells in the peripheral blood of COVID-19 patients and observed that the levels inversely correlated with disease severity. Patients with severe disease had lower numbers of circulating NK cells and displayed increased expression of the inhibitory receptor NKG2A, while patients recovering from the infection had a higher number of circulating NK cells, and these cells had lower NKG2A expression [231]. Since NK cells are the first responders in clearing viral infections, a low number of circulating NK cells may also facilitate the development of secondary infections [226].

Furthermore, the ability of NK cells to exert cytotoxic effects on infected cells and induce interferon production makes them hold potential as cellular therapeutics for COVID-19 treatment. Clinical trials of cryopreserved allogenic NK cell therapy for COVID-19 patients are being conducted by "Celularity." The trial (NCT04365101) includes phase I and phase II studies [226,232]. Chimeric antigen receptor NK cells (CAR-NK cells) have been genetically engineered to express receptor(s) of interest and were originally designed to enhance the ability of NK cells to eliminate cancer cells [233]. A Phase I/II study in early stage COVID-19 patients (within 14 days of illness) employing CAR-NK cell therapy is currently being tested using NK cells derived from human umbilical cord blood and expressing NKG2D and ACE2 (NCT04324996). It is hypothesized that NKG2D-ligands (NKG2DL) are upregulated on virally infected cells, and expressing ACE2 on NK cells will facilitate the elimination of SARS-CoV-2 virions and infected cells by binding to the viral spike protein [226,234]. The potential of NK cell therapy for treating COVID-19 infections is still being researched and warrants efficient clinical trials to vet their efficacy and safety.

Dendritic cells, due to immunotherapy and cell-based vaccination, are the next cell-based therapy candidate for COVID-19 treatment. In the USA, Aivita Biomedical company, as an immuno-oncology company designing personalized vaccines, has recently started an adaptive Phase IB-II trial (Identifier: NCT04386252) of a vaccine consisting of autologous dendritic cells loaded with antigens from SARS-CoV-2, with or without GM-CSF, to prevent COVID-19 in adults.

Vaccines

The high rate of person-to-person transmission of the SARS-COV-2 virus will continue to potentiate the COVID-19 pandemic if left unchallenged. The primary way to control the spread of the virus is the strict adherence to personal protective measures, as recommended by WHO. The development of an effective vaccine is the mainstay in containing the COVID-19 viral pandemic [92]. Various research groups around the globe rose to the urgency of the situation and have worked hard to develop a safe and effective vaccine accessible to the masses at the earliest. They exploited the multitude of technology platforms and scientific data available in the public domain and have primarily focused on the methodologies that had proved successful in formulating a vaccine against SARS [235]. Most of the attempts considered the membrane-bound spike (S) protein of SARS-CoV-2 as the molecule of interest toward vaccine development against COVID-19 [236].

Various COVID-19 vaccine initiatives were instituted by international agencies in collaboration with individual governments and biopharmaceutical companies, namely COVAX, COPTN (COVID-19 Prevention Trials Network), ACTIV, and OWS (Operation Wrap Speed). COVAX is a collaborative alliance between WHO and Gavi—the Vaccine Alliance and the Coalition for Epidemic Preparedness Innovations (CEPI) as part of the World Health Organization's (WHO) Access to COVID-19 Tools (ACT) Accelerator program, aimed at providing low-cost vaccines worldwide [237]. Although the development of a vaccine against RNA viruses is challenging and has inherent risks associated with it [238], more than 300 vaccine projects were initiated as of April 30, 2021. Of these, 14 vaccines are approved for community use, and 60 others are in various phases of the clinical trial pipeline [237]. Most of the candidate vaccines come under four major categories of vaccines, namely, inactivated or attenuated virus vaccine, protein-based vaccines, viral vector vaccines, and gene vaccines. A list of COVID-19 vaccines, from around the world, authorized for administration is given in Table 2.

Of all the vaccine candidates, the first vaccine to get an Emergency Use Listing (EUL) from WHO is the mRNA vaccine, Comirnaty, from Pfizer, on December 31, 2020. It had already procured an Emergency

Table 2 Vaccine candidates currently authorized for administration against SARS-CoV.

Vaccine type	Vaccine name	Manufacturers	Country	Specific details
Inactivated vaccine	Covaxin (BBV152)	Bharat Biotech, ICMR[a]; In US[b]—Ocugen, Bharat Biotech	India	Adjuvanted (Algel-IMDG) whole-virion vaccine of inactivated SARS-CoV-2 [239]; 2 doses 28 days apart; 78% efficiency in ongoing Phase III trial (NCT04471519) [240]
Inactivated vaccine	CoronaVac (PiCoVacc)	Sinovac	China	Formalin-inactivated SARS-CoV-2 adjuvanted with alum [241]; 2 doses 14–28 days apart; varying efficacy reports (above 50%) [242]; ongoing phase III trials (NCT04456595—Brazil, NCT04582344—Turkey, and NCT04508075—Indonesia)
Inactivated vaccine	WIBP-CorV	Wuhan Institute of Biological Products, China National Pharmaceutical Group (Sinopharm)	China	Adjuvanted (alum) vaccine of inactivated SARS-CoV-2 vaccine [243]; 2 doses 21 days apart; 72.5% efficiency [244]; ongoing Phase III trials on Peru, Morocco, and United Arab Emirates
Inactivated vaccine	CoviVac	Chumakov Federal Scientific Center for Research and Development of Immune and Biological Products of the Russian Academy of Sciences	Russia	Inactivated whole-virion SARS-CoV-2 vaccine; 2 doses 14 days apart; Phase I/II trials over, approved for use in Russia [245]; yet to start later-stage trials [237]
Inactivated vaccine	QazVac (QazCovid-in)	Research Institute for Biological Safety Problems	Kazakhstan	Inactivated SARS-CoV-2 vaccine; 2 doses 21 days apart; 96% efficiency [246]; ongoing Phase III trial (NCT04691908) [237]
Recombinant Peptide vaccine	EpiVacCorona	State Research Center of Virology and Biotechnology (VECTOR)	Russia	Three separate short peptide sequences of SARS-CoV-2 spike protein chemically synthesized and conjugated with a larger recombinant carrier protein (fusion of a viral nucleocapsid protein and a bacterial maltose-binding protein [MBP]), adjuvanted with aluminum hydroxide; 2 doses 21–28 days apart; efficacy data not available; ongoing Phase III trial (NCT04780035) [247,248]
Recombinant Peptide vaccine	ZF2001	Anhui Zhifei Longcom Biopharmaceutical Co. Ltd., Institute of Microbiology of the Chinese Academy of Sciences	China, Uzbekistan	Recombinant dimer of the receptor-binding domain (RBD) of the SARS-CoV-2 S protein (RBD-dimer) with aluminum hydroxide as adjuvant; 3 doses 30 days apart; 97% efficacy [249]; ongoing Phase III trials (NCT04646590—China and Uzbekistan)
Recombinant adenovirus vaccine (adenovirus type 5 vector)	Convidicea (Ad5-nCoV)	CanSino Biologics	China	Replication-incompetent adenovirus type 5 (Ad5) vector encoding SARS-CoV-2 spike glycoprotein protein [250]; 1 dose; 65.7% efficiency in ongoing Phase III trial (NCT04526990) [251]

Recombinant adenovirus vaccine (adenovirus type 5 vector)	BBIBP-CorV	Beijing Institute of Biological Products, China National Pharmaceutical Group (Sinopharm)	China	Replication-incompetent adenovirus type 5 (Ad5) vector encoding SARS-CoV-2 spike glycoprotein protein [252]; 2 doses 21–28 days apart; 86% efficiency [253]; ongoing Phase III trials (ChiCTR2000034780—China and NCT04560881—Argentina)
Heterologous recombinant adenovirus vaccine (rAd26 and rAd5)	Sputnik V (Gam-COVID-Vac)	Gamaleya Research Institute of Epidemiology and Microbiology, Acellena Contract Drug Research and Development	Russia	Replication-incompetent recombinant adenovirus (rAd)-based vaccine combining rAd type 26 (Ad26) and rAd type 5 (rAd5) vectors as two separate IM injections carrying full-length SARS-CoV-2 spike protein controlled mutated version of full-length spike protein; 2 doses 21 days apart; 91.6% efficiency in ongoing Phase III trial (NCT04530396) [254]
Nonreplicating Adenovirus vector vaccine	COVID-19 Vaccine AstraZeneca (AZD1222 ChAdOx1); Vaxzevria (Europe) and Covishield (India)	AstraZeneca, Oxford University (In India—AstraZeneca, Serum Institute of India)	United Kingdom	Replication-incompetent chimpanzee adenovirus vector encoding SARS-CoV-2 spike protein; 2 doses 4–12 weeks apart; 76% efficiency in ongoing Phase III trial (NCT04516746) [255,256]; rare and mild thrombotic complications
Nonreplicating Adenovirus vector vaccine	COVID-19 Vaccine Janssen (JNJ-78436735; Ad26. COV2.S)	Janssen Vaccines (Johnson & Johnson)	The Netherlands & US	Replication-incompetent human adenovirus type 26 (Ad26) vector encoding prefusion-controlled mutated version of full-length spike protein; 1 dose; 66% efficiency in Phase III trial (NCT04505722) [257]; rare thrombotic complications
mRNA-based vaccine	Comirnaty (BNT162b2)	Pfizer, BioNTech; Fosun Pharma (in China)	Multinational	LNP[c] -encapsulated modRNA[d] coding for prefusion-controlled mutated version of full-length spike protein; 2 doses 21 days apart; 95% efficiency in Phase III trial (NCT04368728) [258]
mRNA-based vaccine	Moderna COVID-19 Vaccine (mRNA-1273)	Moderna, BARDA[e] , NIAID[f]	US	LNP-encapsulated modRNA coding for prefusion controlled mutated version of full-length spike protein; 2 doses 28 days apart; 94.1% efficiency in Phase III (NCT04470427) [259]

[a] ICMR—Indian Council of Medical Research.
[b] US—United States of America.
[c] LNP—Lipid nanoparticle.
[d] modRNA—nucleoside-modified RNA.
[e] BARDA—Biomedical Advanced Research and Development Authority.
[f] NIAID—National Institute of Allergy and Infectious Diseases.
Modified from Craven J. COVID-19 vaccine tracker. Regulatory FocusTM. https://www.raps.org/news-and-articles/news-articles/2020/3/covid-19-vaccine-tracker. Published 2021. Accessed May 1, 2021.

Use Authorization (EUA) from US FDA for use in people 16 years and older on December 12, 2020. This was followed by EUL for the AstraZeneca/Oxford COVID-19 vaccine, Covishield, on February 15, 2021, and later for the COVID-19 vaccine Janssen by Janssen Vaccines (Johnson & Johnson) on March 12, 2021 [260].

Comirnaty vaccine, formerly designated as BNT162b2, is a nucleoside-modified mRNA (modRNA)-based [261] vaccine developed collaboratively by Pfizer and BioNTech. The vaccine contains an mRNA molecule coding for full-length SARS-CoV-2 spike protein mutated at two proline positions to stabilize it in the prefusion conformation [262] as observed in intact virus under cryo-EM [160]. This lipid nanoparticle (LNP)-encapsulated vaccine was given intramuscularly, 21 days apart, to subjects 16 years and older. While the phase I trial confirmed the safety and immunogenicity of the vaccine [262], the phase III trial with 43,448 participants demonstrated 95% efficiency of the vaccine in preventing COVID-19 infection [258]. The vaccine was also found effective in neutralizing SARS-CoV-2 variants such as B.1.1.7 [263] and B.1.351, though efficacy was two-third less in the case of the latter [264].

Three vaccines that have received sanction for use in the USA are Comirnaty from Pfizer, Janssen from Janssen Vaccines, and Moderna COVID-19 Vaccine from Moderna. While the Janssen vaccine is a single shot nonreplicating viral vector vaccine, the Moderna vaccine is a two-dose mRNA-based vaccine [265]. Moderna vaccine is also a lipid nanoparticle formulated, prefusion-stabilized mRNA vaccine against SARS-CoV-2 spike protein, developed by Moderna and the National Institute of Allergy and Infectious Diseases (NIAID), USA [266]. The phase III clinical trial was carried out at 99 different centers around the USA in a randomized, observer-blinded, placebo-controlled manner, with 30,420 participants receiving two intramuscular injections of the vaccine or the placebo in a 1:1 ratio, 28 days apart. Analysis 14 days after the second injection confirmed the safety and efficacy (94.1%) of the vaccine in preventing even severe COVID-19 illness [259]. The research group reports that the effectiveness of the vaccine in preventing infection from the South African variant, B.1.351, is inconclusive [267].

COVID-19 Vaccine Janssen (formerly called JNJ-78436735, Ad26.COV2.S) is a recombinant viral vector vaccine developed using the Janssen Vaccine's previously established vaccine platform systems for the Ebola vaccine, namely AdVac and PER.C6 [268]. It uses a replication-incompetent human adenovirus type 26 (Ad26) vector, which is designed to encode a prefusion-stabilized version of full-length SARS-CoV-2 spike protein (Ad26.COV2.S). Initial preclinical and clinical experiments established the safety and immunogenicity of the vaccine, with both humoral and cellular responses being stimulated [269]. Interim results from the international phase III clinical trial (ENSEMBLE) involving 43,783 participants, using a single dosage of the vaccine, showed that it is successful in protecting against COVID-19 disease, with 66.9% efficacy for moderate disease and 76.7% efficacy for the severe critical illness after 14 days of administration [257]. By the month of April 2021, there emerged some reports regarding the incidence of cerebral thrombosis and thrombocytopenia after vaccination, prompting FDA and CDC to halt the immunization program. After thorough examinations by the authorities and based on further studies [270], the pause was lifted [271]. European Medicines Agency (EMA) recommended that the vaccine information be updated with a statutory warning of a possible rare complication of "unusual blood clots with low blood platelets." [272]

The COVID-19 Vaccine AstraZeneca (previously denoted as AZD1222, ChAdOx1) was jointly developed by AstraZeneca and the University of Oxford under the trade name Vaxzevria. It is developed under the label Covishield in India by Serum Institute of India and AstraZeneca. It is also an adenoviral vector vaccine based on a non-replicating chimpanzee adenovirus (ChAdOx1) containing the SARS-CoV-2 structural surface glycoprotein, also called spike protein [273]. The results released by the manufacturers about the ongoing phase III clinical trial, via administering two doses of the vaccine 4–12 weeks apart, claim 76% efficiency for the vaccine [255]. The interim results from clinical trials conducted in UK, Brazil, and South Africa, available as preprint in 'The Lancet,' also confirm the observations with an efficacy of 76% and no adverse effects [256]. Scientific reports suggest that while the AstraZeneca vaccine is equally effective against the B.1.1.7 variant [274], it showed only negligible efficiency in protecting against mild-to-moderate cases of COVID-19 caused by the B.1.351 variant [275]. Though some instances of blood clotting complications [276] were reported after vaccine administration, in March 2021, WHO has issued a guidance stating that the benefits of the AstraZeneca vaccine outweigh these mild and rare complications [277].

Covishield was approved for use in India on January 03, 2021, along with another vaccine candidate, Covaxin, by Bharath Biotech. Central Drugs and Standards Committee (CDSCO) has given the sanction for "restricted use in an emergency situation" based on the recommendation by a Subject Expert Committee, who considered the facts such as the comparable efficacy of the vaccines in Indian volunteers and the availability of vaccination options at the wake of the emergence of mutant strains. In the case of Covaxin, approval was granted under "clinical trial mode." Currently, free vaccinations have been given for about three crore COVID-19 frontline warriors, including health care personnel, followed by senior citizens and persons with other health issues above 45 years of age [278].

Covaxin, also designated as BBV152, is a vaccine consisting of whole-virion inactivated SARS-CoV-2, synthesized in India by Bharat Biotech in collaboration with the National Institute of Virology under the Indian Council of Medical Research (ICMR). The inactivated virus is formulated along with an adjuvant, Algel-IMDG, consisting of a toll-like receptor 7/8 agonist imidazoquinoline molecule adsorbed chemically on to alum (Algel) [239]. This adjuvant helps direct the antigen to the draining lymph nodes than to systemic circulation and enthuse cell-mediated immune response [279]. Preliminary reports of the phase III trial from April 21, 2021, which is recruiting 26,000 volunteers, state that Covaxin showed an overall 76% efficacy in protecting against COVID-19 and a 100% efficacy against the disease getting severe [240]. The vaccine is effective against the B.1.1.7 variant [280] and the Indian double mutant variant, B.1.617 [62].

Russia's Sputnik V vaccine received authorization for use in India lately in April 16, 2021 [281], phase III clinical trials and distribution will be managed by Hetero Biopharma and Dr. Reddy's Laboratories. Russia has entered into agreements with enterprises, namely Hetero Biopharma, Panacea Biotech, Virchow Biotech, Stelis Biopharma, and Gland Pharma, for the manufacture of the vaccine [282]. While Johnson & Johnson received approval from CDSCO for the emergency use of their mRNA vaccine for COVID-19 in India [283], the company is also seeking sanction for conducting phase III clinical trials in India [284]. Various other vaccine candidates are also underway in the clinical trial pipeline, such as the nanoparticle vaccine in phase III of the clinical trial, NVX-CoV2373 from Novavax [285], and the 3-dose intradermal DNA vaccine from Zydus Cadila, ZyCoV-D, also in phase III of the clinical trials [286,287].

Because all these COVID-19 vaccine candidates are designed to elicit a comprehensive immune response, it is expected that most of them can provide safety against newly emerging SARS-CoV-2 variants too. Genetic sequences of the viral samples isolated from around the world are continuously monitored for new variant strains, and efforts are underway to improve the vaccine candidates ineffective in handling variants. Hence, the availability of more than a single type of vaccine in addition to the continued practice of social distancing and personal protective measures is essential for reducing further viral mutations and safeguarding the diverse global population [288]. Now that several efficacious and safe vaccination options are available around the globe, it is paramount that regulatory agencies ensure the mass production and equitable distribution of these vaccines. As per the WHO estimate, about one billion and 147 million vaccine doses have been administered globally and in India, respectively, by April 26, 2021 [4]. Supervisory agencies must also warrant that the manufactures of COVID-19 vaccines and repurposed drugs abide by globally accepted regulatory guidelines, such as conducting animal studies prior to clinical trials, to ensure the safety and efficacy of these therapeutics [19].

Future perspectives

Impact of formulations and route of administration

The route of administration can significantly affect the efficiency of drug administration and the efficacy of drugs. It has been observed that to achieve maximum efficacy to relieve pneumonia symptoms, Remdesivir needs to be administered by pulmonary nebulization in addition to the standard mode of intravenous (IV) administration. If given via IV only, due to low cellular permeability and reduced lung distribution, Remdesivir cannot attain therapeutic concentrations and exhibit its complete potential [289]. Hence it is advised to adopt alternative approaches of drug delivery over conventional methods. Drug delivery via inhalation mode, either through oral or nasal route, is advantageous due to its prompt effect and ease of operation. Several earlier reports of local as well as systemic administration of different classes of therapeutics like proteins, nucleic acids, and vaccines stand in favor of pulmonary route of administration [290,291]. Of the various options of nasal formulations available, gels and sprays are easier to use than drops as there won't be any loss of dosage while administering due to the presence of gelling agents. Powder sprays can be delivered deep into the nasal cavity and are more stable [292]. Nanoparticle formulations, due to their applicability to deliver an extensive range of therapeutics, are being employed widely.

The pulmonary route of administration is considered mostly in the case of siRNAs as their local administration helps the clinically significant concentration to deposit directly onto the lung epithelium, thus eliciting a stronger innate immune response. If administered parenterally, a significant dose of siRNA will be lost due to exposure to other organs, serum nucleases, and rapid clearance from the system. Hence, in the case of COVID-19, the inhalation mode of delivery of aerosolized siRNAs by means of a microsprayer can take the drug deep down into the lung tissue [293]. This mode of delivery, though, possess certain challenges such as the need to maintain the particle size of the drug smaller than 1 mm [294] and the inability to use the technique in case of patients with highly inflamed, mucous filled, irritated lung tissue [295].

Due to its potential to be more effective, efforts are underway to develop inhalable siRNAs against COVID-19. The inhalable investigational new drug (IND), VIR2703 (ALN-COV), developed by a company of RNAi expertise, Alnylam, reduced the SARS-CoV-2 replication in vitro and showed 99.9% reactivity with SARS-CoV-2-like sequences [296]. Two other companies, OliX pharmaceuticals and Sirnaomics, are also investing on inhalational siRNAs for COVID-19 [297].

Earlier works in SARS [298] and MERS [299] have shown that nasal administration of vaccines is more effective than intramuscular administration. The nasal mucosa contains all the immuno-components needed to stimulate an antibody-generating immune response and produce IgA in addition to IgG [300]. Also, neuropilin-1 (NRP), the checkpoint protein in T cell memory immune response, is overexpressed in nasal sentinel cells [301]. Hence, various research groups are probing into the possibility of developing a nasal vaccine against COVID-19. An open-label collaborative study conducted by Hubei University of Medicine and Shanghai Jiao Tong University in China examined the prophylactic use of recombinant human interferon alpha (rhIFN-α) nasal drops against COVID-19. It was observed that the rhIFN-α nasal drops protected medical staff from COVID-19 and could possibly be used as a prophylactic in susceptible healthy people [302].

Based on the work carried out by University of Helsinki and the University of Eastern Finland, Rokote Laboratories Finland Ltd has developed a nasal spray vaccine against COVID-19. This adenoviral vaccine, carrying a COVID-19 viral protein to elicit an immune response, showed positive results in animal studies and is about to enter phase I clinical trial [303]. Serum Institute of India (SII) is collaborating with Codagenix, USA, to develop a live-attenuated intranasal vaccine for COVID-19, namely COVI-VAC, and is currently conducting Phase 1 trial (NCT04619628) [304]. Bharath Biotech has formulated a single-dose intranasal vaccine against COVID-19, with the trade name "CoroFlu," [305] in association with University of Wisconsin and FluGen Inc. It is a replication-deficient adenoviral vectored vaccine (BBV154) and is undergoing phase 1 trial (NCT04751682) [304]. If approved after clinical trials, this vaccine will change the immunization scenario in India due to its ease of administration and profound population coverage [306].

Biomaterial and tissue engineering strategies

Regenerative medicine and tissue engineering approaches have also been adopted for designing novel therapeutic strategies for the management of COVID-19. These approaches are centered on the precepts of repair, regeneration, and replacement of diseased or damaged cells, tissues, or organs. The uses of stem cells have been elaborated earlier in section 3.5. In the context of the pandemic, tissue engineering strategies have been proposed to be used to develop drug delivery systems, designing biomaterial-based vaccine platforms and in vitro 3D tissue models for studying disease mechanisms and drug testing (Fig. 3) [307].

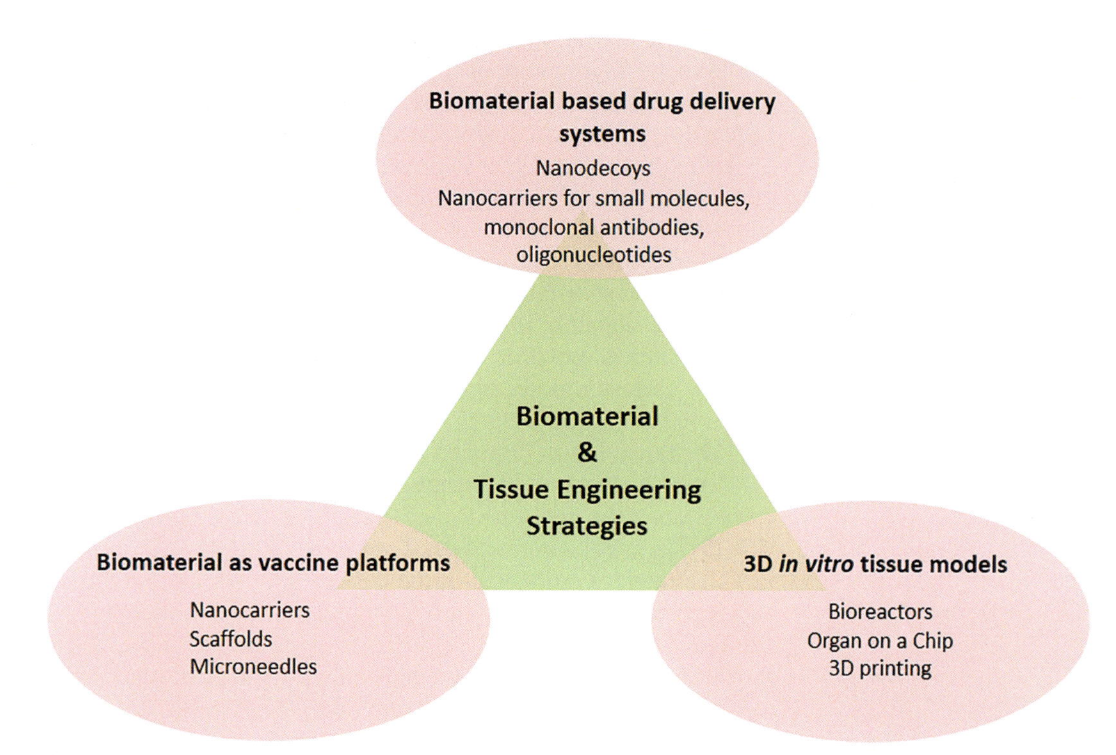

Fig. 3 Outline of different biomaterial and tissue engineering strategies adopted toward mitigating COVID-19.

Biomaterial-based drug delivery systems

Biomaterials are being used to design drug delivery systems, and the most researched vehicles are nanocarriers. They have the advantage of high surface area to volume ratio and small size, which allows for high drug payload and delivery to anatomically privileged sites, respectively. The biomaterial-based delivery systems can also be surface modified to facilitate controlled release, targeted drug delivery, and promote cellular uptake, which can be used to add active targeting ligands or stimulus-responsive moieties [308]. The therapeutic molecules that can be loaded into nanocarriers include small molecules, monoclonal antibodies, and oligonucleotides. Curcumin is a phytochemical that possesses broad-spectrum antiviral activity. However, the low water solubility of curcumin limits its clinical application. Yang et al. improved the solubility of curcumin by loading them into graphene oxide nanoparticles (GSCC). Functionalization with sulfonate groups allowed curcumin-loaded graphene oxide to mimic the cell surface and inhibit viral attachment by a competitive inhibition mechanism. In vitro studies showed that the composite nanoparticles could prevent the respiratory syncytial virus from infecting the host human epithelial type 2 (HEp-2) cells by directly inactivating the virus and inhibiting the viral attachment [309]. Nanoparticles with antibodies combine the physicochemical properties of the nanoparticles with the specific and selective recognition ability of the antibodies to antigens. Hence they find use in diagnosis and therapy.

Arruebo et al. have given a detailed review on antibody conjugated nanoparticles for biomedical application. These precepts can also be adopted for designing therapeutics for COVID-19 [310]. Small interfering RNA (siRNA) degrade the mRNA in a sequence-specific manner at the posttranscriptional level via a process known as posttranscriptional gene silencing (PTGS) [311]. There are several types of siRNA carriers, such as polymer-based carriers (poly lactic-co-glycolic acid (PLGA), chitosan, polyethylenimine (PEI), poly [2 (dimethylamino) ethyl methacrylate] (pDMAEMA), lipoplexes, and peptides such as peptide transduction domains (PTDs) and cell penetrating peptides (CPPs) [312]. Meng et al. successfully designed siRNAs to block the expression of the *envelope E* gene of the SARS virus by approximately 90% [313]. Since sequences are conserved in SARS-CoV and SARS-CoV-2, strategies employed in the designing of siRNA that target the spike and envelope gene for SARS might also have a probability in regulating COVID-19 infection and transmission. SARS-CoV-2 has been shown to target the lung tissue; hence delivery systems such as polymeric particles as inhalants hold potential.

Nanostructures have also been fabricated to mimic living cells; these structures called nanodecoys are designed to incorporate cell membrane-derived materials on the particle surface and serve to trap and sequester the virus. Earlier work by Rao et al. demonstrated the ability of gelatin nanoparticles coated with mosquito-derived cell membranes to act as nanodecoys. These particles effectively adsorb zika virus and thus prevents their replication in susceptible cells. Furthermore, in vivo experiments demonstrated the ability of these nanodecoys to prevent the passage of the zika virus into the fetal mouse brain [314]. With respect to COVID-19, decorating the nanoparticles with the ACE2 protein or related peptide fragments has been explored. Zhang et al. had developed a poly(lactic-co-glycolic acid) (PLGA) nanoparticle covered with cell membranes of human lung epithelial cell ACE2 receptor proteins to act as decoys for SARS-CoV-2. These nanodecoys could neutralize the SARS-CoV-2 infection of Vero E6 cells [315].

Biomaterial-based vaccine platforms

Vaccines prime the immune system to develop adaptive immune responses, which are either T cell mediated, characterized primarily by CD4 + and CD8 + T cells, or B cell mediated, characterized by secreted antibodies generated by B cells. These responses prepare the immune system to rapidly neutralize or eliminate pathogens during a future encounter with the same antigen [316]. Earlier live-attenuated or inactivated whole pathogens were used as vaccines. Subunit vaccines comprise of purified antigens from the pathogens that are antigenic and are able to elicit the desired protective immune response. Subunit vaccines are often formulated with immune-stimulating adjuvants. Recombinant protein vaccines, vectored vaccines, and RNA- and DNA-based vaccines are also currently being used. Ideal vaccines should possess the safety of subunit vaccines, the potency of attenuated or inactivated vaccines, and exert the desire immune response with precise control [316].

Biomaterial-based vaccine platforms include nanocarriers, scaffolds, microneedles, etc., which are an emerging area of science that is directed toward increasing the effectiveness of vaccines against infectious disease. Biomaterial-based vaccines entail the coencapsulation of antigens and adjuvants, effect cargo protection against enzymatic degradation, and improve stability. Additionally, their action can be targeted toward specific cells, mainly antigen-presenting cells, to induce specific immune responses. The combinations and relative concentrations of ligands can be modulated to simultaneously target multiple immune populations and pathways [308]. Vaccinations aim at presenting antigens to the immune system to stimulate recognition and memory in both humoral (B cell) and cell-mediated systems (T cell). These biomaterial-based systems allow for targeted drug delivery and controlled release of the drugs over extended periods and would hence have an advantage over multiple injections [307,316].

Polymeric particles, lipids, self-assembled proteins, and inorganic particles like gold and silica are used as the biomaterial vaccine platforms. Kim et al. had demonstrated that mesoporous silica rods (MSRs) when injected spontaneously assemble in vivo to form macroporous structures to which host immune cells, mainly dendritic cells home. Injection of an MSR-based vaccine formulation was found to enhance systemic helper T cells TH1 and TH2 serum antibody and cytotoxic T cell levels compared to bolus controls [317]. The group had also later conjugated peptide antigens to MSRs via stable thioether and reducible disulfide linkages. This was found to greatly increase the peptide loading and retention compared to passive adsorption. In vivo studies in mice using MSR scaffolds loaded with GM-CSF and CpG-ODN when injected subcutaneously, recruited dendritic cells, and could present antigen in situ with the stable conjugation increasing presentation capacity. The authors propose this system to be a versatile platform to efficiently incorporate peptide antigens in MSR vaccines and potentiate cellular responses [318]. Jiang et al. had developed an oral vaccine comprising of a membrane protein B of *Brachyspira hyodysenteriae* (BmpB), loaded into porous PLGA microparticles, as a vaccine against swine dysentery. The microparticles were further coated with the water-soluble chitosan and conjugated with M (Microfold) cell-homing peptides. Oral immunization of mice with this BmpB vaccine was found to effect increased levels of secretory IgA in mucosal tissue and systemic IgG levels. Microscopic studies had revealed the ability of the vaccine target onto the M cells to enter into the Peyer's patch regions of mouse small intestine [319]. Rungrojcharoenkit had developed trimethyl chitosan nanoparticles (TMC nPs) as the carrier of recombinant influenza hemagglutinin subunit 2 (HA2) and nucleoprotein (NP) with a view of developing a vaccine based on conserved proteins delivered in an adjuvanted nanoparticle system. An in vitro challenge assay demonstrated reduced influenza virus replication in primary human intranasal epithelium cells (HNEpCs) that were earlier primed with the nanoparticles [320]. In relation to SARS-CoV, Pimentel et al. designed a self-assembling polypeptide nanoparticle that repetitively displayed a SARS B cell epitope from the C-terminal heptad repeat of the virus's spike protein. Antisera obtained following immunization in mice exhibited neutralization activity in an in vitro infection inhibition assay using Vero cells [321]. Noh et al. had fabricated injectable hydrogels made from poly(γ-glutamic acid)-based adjuvants, collagens, and viral antigens (iPMH(OVA)). Mice immunized with iPMH(OVA) induced elevated levels of antigen-specific IgG titers and IFN-γ-producing cells and exhibited immunity against intranasal infection of H1N1 and H5N1 [322]. Advancements in the field of biomaterials and medical devices also include the development of microneedle patches as new modes of vaccine delivery. Microneedles are microscale metal or polymer needles which are either coated or loaded with the vaccine. They are used for transdermal delivery of vaccines without penetrating the skin deeply enough to cause pain and target skin-resident immune cells. They are proposed to be alternatives to the conventional needle and syringe for immunization [308]. Microneedles potentially offer improved immunogenicity, simplicity, cost-effectiveness, and safety. The TIV-MNP 2015 study was a randomized, partly blinded, and placebo-controlled, phase 1 clinical trial using dissolvable microneedle patches for influenza vaccination. Results revealed that the vaccines were stable for more than 1 year and generated titers similar to those of existing injectable vaccines. The vaccination was well tolerated and generated robust antibody responses [323]. Polymeric microneedle array-based MERS vaccine using a MERS-CoV S protein subunit trimer as an antigen was developed by Kim et al., and results from in vivo mice immunization studies confirmed a sustained release of antigen-specific antibody responses. The authors have also modified the MERS vaccine to produce a similar microneedle vaccine based on SARS-CoV-2 S protein subunit trimers, which showed similar immune responses in mice [324]. It is to be noted that most of the biomaterial-based vaccine platforms use in vitro cell culture models and in vivo animal models warrant further validation for human use.

3D in vitro tissue models

Development of in vitro models aids in a better understanding of the pathomechanisms of the COVID-19 infection. It would facilitate in-depth studies into the host-pathogen interaction, mechanisms of infection and replication, and help in profiling biomarkers that are pertinent in designing targeted interventions. Furthermore, they would also serve as in vitro drug testing models for high-throughput screening of potential therapeutic agents. Rather than static 2D in vitro cultures, 3D cultures have the added advantage of facilitating the coculture of different cell types promoting three-dimensional cell-cell interactions which emulate the architecture of natural biological tissue. Techniques such as rotating wall vessel bioreactors, organ on a chip, and 3D printing are employed in generating three-dimensional tissue constructs that are either scaffold free (spheroids and organoids) or scaffold based.

The current method of screening therapeutics against the COVID-19 virus adopts the use of 2D cultures of the Vero E6 cell lines. The potential use of human and animal organoids as experimental in vitro platforms is being explored with a view of studying not only SARS-CoV-2 but also other infectious viruses. Suderman et al. had developed a three-dimensional tissue-like assembly of human bronchial-tracheal mesenchymal cells that was overlayed with human bronchial epithelial cells in a 55-mL rotating wall vessel (RWV) bioreactor and used the same to study the infectivity of SARS-CoV [325]. Zhou et al. were able to establish intestinal organoid culture using crypts isolated from the intestines of *R. sinicus* bats. The authors had also developed human intestinal organoids and had used both models for studying the infectivity of SARS-CoV-2 from nasopharyngeal aspirate or sputum of three patients with COVID-19. They observed that both bat and human enteroids developed progressive cytopathic effect after inoculation. The work provided direct evidence of active SARS-CoV-2 replication in human enteroids, an in vitro model of human intestinal epithelium [326]. Monteil et al. generated human capillary and kidney organoids from induced pluripotent stem cells (iPSCs). These organoids were used for evaluating the infectivity of SARS-CoV-2, and the inhibitory effect of human recombinant soluble ACE2 (hrsACE2), a potential therapeutic for COVID-19. The authors demonstrated that both vascular and kidney organoids were infected with SARS-CoV-2, and the supernatant was also capable of infecting monolayers of Vero E6 cells. Furthermore, hrsACE2 was found to decrease infection in a dose-dependent manner [121]. In another study, Youk et al. had used a matrigel-based system to develop feeder-free three-dimensional cultures of human lung alveolar type 2 (hAT2) cells derived from primary human lung tissue. These cultures were used to evaluate cellular response to SARS-CoV-2 by image analysis and single-cell transcriptome profiling. Results obtained show rapid viral replication and the increased expression of interferon-associated genes and proinflammatory genes in infected hAT2 cells, thus demonstrating the potential of 3D in vitro cultures as models for studying the pathomechanisms of SARS-CoV-2 infections [327].

The organ on a chip (OOAC) model employs the precepts of engineering to biology to develop scalable and advanced in vitro models directed toward precision medicine. OOAC is a microfluidic culture device that is designed to simulate the physiological environment of human organs. Microfluidic technology allows the reconstitution of complex cellular interactions on the microscale. The system is designed to regulate key parameters, mainly concentration gradients, shear force, cell patterning, multicellular complexity, tissue boundaries, and interactions [328]. OOACs are being developed for various organs like the liver, lung, kidney, intestines. In order to better mimic the human system, multiorgan on a chip systems that culture cells from various organs and tissues simultaneously and are connected by channels (to achieve multiorgan integration) are being developed and are in their nascent stage [328]. Si et al. had developed a microfluidic device with two microchannels separated by a porous membrane. Basal stem cells and respiratory epithelial cells were cultured on either side of the membrane and then treated with synthetic pseudovirions comprised of the S protein of SARS-CoV-2. The virus entered the lung epithelial cells by binding to the ACE2 receptor expressed on the epithelial cell surface in microchannels. This system was used to screen multiple FDA-approved SARS-CoV-2 inhibitors, including chloroquine, arbidol, toremifene, clomiphene, amodiaquine, verapamil, and amiodarone, by perfusion through the channels. Results showed that amodiaquine and toremifene were able to prevent SARS-CoV-2 infection [329].

The 3D bioprinting is an emerging technology that is used to optimize the conventional 3D cell culture. Through this technology, cells and biomaterials are deposited layer by layer in an organized and precise manner through a computer-aided process. Hence it is possible to fabricate complex architectures that resemble native tissues in a reproducible manner. These fabricated 3D in vitro tissue models hold promise as systems for studying SARS-CoV-2 host cell interactions and as in vitro drug testing models to counteract future pandemics.

Conclusion

The 2019 outbreak of COVID-19 has become one of the largest public health emergencies. The causative agent coronavirus is a zoonotic strain of virus known to infect animals and humans. In this chapter, a brief description of the etiology of the virus, its structure, mode of infection, transmission, and manifestations are provided.

The current clinical scenarios, including the main testing strategies, modes of treatment, and vaccines developed against COVID-19, have been elaborated. Though SARS-CoV-2 (COVID-19 virus) exhibits similarity with SARS-CoV at the molecular level, they differ much in their pathophysiology. The urgency of the pandemic necessitated the repurposing of drugs that were already in use for other diseases. Immunosuppression might be beneficial to reduce mortality in COVID-19 patients with severe symptoms and hyperinflammation. Regenerative medicine has also been adopted to offer various cellular therapeutics against COVID-19, mainly that of mesenchymal stem cell (MSC) therapy, natural killer (NK) cell therapy, and exosomes. Finally, through this chapter, the impact of formulations and their route of administration, along with the potential of biomaterials and tissue engineering strategies to mitigate the COVID-19 pandemic, have been discussed. Tissue engineering approaches have been adopted for designing novel therapeutic modalities for the management of COVID-19, especially biomaterial-based vaccine platforms and in vitro 3D tissue models for studying disease mechanisms and drug testing. The current COVID-19 situation mandates the development and evaluation of therapeutic molecules and effective vaccine candidates, which is key in combating and managing the pandemic.

Acknowledgments

Authors ASR, KAM, BKM, and AJ express their sincere gratitude and appreciation to Dr. Mini S, Director, Advanced Centre for Tissue Engineering (ACTE), and Dr. A. Helen, Head, Department of Biochemistry, University of Kerala, for support and facilities provided. Authors ASR, KAM, BKM, and MS gratefully acknowledge the financial support from University of Kerala under the State Plan fund to ACTE; and AJ for funding from Indian Council of Medical Research (ICMR) Emeritus Scientist Scheme.

References

[1] International Society for Infectious Diseases. Undiagnosed pneumonia—China (HUBEI), https://promedmail.org/promed-post/?id=6864153%20#COVID19; 2021.

[2] Wu F, Zhao S, Yu B, et al. A new coronavirus associated with human respiratory disease in China. Nature 2020;579(7798):265–9.

[3] World Health Organization. WHO Director-General's opening remarks at the media briefing on COVID-19, https://www.who.int/director-general/speeches/detail/who-director-general-s-opening-remarks-at-the-media-briefing-on-covid-19; 2020.

[4] World Health Organization. WHO Coronavirus (COVID-19) Dashboard, https://covid19.who.int/; 2021. [Accessed 1 May 2021].

[5] Google News. Coronavirus (COVID-19), https://news.google.com/covid19/map?hl=en-IN&mid=%2Fm%2F0byh8j&gl=IN&ceid=IN%3Aen; 2021. [Accessed 30 April 2021].

[6] Erkoreka A. Origins of the Spanish Influenza pandemic (1918-1920) and its relation to the First World War. J Mol Genetic Med 2009;3(2):190–4. https://doi.org/10.4172/1747-0862.1000033.
[7] Rajagopal S, Treanor J. Pandemic (avian) influenza. Semin Respir Crit Care Med 2007;28(2):159–70. https://doi.org/10.1055/s-2007-976488.
[8] Rewar S, Mirdha D, Rewar P. Treatment and prevention of pandemic H1N1 influenza. Ann Global Health 2015;81(5):645–53.
[9] Abubakar I, Tillmann T, Banerjee A. Global, regional, and national age-sex specific all-cause and cause-specific mortality for 240 causes of death, 1990-2013: a systematic analysis for the Global Burden of Disease Study 2013. Lancet 2015;385(9963):117–71.
[10] Ifediora OF, Aning K. West Africa's Ebola pandemic: toward effective multilateral responses to health crises. Global Govern 2017;23(2):225–44.
[11] Deeks SG, Overbaugh J, Phillips A, Buchbinder S. HIV infection. Nat Rev Disease Primers 2015;1(1):1–22.
[12] Keele BF, Van Heuverswyn F, Li Y, et al. Chimpanzee reservoirs of pandemic and nonpandemic HIV-1. Science 2006;313(5786):523–6.
[13] Baseler L, Chertow DS, Johnson KM, Feldmann H, Morens DM. The pathogenesis of Ebola virus disease. Annual Rev Pathol: Mechan Disease 2017;12:387–418.
[14] Kruse H, Kirkemo A-M, Handeland K. Wildlife as source of zoonotic infections. Emerg Infect Disease J 2004;10(12):2067. https://doi.org/10.3201/eid1012.040707.
[15] Sood S, Aggarwal V, Aggarwal D, et al. COVID-19 pandemic: from molecular biology, pathogenesis, detection, and treatment to global societal impact. Curr Pharmacol Rep 2020;1–16.
[16] Cheever FS, Daniels JB, Pappenheimer AM, Bailey OT. A murine virus (JHM) causing disseminated encephalomyelitis with extensive destruction of myelin: I. Isolation and biological properties of the virus. J Exp Med 1949;90(3):181–94.
[17] Pappenheimer AM. Pathology of Infection with the JHM Virus1. JNCI 1958;20(5):879–91. https://doi.org/10.1093/jnci/20.5.879.
[18] Bradburne A, Bynoe M, Tyrrell D. Effects of a" new" human respiratory virus in volunteers. Br Med J 1967;3(5568):767.
[19] Kumar M, Al Khodor S. Pathophysiology and treatment strategies for COVID-19. J Transl Med 2020;18(1):353. https://doi.org/10.1186/s12967-020-02520-8.
[20] Gonzalez J, Gomez-Puertas P, Cavanagh D, Gorbalenya A, Enjuanes L. A comparative sequence analysis to revise the current taxonomy of the family Coronaviridae. Arch Virol 2003;148(11):2207–35.
[21] Fehr AR, Perlman S. Coronaviruses: an overview of their replication and pathogenesis. Coronaviruses 2015;1–23.
[22] Weiss SR, Navas-Martin S. Coronavirus pathogenesis and the emerging pathogen severe acute respiratory syndrome coronavirus. Microbiol Mol Biol Rev 2005;69(4):635–64.
[23] Jonassen CM, Kofstad T, Larsen I-L, et al. Molecular identification and characterization of novel coronaviruses infecting graylag geese (Anser anser), feral pigeons (Columbia livia) and mallards (Anas platyrhynchos). J Gen Virol 2005;86(6):1597–607.
[24] Lau SK, Chan JF. Coronaviruses: emerging and re-emerging pathogens in humans and animals. Virol J 2015;12:209. https://doi.org/10.1186/s12985-015-0432-z.
[25] Guo Y-R, Cao Q-D, Hong Z-S, et al. The origin, transmission and clinical therapies on coronavirus disease 2019 (COVID-19) outbreak—an update on the status. Military Med Res 2020;7(1):1–10.
[26] Hussein HA, Hassan RYA, Chino M, Febbraio F. Point-of-care diagnostics of COVID-19: from current work to future perspectives. Sensors 2020;20(15):4289. https://doi.org/10.3390/s2015428.

[27] Chu DK, Pan Y, Cheng SM, et al. Molecular diagnosis of a novel coronavirus (2019-nCoV) causing an outbreak of pneumonia. Clin Chem 2020;66(4):549–55.
[28] Cui J, Li F, Shi Z-L. Origin and evolution of pathogenic coronaviruses. Nat Rev Microbiol 2019;17(3):181–92. https://doi.org/10.1038/s41579-018-0118-9.
[29] Li W, Shi Z, Yu M, et al. Bats are natural reservoirs of SARS-like coronaviruses. Science 2005;310(5748):676–9. https://doi.org/10.1126/science.1118391.
[30] Forni D, Cagliani R, Clerici M, Sironi M. Molecular evolution of human coronavirus genomes. Trends Microbiol 2017;25(1):35–48.
[31] Zhong N, Zheng B, Li Y, et al. Epidemiology and cause of severe acute respiratory syndrome (SARS) in Guangdong, People's Republic of China, in February, 2003. Lancet 2003;362(9393):1353–8.
[32] Gong S, Bao L. The battle against SARS and MERS coronaviruses: reservoirs and animal models. Animal Models Exp Med 2018;1(2):125–33.
[33] Haagmans BL, Al Dhahiry SH, Reusken CB, et al. Middle East respiratory syndrome coronavirus in dromedary camels: an outbreak investigation. Lancet Infect Dis 2014;14(2):140–5.
[34] Khafaie MA, Rahim F. Cross-country comparison of case fatality rates of COVID-19/SARS-COV-2. Osong Public Health Res Persp 2020;11(2):74.
[35] Li Q, Guan X, Wu P, et al. Early transmission dynamics in Wuhan, China, of novel coronavirus–infected pneumonia. N Engl J Med 2020;382.
[36] Zhou P, Yang X-L, Wang X-G, et al. A pneumonia outbreak associated with a new coronavirus of probable bat origin. Nature 2020;579(7798):270–3. https://doi.org/10.1038/s41586-020-2012-7.
[37] Lau SKP, Woo PCY, Li KSM, et al. Severe acute respiratory syndrome coronavirus-like virus in Chinese horseshoe bats. Proc Natl Acad Sci U S A 2005;102(39):14040–5. https://doi.org/10.1073/pnas.0506735102.
[38] Chan JF-W, Kok K-H, Zhu Z, et al. Genomic characterization of the 2019 novel human-pathogenic coronavirus isolated from a patient with atypical pneumonia after visiting Wuhan. Emerg Microbes Infect 2020;9(1):221–36. https://doi.org/10.1080/22221751.2020.1719902.
[39] Zhang T, Wu Q, Zhang Z. Probable pangolin origin of SARS-CoV-2 associated with the COVID-19 outbreak. Curr Biol 2020;30(7):1346–1351.e2.
[40] Lam TT-Y, Jia N, Zhang Y-W, et al. Identifying SARS-CoV-2-related coronaviruses in Malayan pangolins. Nature 2020;583(7815):282–5. https://doi.org/10.1038/s41586-020-2169-0.
[41] Ji W, Wang W, Zhao X, Zai J, Li X. Cross-species transmission of the newly identified coronavirus 2019-nCoV. J Med Virol 2020;92(4):433–40. https://doi.org/10.1002/jmv.25682.
[42] Liu Z, Xiao X, Wei X, et al. Composition and divergence of coronavirus spike proteins and host ACE2 receptors predict potential intermediate hosts of SARS-CoV-2. J Med Virol 2020;92(6):595–601.
[43] Zhao J, Cui W, Tian B-P. The potential intermediate hosts for SARS-CoV-2. Front Microbiol 2020;11:580137. https://doi.org/10.3389/fmicb.2020.580137.
[44] Xiao K, Zhai J, Feng Y, et al. Isolation of SARS-CoV-2-related coronavirus from Malayan pangolins. Nature 2020;583(7815):286–9.
[45] Wong MC, Cregeen SJJ, Ajami NJ, Petrosino JF. Evidence of recombination in coronaviruses implicating pangolin origins of nCoV-2019. BioRxiv 2020.
[46] Wu A, Peng Y, Huang B, et al. Genome composition and divergence of the novel coronavirus (2019-nCoV) originating in China. Cell Host Microbe 2020;27(3):325–8.
[47] Ren L-L, Wang Y-M, Wu Z-Q, et al. Identification of a novel coronavirus causing severe pneumonia in human: a descriptive study. Chin Med J 2020;133.
[48] Petrosillo N, Viceconte G, Ergonul O, Ippolito G, Petersen E. COVID-19, SARS and MERS: are they closely related? Clin Microbiol Infect 2020;26(6):729–34.

[49] Rabaan AA, Al-Ahmed SH, Haque S, et al. SARS-CoV-2, SARS-CoV, and MERS-COV: a comparative overview. Infez Med 2020;28(2):174–84.

[50] Licastro D, Rajasekharan S, Dal Monego S, Segat L, D'Agaro P, Marcello A. Isolation and full-length genome characterization of SARS-CoV-2 from COVID-19 cases in Northern Italy. J Virol 2020;94.

[51] Giri B, Pandey S, Shrestha R, Pokharel K, Ligler FS, Neupane BB. Review of analytical performance of COVID-19 detection methods. Anal Bioanal Chem 2021;1–14.

[52] Boopathi S, Poma AB, Kolandaivel P. Novel 2019 coronavirus structure, mechanism of action, antiviral drug promises and rule out against its treatment. J Biomol Struct Dyn 2021;1–10. https://doi.org/10.1080/07391102.2020.1758788.

[53] Sternberg A, Naujokat C. Structural features of coronavirus SARS-CoV-2 spike protein: targets for vaccination. Life Sci 2020;, 118056.

[54] Xu X, Chen P, Wang J, et al. Evolution of the novel coronavirus from the ongoing Wuhan outbreak and modeling of its spike protein for risk of human transmission. Sci China Life Sci 2020;63(3):457–60.

[55] Zhou F, Yu T, Du R, et al. Clinical course and risk factors for mortality of adult inpatients with COVID-19 in Wuhan, China: a retrospective cohort study. Lancet 2020;395(10229):1054–62.

[56] Shang J, Ye G, Shi K, et al. Structural basis of receptor recognition by SARS-CoV-2. Nature 2020;581(7807):221–4. https://doi.org/10.1038/s41586-020-2179-y.

[57] Coutard B, Valle C, de Lamballerie X, Canard B, Seidah NG, Decroly E. The spike glycoprotein of the new coronavirus 2019-nCoV contains a furin-like cleavage site absent in CoV of the same clade. Antivir Res 2020;176. https://doi.org/10.1016/j.antiviral.2020.104742, 104742.

[58] Johnson BA, Xie X, Kalveram B, et al. Furin cleavage site is key to SARS-CoV-2 pathogenesis. BioRxiv 2020.

[59] Rahman MS, Islam MR, ASMRU A, et al. Evolutionary dynamics of SARS-CoV-2 nucleocapsid protein and its consequences. J Med Virol 2021;93(4):2177–95. https://doi.org/10.1002/jmv.26626.

[60] Tang JW, Tambyah PA, Hui DS. Emergence of a new SARS-CoV-2 variant in the UK. J Infect 2021;82.

[61] World Health Organization. COVID-19 weekly epidemiological update—27 April 2021, https://www.who.int/publications/m/item/weekly-epidemiological-update-on-covid-19- - -27-april-2021#:~:text=Globally%2C%20new%20COVID%2D19%20cases,87%20000%20new%20deaths%20reported; 2021.

[62] Yadav P, Sapkal GN, Abraham P, et al. Neutralization of variant under investigation B.1.617 with sera of BBV152 vaccinees. bioRxiv 2021;ciab411.

[63] Chan JF-W, Yuan S, Kok K-H, et al. A familial cluster of pneumonia associated with the 2019 novel coronavirus indicating person-to-person transmission: a study of a family cluster. Lancet 2020;395(10223):514–23.

[64] Liu J, Liao X, Qian S, et al. Community transmission of severe acute respiratory syndrome Coronavirus 2, Shenzhen, China, 2020. Emerg Infect Dis 2020;26(6):1320–3. https://doi.org/10.3201/eid2606.200239.

[65] Huang C, Wang Y, Li X, et al. Clinical features of patients infected with 2019 novel coronavirus in Wuhan, China. Lancet 2020;395(10223):497–506.

[66] World Health Organization. Infection prevention and control of epidemic- and pandemic-prone acute respiratory infections in health care WHO Guidelines, https://apps.who.int/iris/bitstream/handle/10665/112656/9789241507134_eng.pdf;jsessionid=41AA684FB64571CE8D8A453C4F2B2096?sequence=1; 2021.

[67] World Health Organization. Mask use in the context of COVID-19 Interim guidance, https://www.who.int/publications/i/item/advice-on-the-use-of-masks-in-the-community-during-home-care-and-in-healthcare-settings-in-the-context-of-the-novel-coronavirus-(2019-ncov)-outbreak; 2021.

[68] Van Doremalen N, Bushmaker T, Morris DH, et al. Aerosol and surface stability of SARS-CoV-2 as compared with SARS-CoV-1. N Engl J Med 2020;382(16):1564–7.
[69] Chia PY, Coleman KK, Tan YK, et al. Detection of air and surface contamination by SARS-CoV-2 in hospital rooms of infected patients. Nat Commun 2020;11(1):1–7.
[70] Wang W, Xu Y, Gao R, et al. Detection of SARS-CoV-2 in different types of clinical specimens. JAMA 2020;323(18):1843–4.
[71] Le Chang LZ, Gong H, Wang L, Wang L. Severe acute respiratory syndrome coronavirus 2 RNA detected in blood donations. Emerg Infect Dis 2020;26(7):1631.
[72] Centers for Disease Control and Prevention. Improve how your mask protects you, https://www.cdc.gov/coronavirus/2019-ncov/your-health/effective-masks.html; 2021. [Accessed 1 May 2021].
[73] Chau CH, Strope JD, Figg WD. COVID-19 Clinical Diagnostics and Testing Technology. Pharmacotherapy 2020;40(8):857–68. https://doi.org/10.1002/phar.2439.
[74] Manabe YC, Sharfstein JS, Armstrong K. The need for more and better testing for COVID-19. JAMA 2020;324(21):2153–4.
[75] Pascarella G, Strumia A, Piliego C, et al. COVID-19 diagnosis and management: a comprehensive review. J Intern Med 2020;288(2):192–206.
[76] Itou T, Markotter W, Nel LH. Reverse transcription-loop-mediated isothermal amplification system for the detection of rabies virus. In: Current laboratory techniques in rabies diagnosis, research and prevention. Elsevier; 2014. p. 85–95.
[77] Kellner MJ, Koob JG, Gootenberg JS, Abudayyeh OO, Zhang F. SHERLOCK: nucleic acid detection with CRISPR nucleases. Nat Protoc 2019;14(10):2986–3012.
[78] Hou T, Zeng W, Yang M, et al. Development and evaluation of a rapid CRISPR-based diagnostic for COVID-19. PLoS Pathog 2020;16(8), e1008705.
[79] Guo X, Guo Z, Duan C, et al. Long-term persistence of IgG antibodies in SARS-CoV infected healthcare workers. MedRxiv 2020.
[80] Kontou PI, Braliou GG, Dimou NL, Nikolopoulos G, Bagos PG. Antibody tests in detecting SARS-CoV-2 infection: a meta-analysis. Diagnostics 2020;10(5):319.
[81] Raptis CA, Hammer MM, Short RG, et al. Chest CT and coronavirus disease (COVID-19): a critical review of the literature to date. Am J Roentgenol 2020;215(4):839–42.
[82] Salian VS, Wright JA, Vedell PT, et al. COVID-19 transmission, current treatment, and future therapeutic strategies. Mol Pharm 2021;18(3):754–71.
[83] Wilson NM, Norton A, Young FP, Collins DW. Airborne transmission of severe acute respiratory syndrome coronavirus-2 to healthcare workers: a narrative review. Anaesthesia 2020;75(8):1086–95. https://doi.org/10.1111/anae.15093.
[84] Bikdeli B, Madhavan MV, Jimenez D, et al. COVID-19 and thrombotic or thromboembolic disease: implications for prevention, antithrombotic therapy, and follow-up: JACC state-of-the-art review. J Am Coll Cardiol 2020;75(23):2950–73. https://doi.org/10.1016/j.jacc.2020.04.031.
[85] Moores LK, Tritschler T, Brosnahan S, et al. Prevention, diagnosis, and treatment of VTE in patients with coronavirus disease 2019: CHEST guideline and expert panel report. Chest 2020;158(3):1143–63.
[86] Cumhur Cure M, Kucuk A, Cure E. NSAIDs may increase the risk of thrombosis and acute renal failure in patients with COVID-19 infection. Therapies 2020;75(4):387–8. https://doi.org/10.1016/j.therap.2020.06.012.
[87] Singh AK, Majumdar S, Singh R, Misra A. Role of corticosteroid in the management of COVID-19: a systemic review and a Clinician's perspective. Diabetes Metab Syndr Clin Res Rev 2020;14(5):971–8. https://doi.org/10.1016/j.dsx.2020.06.054.
[88] Attaway A. Management of patients with COPD during the COVID-19 pandemic. Cleve Clin J Med 2020.

[89] Krishna G, Pillai VS, Veettil MV. Approaches and advances in the development of potential therapeutic targets and antiviral agents for the management of SARS-CoV-2 infection. Eur J Pharmacol 2020;885, 173450.
[90] World Health Organization. COVID-19 trials updated on 29April 2021, https://worldhealthorg-my.sharepoint.com/:x:/g/personal/karamg_who_int/EReltiDVLS5HuSNOVB21VRAByz3zAv-jxPoNS3cgtTywkA?e=5NWl6s; 2021. [Accessed 2 May 2021].
[91] Ko M, Jeon S, Ryu W-S, Kim S. Comparative analysis of antiviral efficacy of FDA-approved drugs against SARS-CoV-2 in human lung cells. J Med Virol 2021;93(3):1403–8. https://doi.org/10.1002/jmv.26397.
[92] Jamshaid H, Zahid F, Ud Din I, et al. Diagnostic and treatment strategies for COVID-19. AAPS PharmSciTech 2020;21(6):1–14.
[93] World Health Organization. WHO Director-General's opening remarks at the media briefing on COVID-19—18 March 2020, https://www.who.int/director-general/speeches/detail/who-director-general-s-opening-remarks-at-the-media-briefing-on-covid-19- - -18-march-2020; 2020.
[94] Warren TK, Jordan R, Lo MK, et al. Therapeutic efficacy of the small molecule GS-5734 against Ebola virus in rhesus monkeys. Nature 2016;531(7594):381–5.
[95] Elfiky AA. Anti-HCV, nucleotide inhibitors, repurposing against COVID-19. Life Sci 2020;248, 117477.
[96] Silverman E, Feuerstein A, Herper M. New data on Gilead's Remdesivir, released by accident, show no benefit for coronavirus patients. Company still sees reason for hope. STAT; 2020. https://www.statnews.com/2020/04/23/data-on-gileads-remdesivir-released-by-accident-show-no-benefit-for-coronavirus-patients/. [Accessed 1 May 2021].
[97] U.S. Food and Drug Administration. Coronavirus (COVID-19) update: FDA issues emergency use authorization for potential COVID-19 treatment, https://www.fda.gov/news-events/press-announcements/coronavirus-covid-19-update-fda-issues-emergency-use-authorization-potential-covid-19-treatment; 2021.
[98] Savarino A, Di Trani L, Donatelli I, Cauda R, Cassone A. New insights into the antiviral effects of chloroquine. Lancet Infect Dis 2006;6(2):67–9.
[99] Chandwani A, Shuter J. Lopinavir/ritonavir in the treatment of HIV-1 infection: a review. Ther Clin Risk Manag 2008;4(5):1023.
[100] Zhang L, Lin D, Kusov Y, et al. α-Ketoamides as broad-spectrum inhibitors of coronavirus and enterovirus replication: structure-based design, synthesis, and activity assessment. J Med Chem 2020;63(9):4562–78.
[101] World Health Organization. "Solidarity" clinical trial for COVID-19 treatments, https://www.who.int/emergencies/diseases/novel-coronavirus-2019/global-research-on-novel-coronavirus-2019-ncov/solidarity-clinical-trial-for-covid-19-treatments; 2021. [Accessed 1 May 2021].
[102] Pan H, Peto R, Karim QA, et al. Repurposed antiviral drugs for COVID-19—interim WHO SOLIDARITY trial results. MedRxiv 2020. https://doi.org/10.1101/2020.10.15.20209817.
[103] Furuta Y, Komeno T, Nakamura T. Favipiravir (T-705), a broad spectrum inhibitor of viral RNA polymerase. Proc Jpn Acad Ser B 2017;93(7):449–63. https://doi.org/10.2183/pjab.93.027.
[104] Mifsud EJ, Hayden FG, Hurt AC. Antivirals targeting the polymerase complex of influenza viruses. Antivir Res 2019;169. https://doi.org/10.1016/j.antiviral.2019.104545, 104545.
[105] Chan KW, Wong VT, Tang SCW. COVID-19: an update on the epidemiological, clinical, preventive and therapeutic evidence and guidelines of integrative Chinese-Western medicine for the management of 2019 novel coronavirus disease. Am J Chin Med 2020;48(3):737–62. https://doi.org/10.1142/s0192415x20500378.

[106] Udwadia ZF, Singh P, Barkate H, et al. Efficacy and safety of favipiravir, an oral RNA-dependent RNA polymerase inhibitor, in mild-to-moderate COVID-19: a randomized, comparative, open-label, multicenter, phase 3 clinical trial. Int J Infect Dis 2021;103:62–71. https://doi.org/10.1016/j.ijid.2020.11.142.
[107] Chen H, Zhang Z, Wang L, et al. First clinical study using HCV protease inhibitor danoprevir to treat COVID-19 patients. Medicine 2020;99(48).
[108] Wang X, Cao R, Zhang H, et al. The anti-influenza virus drug, arbidol is an efficient inhibitor of SARS-CoV-2 in vitro. Cell Discov 2020;6(1):28. https://doi.org/10.1038/s41421-020-0169-8.
[109] Nojomi M, Yassin Z, Keyvani H, et al. Effect of Arbidol (Umifenovir) on COVID-19: a randomized controlled trial. BMC Infect Dis 2020;20(1):954. https://doi.org/10.1186/s12879-020-05698-w.
[110] Rappe JCF, de Wilde A, Di H, et al. Antiviral activity of K22 against members of the order Nidovirales. Virus Res 2018;246:28–34. https://doi.org/10.1016/j.virusres.2018.01.002.
[111] Lundin A, Dijkman R, Bergström T, et al. Targeting membrane-bound viral RNA synthesis reveals potent inhibition of diverse coronaviruses including the middle East respiratory syndrome virus. PLoS Pathog 2014;10(5). https://doi.org/10.1371/journal.ppat.1004166, e1004166.
[112] Jamiu AT, Aruwa CE, Abdulakeem IA, Ajao AA, Sabiu S. Phytotherapeutic Evidence Against Coronaviruses and Prospects for COVID-19. Pharm J 2020;12(6).
[113] Kawase M, Shirato K, van der Hoek L, Taguchi F, Matsuyama S. Simultaneous treatment of human bronchial epithelial cells with serine and cysteine protease inhibitors prevents severe acute respiratory syndrome coronavirus entry. J Virol 2012;86(12):6537–45.
[114] Yamamoto M, Matsuyama S, Li X, et al. Identification of nafamostat as a potent inhibitor of Middle East respiratory syndrome coronavirus S protein-mediated membrane fusion using the split-protein-based cell-cell fusion assay. Antimicrob Agents Chemother 2016;60(11):6532–9.
[115] Zhu H, Du W, Song M, Liu Q, Herrmann A, Huang Q. Spontaneous binding of potential COVID-19 drugs (Camostat and Nafamostat) to human serine protease TMPRSS2. Comput Struct Biotechnol J 2020;19:467–76. https://doi.org/10.1016/j.csbj.2020.12.035.
[116] Hoffmann M, Kleine-Weber H, Schroeder S, et al. SARS-CoV-2 Cell Entry Depends on ACE2 and TMPRSS2 and Is Blocked by a Clinically Proven Protease Inhibitor. Cell 2020;181(2):271–280.e8. https://doi.org/10.1016/j.cell.2020.02.052.
[117] Breining P, Frølund AL, Højen JF, et al. Camostat mesylate against SARS-CoV-2 and COVID-19—Rationale, dosing and safety. Basic Clin Pharmacol Toxicol 2021;128(2):204–12. https://doi.org/10.1111/bcpt.13533.
[118] Hoffmann M, Hofmann-Winkler H, Smith JC, et al. Camostat mesylate inhibits SARS-CoV-2 activation by TMPRSS2-related proteases and its metabolite GBPA exerts antiviral activity. EBioMed 2021;65, 103255.
[119] Hoffmann M, Schroeder S, Kleine-Weber H, Müller MA, Drosten C, Pöhlmann S. Nafamostat mesylate blocks activation of SARS-CoV-2: new treatment option for COVID-19. Antimicrob Agents Chemother 2020;64(6). https://doi.org/10.1128/aac.00754-20, e00754-20.
[120] Haschke M, Schuster M, Poglitsch M, et al. Pharmacokinetics and pharmacodynamics of recombinant human angiotensin-converting enzyme 2 in healthy human subjects. Clin Pharmacokinet 2013;52(9):783–92. https://doi.org/10.1007/s40262-013-0072-7.
[121] Monteil V, Kwon H, Prado P, et al. Inhibition of SARS-CoV-2 infections in engineered human tissues using clinical-grade soluble human ACE2. Cell 2020;181(4):905–913.e7.

[122] Khan A, Benthin C, Zeno B, et al. A pilot clinical trial of recombinant human angiotensin-converting enzyme 2 in acute respiratory distress syndrome. Crit Care 2017;21(1):234. https://doi.org/10.1186/s13054-017-1823-x.

[123] Zoufaly A, Poglitsch M, Aberle JH, et al. Human recombinant soluble ACE2 in severe COVID-19. Lancet Respir Med 2020;8(11):1154–8. https://doi.org/10.1016/s2213-2600(20)30418-5.

[124] Apeiron Biologics. Recombinant human angiotensin-converting enzyme 2 (rhACE2) as a treatment for patients with COVID-19 (APN01-COVID-19), https://clinicaltrials.gov/ct2/show/NCT04335136; 2020. [Accessed 1 May 2021].

[125] Technology Networks. APN01 shows clinical benefit for severely Ill COVID-19 patients in phase II trial, https://www.technologynetworks.com/biopharma/product-news/apn01-shows-clinical-benefit-for-severely-ill-covid-19-patients-in-phase-ii-trial-346670; 2021.

[126] National Institute of Health. COVID-19 treatment guidelines—Corticosteroids, https://www.covid19treatmentguidelines.nih.gov/immunomodulators/corticosteroids/; 2021.

[127] Horby P, Lim WS, Emberson JR, et al. Dexamethasone in Hospitalized Patients with Covid-19. N Engl J Med 2021;384(8):693–704. https://doi.org/10.1056/NEJMoa2021436.

[128] U.S. National Library of Medicine. Efficacy and safety of IFN-α2β in the treatment of novel coronavirus patients, https://clinicaltrials.gov/ct2/show/NCT04293887; 2020.

[129] Chinese Clinical Trial Registry. Comparative effectiveness and safety of ribavirin plus interferon-alpha, lopinavir/ritonavir plus interferon-alpha and ribavirin plus lopinavir/ritonavir plus interferon-alphain in patients with mild to moderate novel coronavirus pneumonia, http://www.chictr.org.cn/showprojen.aspx?proj=48782; 2021. Accessed 1 May 2021.

[130] Kang JH, Kao LT, Lin HC, Wang TJ, Yang TY. Do outpatient statins and ACEIs/ARBs have synergistic effects in reducing the risk of pneumonia? A population-based case-control study. PLoS One 2018;13(6). https://doi.org/10.1371/journal.pone.0199981, e0199981.

[131] Piyush R, Rajarshi K, Chatterjee A, Khan R, Ray S. Nucleic acid-based therapy for coronavirus disease 2019. Heliyon 2020;6(9). https://doi.org/10.1016/j.heliyon.2020.e05007, e05007.

[132] Wu CJ, Huang HW, Liu CY, Hong CF, Chan YL. Inhibition of SARS-CoV replication by siRNA. Antivir Res 2005;65(1):45–8. https://doi.org/10.1016/j.antiviral.2004.09.005.

[133] Kalhori MR, Saadatpour F, Arefian E, et al. The potential therapeutic effect of RNA interference and natural products on COVID-19: a review of the coronaviruses infection. Front Pharmacol 2021;12(116). https://doi.org/10.3389/fphar.2021.616993.

[134] Uludağ H, Parent K, Aliabadi HM, Haddadi A. Prospects for RNAi therapy of COVID-19. Front Bioeng Biotechnol 2020;8(916). https://doi.org/10.3389/fbioe.2020.00916.

[135] Fukushima A, Fukuda N, Lai Y, et al. Development of a chimeric DNA-RNA hammerhead ribozyme targeting SARS virus. Intervirology 2009;52(2):92–9. https://doi.org/10.1159/000215946.

[136] Dönmüş B, Ünal S, Kirmizitaş FC, Türkoğlu LN. Virus-associated ribozymes and nano carriers against COVID-19. Artif Cells Nanomed Biotechnol 2021;49(1):204–18. https://doi.org/10.1080/21691401.2021.1890103.

[137] Sharti M, Esmaeili Gouvarchin Ghaleh H, Dorostkar R, Jalali KB. Double-stranded RNA activated caspase oligomerizer (DRACO): design, subcloning, and antiviral investigation. J Appl Biotechnol Rep 2021;8(1). https://doi.org/10.30491/jabr.2020.111083.

[138] Zimmermann P, Curtis N. Coronavirus infections in children including COVID-19: an overview of the epidemiology, clinical features, diagnosis, treatment and prevention options in children. Pediatr Infect Dis J 2020;39(5):355–68. https://doi.org/10.1097/inf.0000000000002660.

[139] Arshad S, Kilgore P, Chaudhry ZS, et al. Treatment with hydroxychloroquine, azithromycin, and combination in patients hospitalized with COVID-19. Int J Infect Dis 2020;97:396–403. https://doi.org/10.1016/j.ijid.2020.06.099.

[140] Fredeking TM, Zavala-Castro JE, González-Martínez P, et al. Dengue patients treated with doxycycline showed lower mortality associated to a reduction in IL-6 and TNF levels. Recent Pat Antiinfect Drug Discov 2015;10(1):51–8. https://doi.org/10.2174/1574891x10666150410153839.

[141] Rothan HA, Bahrani H, Mohamed Z, et al. A combination of doxycycline and ribavirin alleviated chikungunya infection. PLoS One 2015;10(5). https://doi.org/10.1371/journal.pone.0126360, e0126360.

[142] Griffin MO, Fricovsky E, Ceballos G, Villarreal F. Tetracyclines: a pleitropic family of compounds with promising therapeutic properties. Review of the literature. Am J Phys Cell Phys 2010;299(3):C539–48. https://doi.org/10.1152/ajpcell.00047.2010.

[143] Yates PA, Newman SA, Oshry LJ, Glassman RH, Leone AM, Reichel E. Doxycycline treatment of high-risk COVID-19-positive patients with comorbid pulmonary disease. Ther Adv Respir Dis 2020;14. https://doi.org/10.1177/1753466620951053.

[144] Arabi YM, Deeb AM, Al-Hameed F, et al. Macrolides in critically ill patients with Middle East Respiratory Syndrome. Int J Infect Dis 2019;81:184–90. https://doi.org/10.1016/j.ijid.2019.01.041.

[145] PRINCIPLE Trial Collaborative Group. Azithromycin for community treatment of suspected COVID-19 in people at increased risk of an adverse clinical course in the UK (PRINCIPLE): a randomised, controlled, open-label, adaptive platform trial. Lancet 2021;397(10279):1063–74. https://doi.org/10.1016/s0140-6736(21)00461-x.

[146] U.S. National Library of Medicine. Ivermectin-Azithromycin-Cholecalciferol (IvAzCol) combination therapy for COVID-19 (IvAzCol), https://clinicaltrials.gov/ct2/show/NCT04399746; 2020.

[147] Yan H, Sun J, Wang K, et al. Repurposing CFDA-approved drug carrimycin as an antiviral agent against human coronaviruses, including the currently pandemic SARS-CoV-2. Acta Pharm Sin B 2021. https://doi.org/10.1016/j.apsb.2021.02.024.

[148] Balfour H. FDA approves phase III trial of synthetic biological COVID-19 treatment. European Pharmaceutical Review; 2021. https://www.europeanpharmaceuticalreview.com/news/138938/fda-approves-phase-iii-trial-of-synthetic-biological-covid-19-treatment/.

[149] Leis JA, Born KB, Theriault G, Ostrow O, Grill A, Johnston KB. Using antibiotics wisely for respiratory tract infection in the era of covid-19. BMJ 2020;371. https://doi.org/10.1136/bmj.m4125, m4125.

[150] Yang Y, Islam MS, Wang J, Li Y, Chen X. Traditional Chinese medicine in the treatment of patients infected with 2019-new coronavirus (SARS-CoV-2): a review and perspective. Int J Biol Sci 2020;16(10):1708–17. https://doi.org/10.7150/ijbs.45538.

[151] Yi L, Li Z, Yuan K, et al. Small molecules blocking the entry of severe acute respiratory syndrome coronavirus into host cells. J Virol 2004;78(20):11334–9. https://doi.org/10.1128/jvi.78.20.11334-11339.2004.

[152] Elshabrawy HA. SARS-CoV-2: an update on potential antivirals in light of SARS-CoV antiviral drug discoveries. Vaccines (Basel) 2020;8(2). https://doi.org/10.3390/vaccines8020335.

[153] Theoharides TC, Cholevas C, Polyzoidis K, Politis A. Long-COVID syndrome-associated brain fog and chemofog: luteolin to the rescue. Biofactors 2021;47(2):232–41. https://doi.org/10.1002/biof.1726.

[154] Wanjarkhedkar P, Sarade G, Purandare B, Kelkar D. A prospective clinical study of an Ayurveda regimen in COVID 19 patients. J Ayurveda Integr Med 2020. https://doi.org/10.1016/j.jaim.2020.10.008.
[155] Gautam S, Gautam A, Chhetri S, Bhattarai U. Immunity against COVID-19: potential role of Ayush Kwath. J Ayurveda Integr Med 2020. https://doi.org/10.1016/j.jaim.2020.08.003.
[156] Ram TS, Munikumar M, Raju VN, et al. In silico evaluation of the compounds of the ayurvedic drug, AYUSH-64, for the action against the SARS-CoV-2 main protease. J Ayurveda Integr Med 2021. https://doi.org/10.1016/j.jaim.2021.02.004.
[157] Gundeti MS, Bhurke LW, Mundada PS, et al. AYUSH 64, a polyherbal Ayurvedic formulation in Influenza-like illness—results of a pilot study. J Ayurveda Integr Med 2020. https://doi.org/10.1016/j.jaim.2020.05.010.
[158] Reddy RG, Gosavi RV, Yadav B, et al. AYUSH-64 as add-on to standard care in asymptomatic and mild cases of COVID-19: a randomized controlled trial. OSFPreprints 2020.
[159] Rao A, Ranganatha R, Vikneswaran G, et al. AYUSH medicine as add-on therapy for mild category COVID-19; an open label randomised, controlled clinical trial. MedRxiv 2020. https://doi.org/10.1101/2020.12.06.20245019.
[160] Wrapp D, Wang N, Corbett KS, et al. Cryo-EM structure of the 2019-nCoV spike in the prefusion conformation. Science 2020;367(6483):1260–3.
[161] Wan S, Yi Q, Fan S, et al. Characteristics of lymphocyte subsets and cytokines in peripheral blood of 123 hospitalized patients with 2019 novel coronavirus pneumonia (NCP). MedRxiv 2020.
[162] Wang D, Hu B, Hu C, et al. Clinical characteristics of 138 hospitalized patients with 2019 novel coronavirus-infected pneumonia in Wuhan, China. JAMA 2020;323(11):1061–9.
[163] Xu Z, Shi L, Wang Y, et al. Pathological findings of COVID-19 associated with acute respiratory distress syndrome. Lancet Respir Med 2020;8(4):420–2.
[164] Dandekar AA, Perlman S. Immunopathogenesis of coronavirus infections: implications for SARS. Nat Rev Immunol 2005;5(12):917–27.
[165] Wong CK, Lam CW, Wu AK, et al. Plasma inflammatory cytokines and chemokines in severe acute respiratory syndrome. Clin Exp Immunol 2004;136(1):95–103. https://doi.org/10.1111/j.1365-2249.2004.02415.x.
[166] Wu S, Wang Y, Jin X, Tian J, Liu J, Mao Y. Environmental contamination by SARS-CoV-2 in a designated hospital for coronavirus disease 2019. Am J Infect Control 2020;48(8):910–4. https://doi.org/10.1016/j.ajic.2020.05.003.
[167] Mehta P, McAuley DF, Brown M, Sanchez E, Tattersall RS, Manson JJ. COVID-19: consider cytokine storm syndromes and immunosuppression. Lancet 2020;395(10229):1033–4. https://doi.org/10.1016/s0140-6736(20)30628-0.
[168] Hutchinson M, Tattersall RS, Manson JJ. Haemophagocytic lymphohisticytosis—an underrecognized hyperinflammatory syndrome. Rheumatology 2019;58(Suppl_6):vi23–30.
[169] Farquhar JW, Claireaux AE. Familial haemophagocytic reticulosis. Arch Dis Child 1952;27(136):519–25. https://doi.org/10.1136/adc.27.136.519.
[170] Halyabar O, Chang MH, Schoettler ML, et al. Calm in the midst of cytokine storm: a collaborative approach to the diagnosis and treatment of hemophagocytic lymphohistiocytosis and macrophage activation syndrome. Pediatr Rheumatol Online J 2019;17(1):7. https://doi.org/10.1186/s12969-019-0309-6.
[171] Guan W, Ni Z, Hu Y, et al. Clinical characteristics of coronavirus disease 2019 in China. N Engl J Med 2020;382(18):1708–20.
[172] Fang Y, Zhang H, Xu Y, Xie J, Pang P, Ji W. CT manifestations of two cases of 2019 novel coronavirus (2019-nCoV) pneumonia. Radiology 2020;295(1):208–9.
[173] Choi IA, Lee SJ, Park W, et al. Effects of tocilizumab therapy on serum interleukin-33 and interleukin-6 levels in patients with rheumatoid arthritis. Arch Rheumatol 2018;33(4):389–94. https://doi.org/10.5606/ArchRheumatol.2018.6753.

[174] Zhou Y, Fu B, Zheng X, et al. Aberrant pathogenic GM-CSF+ T cells and inflammatory CD14+ CD16+ monocytes in severe pulmonary syndrome patients of a new coronavirus. BioRxiv 2020.
[175] Riegler LL, Jones GP, Lee DW. Current approaches in the grading and management of cytokine release syndrome after chimeric antigen receptor T-cell therapy. Ther Clin Risk Manag 2019;15:323.
[176] Shimabukuro-Vornhagen A, Gödel P, Subklewe M, et al. Cytokine release syndrome. J Immunother Cancer 2018;6(1):56. https://doi.org/10.1186/s40425-018-0343-9.
[177] Tanaka T, Narazaki M, Kishimoto T. Immunotherapeutic implications of IL-6 blockade for cytokine storm. Immunotherapy 2016;8(8):959–70. https://doi.org/10.2217/imt-2016-0020.
[178] Xu X, Han M, Li T, et al. Effective treatment of severe COVID-19 patients with tocilizumab. Proc Natl Acad Sci 2020;117(20):10970–5.
[179] Shakoory B, Carcillo JA, Chatham WW, et al. Interleukin-1 receptor blockade is associated with reduced mortality in sepsis patients with features of macrophage activation syndrome: reanalysis of a prior phase III trial. Crit Care Med 2016;44(2):275–81. https://doi.org/10.1097/ccm.0000000000001402.
[180] Eloseily EM, Weiser P, Crayne CB, et al. Benefit of anakinra in treating pediatric secondary hemophagocytic lymphohistiocytosis. Arthritis Rheum 2020;72(2):326–34.
[181] Zaki AM, van Boheemen S, Bestebroer TM, Osterhaus AD, Fouchier RA. Isolation of a novel coronavirus from a man with pneumonia in Saudi Arabia. N Engl J Med 2012;367(19):1814–20. https://doi.org/10.1056/NEJMoa1211721.
[182] Taylor PC. Clinical efficacy of launched JAK inhibitors in rheumatoid arthritis. Rheumatology (Oxford) 2019;58(Suppl 1):i17–26. https://doi.org/10.1093/rheumatology/key225.
[183] Witte T. JAK Inhibitors in Rheumatology. Dtsch Med Wochenschr 2019;144(11):748–52. https://doi.org/10.1055/a-0652-2731.
[184] Zhang X, Zhang Y, Qiao W, Zhang J, Qi Z. Baricitinib, a drug with potential effect to prevent SARS-COV-2 from entering target cells and control cytokine storm induced by COVID-19. Int Immunopharmacol 2020;106749.
[185] Sin JH, Zangardi ML. Ruxolitinib for secondary hemophagocytic lymphohistiocytosis: first case report. Hematol Oncol Stem Cell Ther 2019;12(3):166–70. https://doi.org/10.1016/j.hemonc.2017.07.002.
[186] Broglie L, Pommert L, Rao S, et al. Ruxolitinib for treatment of refractory hemophagocytic lymphohistiocytosis. Blood Adv 2017;1(19):1533–6.
[187] Ahmed A, Merrill SA, Alsawah F, et al. Ruxolitinib in adult patients with secondary haemophagocytic lymphohistiocytosis: an open-label, single-centre, pilot trial. Lancet Haematol 2019;6(12):e630–7. https://doi.org/10.1016/s2352-3026(19)30156-5.
[188] Das R, Guan P, Sprague L, et al. Janus kinase inhibition lessens inflammation and ameliorates disease in murine models of hemophagocytic lymphohistiocytosis. Blood 2016;127(13):1666–75.
[189] Richardson P, Griffin I, Tucker C, et al. Baricitinib as potential treatment for 2019-nCoV acute respiratory disease. Lancet 2020;395(10223):e30–1. https://doi.org/10.1016/s0140-6736(20)30304-4.
[190] Romaguera R, Cruz-González I, Jurado-Román A. Considerations on the invasive management of ischemic and structural heart disease during the COVID-19 coronavirus outbreak. Consensus statement of the interventional cardiology association and the ischemic heart disease and acute cardiac Care Association of the Spanish Society of cardiology. REC Interv Cardiol 2020;2:112–7.
[191] Valgimigli M, Bueno H, Byrne R. ESC focused update on dual antiplatelet therapy in coronary artery disease developed in collaboration with EACTS. EurJCardiothoracSurg 2018;53:34–78.

[192] Li X, Xu S, Yu M, et al. Risk factors for severity and mortality in adult COVID-19 inpatients in Wuhan. J Allergy Clin Immunol 2020;146(1):110–8.
[193] Liao JK, Laufs U. Pleiotropic effects of statins. Annu Rev Pharmacol Toxicol 2005;45:89–118.
[194] Ferrara F, Granata G, Pelliccia C, La Porta R, Vitiello A. The added value of pirfenidone to fight inflammation and fibrotic state induced by SARS-CoV-2. Eur J Clin Pharmacol 2020;76(11):1615–8.
[195] Yildirim M, Kayalar O, Atahan E, Oztay F. Anti-fibrotic effect of Atorvastatin on the lung fibroblasts and myofibroblasts. Eur Respir J 2018;52(Suppl 62):PA991. https://doi.org/10.1183/13993003.congress-2018.PA991.
[196] Castiglione V, Chiriacò M, Emdin M, Taddei S, Vergaro G. Statin therapy in COVID-19 infection. Eur Heart J-Cardiovasc Pharmacother 2020;6(4):258–9.
[197] Mount Auburn Hospital. Atorvastatin as adjunctive therapy in COVID-19 (STATCO19), https://clinicaltrials.gov/ct2/show/NCT04380402; 2021.
[198] Bloch EM, Shoham S, Casadevall A, et al. Deployment of convalescent plasma for the prevention and treatment of COVID-19. J Clin Invest 2020;130(6):2757–65.
[199] Mair-Jenkins J, Saavedra-Campos M, Baillie JK, et al. The effectiveness of convalescent plasma and hyperimmune immunoglobulin for the treatment of severe acute respiratory infections of viral etiology: a systematic review and exploratory meta-analysis. J Infect Dis 2015;211(1):80–90.
[200] Luke TC, Casadevall A, Watowich SJ, Hoffman SL, Beigel JH, Burgess TH. Hark back: passive immunotherapy for influenza and other serious infections. Crit Care Med 2010;38:e66–73.
[201] van Griensven J, De Weiggheleire A, Delamou A, et al. The use of Ebola convalescent plasma to treat Ebola virus disease in resource-constrained settings: a perspective from the field. Clin Infect Dis 2016;62(1):69–74.
[202] Planitzer CB, Modrof J, Kreil TR. West Nile virus neutralization by US plasma-derived immunoglobulin products. J Infect Dis 2007;196(3):435–40.
[203] Owji H, Negahdaripour M, Hajighahramani N. Immunotherapeutic approaches to curtail COVID-19. Int Immunopharmacol 2020;, 106924.
[204] Rojas M, Rodríguez Y, Monsalve DM, et al. Convalescent plasma in Covid-19: possible mechanisms of action. Autoimmun Rev 2020;, 102554.
[205] Tiberghien P, de Lamballerie X, Morel P, Gallian P, Lacombe K, Yazdanpanah Y. Collecting and evaluating convalescent plasma for COVID-19 treatment: why and how? Vox Sang 2020;115(6):488–94.
[206] Li L, Zhang W, Hu Y, et al. Effect of convalescent plasma therapy on time to clinical improvement in patients with severe and life-threatening COVID-19: a randomized clinical trial. JAMA 2020;324(5):460–70.
[207] Joyner MJ, Senefeld JW, Klassen SA, et al. Effect of convalescent plasma on mortality among hospitalized patients with COVID-19: initial three-month experience. MedRxiv 2020. https://doi.org/10.1101/2020.08.12.20169359.
[208] Clinical Trials Registry India. Study to assess the efficacy and safety of convalescent plasma in moderate COVID-19 disease; 2021. http://ctri.nic.in/Clinicaltrials/advsearch.php. Accessed 1 May 2021.
[209] Agarwal A, Mukherjee A, Kumar G, Chatterjee P, Bhatnagar T, Malhotra P. Convalescent plasma in the management of moderate covid-19 in adults in India: open label phase II multicentre randomised controlled trial (PLACID trial). BMJ 2020;371.
[210] Basiri A, Mansouri F, Azari A, et al. Stem cell therapy potency in personalizing severe COVID-19 treatment. Stem Cell Rev Rep 2021;1-21.
[211] Du J, Li H, Lian J, Zhu X, Qiao L, Lin J. Stem cell therapy: a potential approach for treatment of influenza virus and coronavirus-induced acute lung injury. Stem Cell Res Ther 2020;11(1):192. https://doi.org/10.1186/s13287-020-01699-3.
[212] Uccelli A, de Rosbo NK. The immunomodulatory function of mesenchymal stem cells: mode of action and pathways. Ann N Y Acad Sci 2015;1351(1):114–26.

[213] Li Y, Xu J, Shi W, et al. Mesenchymal stromal cell treatment prevents H9N2 avian influenza virus-induced acute lung injury in mice. Stem Cell Res Ther 2016;7(1):1–11.
[214] Chan MC, Kuok DI, Leung CY, et al. Human mesenchymal stromal cells reduce influenza A H5N1-associated acute lung injury in vitro and in vivo. Proc Natl Acad Sci U S A 2016;113(13):3621–6. https://doi.org/10.1073/pnas.1601911113.
[215] Zheng G, Huang L, Tong H, et al. Treatment of acute respiratory distress syndrome with allogeneic adipose-derived mesenchymal stem cells: a randomized, placebo-controlled pilot study. Respir Res 2014;15(1):1–10.
[216] Liang B, Chen J, Li T, et al. Clinical remission of a critically ill COVID-19 patient treated by human umbilical cord mesenchymal stem cells: a case report. Medicine 2020;99(31).
[217] Leng Z, Zhu R, Hou W, et al. Transplantation of ACE2-mesenchymal stem cells improves the outcome of patients with COVID-19 pneumonia. Aging Dis 2020;11(2):216.
[218] Gorman E, Millar J, McAuley D, O'Kane C. Mesenchymal stromal cells for acute respiratory distress syndrome (ARDS), sepsis, and COVID-19 infection: optimizing the therapeutic potential. Expert Rev Respir Med 2021;15(3):301–24. https://doi.org/10.1080/17476348.2021.1848555.
[219] Musial C, Gorska-Ponikowska M. Medical progress: stem cells as a new therapeutic strategy for COVID-19. Stem Cell Res 2021;, 102239.
[220] Park K-S, Svennerholm K, Shelke GV, et al. Mesenchymal stromal cell-derived nanovesicles ameliorate bacterial outer membrane vesicle-induced sepsis via IL-10. Stem Cell Res Ther 2019;10(1):1–14.
[221] Rezakhani L, Kelishadrokhi AF, Soleimanizadeh A, Rahmati S. Mesenchymal stem cell (MSC)-derived exosomes as a cell-free therapy for patients Infected with COVID-19: real opportunities and range of promises. Chem Phys Lipids 2021;234. https://doi.org/10.1016/j.chemphyslip.2020.105009, 105009.
[222] Wang J, Huang R, Xu Q, et al. Mesenchymal stem cell-derived extracellular vesicles alleviate acute lung injury via transfer of miR-27a-3p. Crit Care Med 2020;48(7):e599–610. https://doi.org/10.1097/ccm.0000000000004315.
[223] Sengupta V, Sengupta S, Lazo A, Woods P, Nolan A, Bremer N. Exosomes derived from bone marrow mesenchymal stem cells as treatment for severe COVID-19. Stem Cells Dev 2020;29(12):747–54.
[224] Gupta A, Kashte S, Gupta M, Rodriguez HC, Gautam SS, Kadam S. Mesenchymal stem cells and exosome therapy for COVID-19: current status and future perspective. Hum Cell 2020;33(4):907–18.
[225] Golchin A. Cell-based therapy for severe COVID-19 patients: clinical trials and cost-utility. Stem Cell Rev Rep 2021;17(1):56–62. https://doi.org/10.1007/s12015-020-10046-1.
[226] Market M, Angka L, Martel AB, et al. Flattening the COVID-19 curve with natural killer cell based immunotherapies. Front Immunol 2020;11.
[227] Basiri A, Pazhouhnia Z, Beheshtizadeh N, Hoseinpour M, Saghazadeh A, Rezaei N. Regenerative medicine in COVID-19 treatment: real opportunities and range of promises. Stem Cell Rev Rep 2021;1–13.
[228] Orange JS. Natural killer cell deficiency. J Allergy Clin Immunol 2013;132(3):515–25.
[229] Dong Z, Wei H, Sun R, Hu Z, Gao B, Tian Z. Involvement of natural killer cells in PolyI: C-induced liver injury. J Hepatol 2004;41(6):966–73.
[230] Al-Tawfiq JA, Hinedi K, Abbasi S, Babiker M, Sunji A, Eltigani M. Hematologic, hepatic, and renal function changes in hospitalized patients with Middle East respiratory syndrome coronavirus. Int J Lab Hematol 2017;39(3):272–8. https://doi.org/10.1111/ijlh.12620.
[231] Zheng M, Gao Y, Wang G, et al. Functional exhaustion of antiviral lymphocytes in COVID-19 patients. Cell Mol Immunol 2020;17(5):533–5.

[232] Celularity Incorporated. Natural killer cell (CYNK-001) infusions in adults with COVID-19 (CYNKCOVID), https://clinicaltrials.gov/ct2/show/NCT04365101; 2021. [Accessed 1 May 2021].
[233] Liu E, Marin D, Banerjee P, et al. Use of CAR-transduced natural killer cells in CD19-positive lymphoid tumors. N Engl J Med 2020;382(6):545–53.
[234] Hammer Q, Rückert T, Romagnani C. Natural killer cell specificity for viral infections. Nat Immunol 2018;19(8):800–8.
[235] Graham BS. Rapid COVID-19 vaccine development. Science 2020;368(6494):945–6.
[236] Liu C, Zhou Q, Li Y, et al. Research and development on therapeutic agents and vaccines for COVID-19 and related human coronavirus diseases. ACS Central Sci 2020;6(3):315–31. https://doi.org/10.1021/acscentsci.0c00272.
[237] Craven J. COVID-19 vaccine tracker. Regulatory focus, https://www.raps.org/news-and-articles/news-articles/2020/3/covid-19-vaccine-tracker; 2021. [Accessed 1 May 2021].
[238] Jiang S. Don't rush to deploy COVID-19 vaccines and drugs without sufficient safety guarantees. Nature 2020;321-321.
[239] Ella R, Vadrevu KM, Jogdand H, et al. Safety and immunogenicity of an inactivated SARS-CoV-2 vaccine, BBV152: a double-blind, randomised, phase 1 trial. Lancet Infect Dis 2021.
[240] Ocugen. Ocugen's COVID-19 vaccine co-development partner, Bharat Biotech, shares second interim results demonstrating 100% protection against severe disease including hospitalization, https://ocugen.gcs-web.com/news-releases/news-release-details/ocugens-covid-19-vaccine-co-development-partner-bharat-biotech-0; 2021.
[241] Gao Q, et al. Development of an inactivated vaccine candidate for SARS-CoV-2. Science 2020;369(6499):77–81.
[242] Pearson S, Magalhaes L. Sinovac's Covid-19 vaccine shown effective in Brazil trials. The Wall Street Journal; 2020. Available from: https://www.wsj.com/articles/sinovacs-covid-19-vaccine-shown-to-be-effective-in-brazil-trials-11608581330.
[243] Xia S, et al. Effect of an inactivated vaccine against SARS-CoV-2 on safety and immunogenicity outcomes: interim analysis of 2 randomized clinical trials. JAMA 2020;324(10):951–60.
[244] Reuters. Sinopharm's Wuhan unit reports 72.5% efficacy for COVID shot, seeks approval in China. Available from: https://www.reuters.com/business/healthcare-pharmaceuticals/sinopharms-wuhan-unit-reports-725-efficacy-covid-shot-seeks-approval-china-2021-02-24/; 2021.
[245] Interfax. The participants in the vaccine trials of the Chumakov Center did not reveal any adverse reactions. Available from: https://www.interfax.ru/russia/738613; 2021.
[246] Meyer D. A new vaccine on the scene: Kazakhstan begins rollout of Homegrown QazVac. Fortune; 2021. Available from: https://fortune.com/2021/04/26/new-covid-19-vaccine-kazakhstan-qazvac/.
[247] U.S. National Library of Medicine. Study of the tolerability, safety, immunogenicity and preventive efficacy of the EpiVacCorona vaccine for the prevention of COVID-19. Available from: https://clinicaltrials.gov/ct2/show/NCT04780035; 2021.
[248] Reuters. Russia completes clinical trials of second potential COVID-19 vaccine: RIA. Available from: https://www.reuters.com/article/uk-health-coronavirus-russia-vaccine-vec/russia-completes-clinical-trials-of-second-potential-covid-19-vaccine-ria-idUSKBN26L13B; 2020.
[249] Yang S, et al. Safety and immunogenicity of a recombinant tandem-repeat dimeric RBD-based protein subunit vaccine (ZF2001) against COVID-19 in adults: two randomised, double-blind, placebo-controlled, phase 1 and 2 trials. Lancet Infect Dis 2021.

[250] Zhu F-C, et al. Safety, tolerability, and immunogenicity of a recombinant adenovirus type-5 vectored COVID-19 vaccine: a dose-escalation, open-label, non-randomised, first-in-human trial. Lancet 2020;395(10240):1845–54.
[251] Peshimam GN, Farooq U. CanSinoBIO's COVID-19 vaccine 65.7% effective in global trials, Pakistan official says. Available from: https://www.reuters.com/article/us-health-coronavirus-vaccine-pakistan/cansinobios-covid-19-vaccine-65-7-effective-in-global-trials-pakistan-official-says-idUSKBN2A81N0; 2021.
[252] Xia S, et al. Safety and immunogenicity of an inactivated SARS-CoV-2 vaccine, BBIBP-CorV: a randomised, double-blind, placebo-controlled, phase 1/2 trial. Lancet Infect Dis 2021;21(1):39–51.
[253] Reuters. UAE says Sinopharm vaccine has 86% efficacy against COVID-19. Available from: https://www.reuters.com/business/healthcare-pharmaceuticals/uae-says-sinopharm-vaccine-has-86-efficacy-against-covid-19-2020-12-11/; 2020.
[254] Logunov DY, et al. Safety and efficacy of an rAd26 and rAd5 vector-based heterologous prime-boost COVID-19 vaccine: an interim analysis of a randomised controlled phase 3 trial in Russia. Lancet 2021;397(10275):671–81.
[255] AstraZeneca. AZD1222 US Phase III primary analysis confirms safety and efficacy, https://www.astrazeneca.com/content/astraz/media-centre/press-releases/2021/azd1222-us-phase-iii-primary-analysis-confirms-safety-and-efficacy.html; 2021. [Accessed 1 May 2021].
[256] Voysey M, Clemens SAC, Madhi SA, et al. Single-dose administration and the influence of the timing of the booster dose on immunogenicity and efficacy of ChAdOx1 nCoV-19 (AZD1222) vaccine: a pooled analysis of four randomised trials. Lancet 2021;397(10277):881–91.
[257] Sadoff J, Gray G, Vandebosch A, et al. Safety and efficacy of single-dose Ad26. COV2. S vaccine against COVID-19. N Engl J Med 2021.
[258] Polack FP, Thomas SJ, Kitchin N, et al. Safety and efficacy of the BNT162b2 mRNA Covid-19 vaccine. N Engl J Med 2020;383(27):2603–15.
[259] Baden LR, El Sahly HM, Essink B, et al. Efficacy and safety of the mRNA-1273 SARS-CoV-2 vaccine. N Engl J Med 2021;384(5):403–16.
[260] World Health Organization. Coronavirus disease (COVID-19): Vaccines, https://www.who.int/news-room/q-a-detail/coronavirus-disease-(covid-19)-vaccines; 2021. [Accessed 1 May 2021].
[261] Karikó K, Muramatsu H, Welsh FA, et al. Incorporation of pseudouridine into mRNA yields superior nonimmunogenic vector with increased translational capacity and biological stability. Mol Ther 2008;16(11):1833–40.
[262] Walsh EE, Frenck Jr RW, Falsey AR, et al. Safety and immunogenicity of two RNA-based Covid-19 vaccine candidates. N Engl J Med 2020;383(25):2439–50.
[263] Muik A, Wallisch A-K, Sänger B, et al. Neutralization of SARS-CoV-2 lineage B. 1.1. 7 pseudovirus by BNT162b2 vaccine–elicited human sera. Science 2021;371(6534):1152–3.
[264] Liu Y, Liu J, Xia H, et al. Neutralizing activity of BNT162b2-elicited serum. N Engl J Med 2021;384(15):1466–8.
[265] Centers for Disease Control and Prevention. Different COVID-19 vaccines, https://www.cdc.gov/coronavirus/2019-ncov/vaccines/different-vaccines.html; 2021. [Accessed 1 May 2021].
[266] Corbett KS, Edwards DK, Leist SR, et al. SARS-CoV-2 mRNA vaccine design enabled by prototype pathogen preparedness. Nature 2020;586(7830):567–71.
[267] Wu K, Werner AP, Koch M, et al. Serum neutralizing activity elicited by mRNA-1273 vaccine. N Engl J Med 2021;384(15):1468–70.
[268] Custers J, Kim D, Leyssen M, et al. Vaccines based on replication incompetent Ad26 viral vectors: standardized template with key considerations for a risk/benefit assessment. Vaccine 2021.

[269] Bos R, Rutten L, van der Lubbe JEM, et al. Ad26 vector-based COVID-19 vaccine encoding a prefusion-stabilized SARS-CoV-2 Spike immunogen induces potent humoral and cellular immune responses. NPJ Vaccines 2020;5(1):91. https://doi.org/10.1038/s41541-020-00243-x.
[270] Sadoff J, Davis K, Douoguih M. Thrombotic thrombocytopenia after Ad26.COV2. S vaccination—response from the manufacturer. N Engl J Med 2021.
[271] U.S. Food and Drug Administration. FDA and CDC lift recommended pause on Johnson & Johnson (Janssen) COVID-19 vaccine use following thorough safety review, https://www.fda.gov/news-events/press-announcements/fda-and-cdc-lift-recommended-pause-johnson-johnson-janssen-covid-19-vaccine-use-following-thorough; 2021.
[272] European Medicines Agency. COVID-19 vaccine Janssen: EMA finds possible link to very rare cases of unusual blood clots with low blood platelets, https://www.ema.europa.eu/en/news/covid-19-vaccine-janssen-ema-finds-possible-link-very-rare-cases-unusual-blood-clots-low-blood; 2021.
[273] Voysey M, Clemens SAC, Madhi SA, et al. Safety and efficacy of the ChAdOx1 nCoV-19 vaccine (AZD1222) against SARS-CoV-2: an interim analysis of four randomised controlled trials in Brazil, South Africa, and the UK. Lancet 2021;397(10269):99–111.
[274] Emary KR, Golubchik T, Aley PK, et al. Efficacy of ChAdOx1 nCoV-19 (AZD1222) vaccine against SARS-CoV-2 variant of concern 202012/01 (B. 1.1. 7): an exploratory analysis of a randomised controlled trial. Lancet 2021.
[275] Madhi SA, Baillie V, Cutland CL, et al. Efficacy of the ChAdOx1 nCoV-19 Covid-19 vaccine against the B. 1.351 variant. N Engl J Med 2021.
[276] Paul-Ehrlich-Institut. The Paul-Ehrlich-Institut informs—Temporary Suspension of Vaccination with COVID-19 Vaccine AstraZeneca, https://www.pei.de/EN/newsroom/hp-news/2021/210315-pei-informs-temporary-suspension-vaccination-astra-zeneca.html; 2021.
[277] World Health Organization. Statement of the Strategic Advisory Group of Experts (SAGE) on Immunization: Continued Review of Emerging Evidence on AstraZeneca COVID-19 Vaccines, https://www.who.int/news/item/22-04-2021-statement-of-the-strategic-advisory-group-of-experts-(sage)-on-immunization-continued-review-of-emerging-evidence-on-astrazeneca-covid-19-vaccines; 2021.
[278] The Hindu. Coronavirus | India approves COVID-19 Vaccines Covishield and Covaxin for emergency use, https://www.thehindu.com/news/national/drug-controller-general-approves-covishield-and-covaxin-in-india-for-emergency-use/article33485539.ece; 2021.
[279] Shukla NM, Salunke DB, Balakrishna R, Mutz CA, Malladi SS, David SA. Potent adjuvanticity of a pure TLR7-agonistic imidazoquinoline dendrimer. PLoS One 2012;7(8), e43612.
[280] Sapkal GN, Yadav P, Ella R, et al. Neutralization of UK-variant VUI-202012/01 with COVAXIN vaccinated human serum. BioRxiv 2021.
[281] Sputnik V. Sputnik V vaccine authorized in India, https://sputnikvaccine.com/newsroom/pressreleases/sputnik-v-authorized-for-use-in-india/; 2021.
[282] The Hindu. Panel nod for Sputnik trials by Hetero Pharma, https://www.thehindu.com/news/national/panel-nod-for-sputnik-trials-by-hetero-pharma/article34444888.ece; 2021.
[283] Central Drugs Standard Control Organization. Notice regarding guidance for approval Covid-19 vaccine in India for restricted use in emergency situation which are already approved for restricted use by US FDA EMA, UK MHRA, PMDA Japan or which are listed in WHO emergency use listing, https://cdsco.gov.in/opencms/export/sites/CDSCO_WEB/Pdf-documents/notice15april21.pdf; 2021.

[284] Times of India. J&J seeks permission for phase-3 trial of its Single-Sho, http://timesofindia.indiatimes.com/articleshow/82150152.cms?utm_source=contentofinterest&utm_medium=text&utm_campaign=cppst; 2021.
[285] Keech C, Albert G, Cho I, et al. Phase 1–2 trial of a SARS-CoV-2 recombinant spike protein nanoparticle vaccine. N Engl J Med 2020;383(24):2320–32.
[286] Clinical Trials Registry India. A phase III, randomized, multi-centre, double blind, placebo controlled, study to evaluate efficacy, safety and immunogenicity of novel corona virus-2019-nCov vaccine candidate of M/s Cadila Healthcare Limited; 2021. http://ctri.nic.in/Clinicaltrials/showallp.php?mid1=51254&EncHid=&userName=ZyCoV-D. Accessed 1 May 2021.
[287] Deccan Herald. Zydus Cadila gets DCGI nod for phase-3 trials of its Covid-19 vaccine candidate, https://www.deccanherald.com/national/zydus-cadila-gets-dcgi-nod-for-phase-3-trials-of-its-covid-19-vaccine-candidate-934751.html; 2021.
[288] Forni G, Mantovani A. COVID-19 vaccines: where we stand and challenges ahead. Cell Death Differ 2021;28(2):626–39.
[289] Sun D. Remdesivir with IV Administration Alone is Unlikely to Achieve Adequate Efficacy and Pulmonary Delivery should be Investigated in COVID-19 Patients. Ann Arbor 2020;1001:48109.
[290] Bhavane R, Karathanasis E, Annapragada AV. Agglomerated vesicle technology: a new class of particles for controlled and modulated pulmonary drug delivery. J Control Release 2003;93(1):15–28.
[291] Velasquez LS, Shira S, Berta AN, et al. Intranasal delivery of Norwalk virus-like particles formulated in an in situ gelling, dry powder vaccine. Vaccine 2011;29(32):5221–31.
[292] Yusuf H, Kett V. Current prospects and future challenges for nasal vaccine delivery. Human Vaccines Immunother 2017;13(1):34–45.
[293] Gatta AK, Josyula VR. Small interfering RNA: a tailored approach to explore the therapeutic potential in COVID-19. Mol Ther-Nucleic Acids 2021;23:640–2.
[294] Kandil R, Merkel OM. Pulmonary delivery of siRNA as a novel treatment for lung diseases. Ther Deliv 2019;10(4):203–6. https://doi.org/10.4155/tde-2019-0009.
[295] Youngren-Ortiz SR, Gandhi NS, España-Serrano L, Chougule MB. Aerosol delivery of siRNA to the lungs. Part 1: rationale for gene delivery systems. KONA Powder Particle J 2016;, 2016014.
[296] Akinc A. ALN-COV: an investigational RNAi therapeutic for COVID-19. In: The Oligonucleotide Therapeutic Society Meeting; 2020. https://www.alnylam.com/wp-content/uploads/2020/09/OTS-2020_Akinc.pdf. [Accessed 1 May 2021].
[297] Hodgson J. The pandemic pipeline. Nat Biotechnol 2020;38(5):523–32. https://doi.org/10.1038/d41587-020-00005-z.
[298] Du L, Zhao G, Lin Y, et al. Intranasal vaccination of recombinant adeno-associated virus encoding receptor-binding domain of severe acute respiratory syndrome coronavirus (SARS-CoV) spike protein induces strong mucosal immune responses and provides long-term protection against SARS-CoV infection. J Immunol 2008;180(2):948–56. https://doi.org/10.4049/jimmunol.180.2.948.
[299] Kim MH, Kim HJ, Chang J. Superior immune responses induced by intranasal immunization with recombinant adenovirus-based vaccine expressing full-length Spike protein of Middle East respiratory syndrome coronavirus. PLoS One 2019;14(7), e0220196.
[300] Travis CR. As plain as the nose on your face: the case for a nasal (mucosal) route of vaccine administration for Covid-19 disease prevention. Front Immunol 2020;11(2611). https://doi.org/10.3389/fimmu.2020.591897.
[301] Mayi BS, Leibowitz JA, Woods AT, Ammon KA, Liu AE, Raja A. The role of Neuropilin-1 in COVID-19. PLoS Pathog 2021;17(1). https://doi.org/10.1371/journal.ppat.1009153, e1009153.

[302] Meng Z, Wang T, Chen L, et al. An experimental trial of recombinant human interferon alpha nasal drops to prevent COVID-19 in medical staff in an epidemic area. medRxiv 2020. https://doi.org/10.1101/2020.04.11.20061473.
[303] Pharmaceutical Technology. Rokote Laboratories Finland to unveil New Nasal Vaccine for Covid-19, https://www.pharmaceutical-technology.com/news/rokote-laboratories-nasal-vaccine/; 2021.
[304] Sharun K, Dhama K. India's role in COVID-19 vaccine diplomacy. J Travel Med 2021.
[305] Chakraborty C, Agoramoorthy G. India's cost-effective COVID-19 vaccine development initiatives. Vaccine 2020;38(50):7883–4. https://doi.org/10.1016/j.vaccine.2020.10.056.
[306] Mordani S. Bharat Biotech's Nasal Vaccine for Covid-19 likely to be a game-changer. Here's why. India Today; 2021. https://www.indiatoday.in/coronavirus-outbreak/vaccine-updates/story/nasal-vaccine-clinical-trial-bharat-biotech-1775710-2021-03-04.
[307] Tatara AM. Role of tissue engineering in COVID-19 and future viral outbreaks. Tissue Eng A 2020;26(9-10):468–74.
[308] Bookstaver ML, Tsai SJ, Bromberg JS, Jewell CM. Improving vaccine and immunotherapy design using biomaterials. Trends Immunol 2018;39(2):135–50.
[309] Yang XX, Li CM, Li YF, Wang J, Huang CZ. Synergistic antiviral effect of curcumin functionalized graphene oxide against respiratory syncytial virus infection. Nanoscale 2017;9(41):16086–92.
[310] Arruebo M, Valladares M, González-Fernández Á. Antibody-conjugated nanoparticles for biomedical applications. J Nanomater 2009.
[311] Gatta AK, Philip NV, Udupa N, Reddy MS, Mutalik S, Josyula VR. Strategic design of potential siRNA molecules for in vitro Evaluation in ABCG2 resistant Breast cancer and in vivo toxicity Determination. Res J Pharm Technol 2021;14(1):55–61.
[312] Merkel OM, Rubinstein I, Kissel T. siRNA delivery to the lung: what's new? Adv Drug Deliv Rev 2014;75:112–28.
[313] Meng B, Lui Y, Meng S, Cao C, Hu Y. Identification of effective siRNA blocking the expression of SARS viral envelope E and RDRP genes. Mol Biotechnol 2006;33(2):141–8.
[314] Rao L, Wang W, Meng Q-F, et al. A Biomimetic nanodecoy traps zika virus to prevent viral infection and fetal microcephaly development. Nano Lett 2019;19(4):2215–22. https://doi.org/10.1021/acs.nanolett.8b03913.
[315] Zhang Q, Honko A, Zhou J, et al. Cellular nanosponges inhibit SARS-CoV-2 infectivity. Nano Lett 2020;20(7):5570–4. https://doi.org/10.1021/acs.nanolett.0c02278.
[316] Yenkoidiok-Douti L, Jewell CM. Integrating biomaterials and immunology to improve vaccines against infectious diseases. ACS Biomater Sci Eng 2020;6(2):759–78.
[317] Kim J, Li WA, Choi Y, et al. Injectable, spontaneously assembling, inorganic scaffolds modulate immune cells in vivo and increase vaccine efficacy. Nat Biotechnol 2015;33(1):64–72.
[318] Dellacherie MO, Li AW, Lu BY, Mooney DJ. Covalent conjugation of peptide antigen to mesoporous silica rods to enhance cellular responses. Bioconjug Chem 2018;29(3):733–41.
[319] Jiang T, Singh B, Li H-S, et al. Targeted oral delivery of BmpB vaccine using porous PLGA microparticles coated with M cell homing peptide-coupled chitosan. Biomaterials 2014;35(7):2365–73.
[320] Rungrojcharoenkit K, Sunintaboon P, Ellison D, et al. Development of an adjuvanted nanoparticle vaccine against influenza virus, an in vitro study. PLoS One 2020;15(8), e0237218.

[321] Pimentel TA, Yan Z, Jeffers SA, Holmes KV, Hodges RS, Burkhard P. Peptide nanoparticles as novel immunogens: design and analysis of a prototypic severe acute respiratory syndrome vaccine. Chem Biol Drug Des 2009;73(1):53–61.
[322] Noh HJ, Noh YW, Heo MB, et al. Injectable and pathogen-mimicking hydrogels for enhanced protective immunity against emerging and highly pathogenic influenza virus. Small 2016;12(45):6279–88. https://doi.org/10.1002/smll.201602344.
[323] Rouphael NG, Paine M, Mosley R, et al. The safety, immunogenicity, and acceptability of inactivated influenza vaccine delivered by microneedle patch (TIV-MNP 2015): a randomised, partly blinded, placebo-controlled, phase 1 trial. Lancet 2017;390(10095):649–58.
[324] Kim E, Erdos G, Huang S, et al. Microneedle array delivered recombinant coronavirus vaccines: immunogenicity and rapid translational development. EBioMed 2020;55, 102743.
[325] Suderman M, McCarthy M, Mossell E, et al. Three-dimensional human bronchial-tracheal epithelial tissue-like assemblies (TLAs) as hosts for severe acute respiratory syndrome (SARS)-CoV infection; 2006.
[326] Zhou J, Li C, Liu X, et al. Infection of bat and human intestinal organoids by SARS-CoV-2. Nat Med 2020;26(7):1077–83.
[327] Youk J, Kim T, Evans KV, et al. Three-dimensional human alveolar stem cell culture models reveal infection response to SARS-CoV-2. Cell Stem Cell 2020;27(6):905–919.e10.
[328] Wu Y-T, Yang C-E, Ko C-H, Wang Y-N, Liu C-C, Fu L-M. Microfluidic detection platform with integrated micro-spectrometer system. Chem Eng J 2020;393, 124700.
[329] Si H, Xu G, Jing F, Sun P, Zhao D, Wu D. A multi-volume microfluidic device with no reagent loss for low-cost digital PCR application. Sensors Actuators B Chem 2020;318, 128197.

Index

Note: Page numbers followed by *f* indicate figures and *t* indicate tables.

D

E

Printed in the United States
by Baker & Taylor Publisher Services